TO RENEW PLEASE
QUOTE BORROWER
NUMBER.

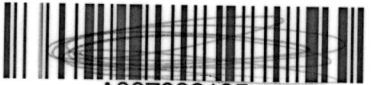

A007092105

D1743250

WITHDRAWN
FOR
SALE

Linear Circuit Analysis

Linear Circuit Analysis

Chi Kong Tse
Hong Kong Polytechnic University

CHESHIRE LIBRARIES	
Morley Books	11.11.98
621·38150285	1595p

Addison-Wesley

Harlow, England ● Reading, Massachusetts ● Menlo Park, California ● New York
Don Mills, Ontario ● Amsterdam ● Bonn ● Sydney ● Singapore
Tokyo ● Madrid ● San Juan ● Milan ● Mexico City ● Seoul ● Taipei

© Addison Wesley Longman Limited 1998

Addison Wesley Longman Limited
Edinburgh Gate
Harlow
Essex CM20 2JE
England

and Associated Companies throughout the world.

The right of Chi Kong Tse to be identified as author of this Work has been asserted by him in accordance with the Copyright, Designs and Patents Act 1988.

All rights reserved. No part of this publication may be reproduced, stored in a retrieval system, or transmitted in any form or by any means, electronic, mechanical, photocopying, recording or otherwise, without either the prior written permission of the publisher or a licence permitting restricted copying in the United Kingdom issued by the Copyright Licensing Agency Ltd, 90 Tottenham Court Road, London W1P 9HE.

The programs in this book have been included for their instructional value. They have been tested with care but are not guaranteed for any particular purpose. The publisher does not offer any warranties or representations nor does it accept any liabilities with respect to the programs.

Many of the designations used by manufacturers and sellers to distinguish their products are claimed as trademarks. Addison Wesley Longman Limited has made every attempt to supply trademark information about manufacturers and their products mentioned in this book. A list of trademark designations and their owners appears on this page.

Cover designed by Designers & Partners, Oxford
Cover illustration by the author
Produced by Addison Wesley Longman China Limited, Hong Kong

First printed 1998

ISBN 0-201-34296-0

British Library Cataloging-in-Publication Data
A catalogue record for this book is available from the British Library

Trademark notice
MicroSim and PSpice are registered trademarks of MicroSim
Microsoft is a registered trademark of Microsoft Corporation

To Belinda

Preface

This book is designed to serve as a circuit analysis text in an electrical engineering curriculum. The material covered in this book provides a comprehensive, yet concise, treatment of techniques for the analysis of linear electric circuits. Suitable for a one or two-semester introductory course in circuit analysis at the first-year level, this book primarily aims to provide a firm electric circuits foundation for students studying electrical engineering as well as other engineering disciplines. Covering also advanced topological approaches to circuit analysis, this book provides useful supplementary reference for students taking an advanced course in circuit analysis at final-year undergraduate or postgraduate levels.

In the discussion of conventional topics such as basic circuit theorems, first-order transient responses and frequency responses, the treatments described in this book emphasize fast inspection of circuits with minimal use of mathematics. This kind of to-the-point treatment, coupled with a concise style, makes this book suitable for use in curricula that allow very limited time for studying circuit analysis. In the discussion of circuit topologies and their applications in formulating efficient analysis methods, this book, unlike many advanced circuit theory texts or research monographs which involve advanced use of linear algebra and set theory, presents the essential topological techniques at a level suitable for an undergraduate course in electrical engineering. The pre-requisites are elementary algebra and basic calculus. Many of the basic mathematical skills have been included in self-contained appendices at the end of the book.

The use of computer-aided analysis programmes for the study of electric circuits has become increasingly popular. The obvious advantage of using computers in circuit analysis is the great saving in time, especially when large circuits are analyzed. In this book, the PSPICE analysis programme is used to illustrate how the computer helps solve circuit problems. Common features of the Windows version of PSPICE (version 6.0 or above) will be introduced. Although the emphasis is on circuits containing linear passive elements, the analysis procedures described in this book are applicable to circuits containing active devices.

As mentioned before, this book permits a two-fold usage. First, Chapters 1 through 8 can be used in a one or two-semester course in linear circuit analysis, with optional inclusion of Chapter 12 if it has not already been covered in an analogue circuits course. Second, Chapters 9 through 11 can form part of the material for an advanced course in circuit theory.

This book begins with a review, in Chapter 1, of some fundamental concepts of electric circuits, starting from the definitions of voltage and current, and the Ohm's law relation. Some basic circuit analysis methods, which are applicable to circuits containing a relatively small number of components connected in some relatively simple fashions, are discussed in this introductory chapter. These methods may be regarded as *ad hoc,* in that they are not generally applicable to circuits of arbitrary configurations.

Chapter 2 discusses some important circuit theorems which, when suitably applied, can drastically simplify a circuit problem. The emphases are on the underlying concepts of the theorems as well as the ways in which the theorems are applied to solve circuit problems.

Chapter 3 covers the nodal and mesh methods which are stereotypically adopted in most standard circuit theory texts as the systematic analysis methods. The problems are categorized according to the presence of the type of voltage and current sources. The principle of superposition is introduced as a consequence of linearity, and is made use of in the solution of circuits containing more than one independent sources. To complete the solution, the resulting linear system of equations needs to be solved. For students who have little or no background in linear equations, Appendix B provides a quick reference to the essential techniques for solving linear systems of equations.

Chapter 4 introduces the two basic energy storage elements, namely the inductor and the capacitor. The typical time-domain transient response is derived for simple first-order circuits. The emphases are on physical inspection of circuits and fast determination of time constants, initial states and final states, without solving differential equations. The continuity of capacitor voltages and inductor currents in circuits containing switches is also discussed in the light of circuit configurations.

Chapter 5 studies mutual inductance and the transformer action, with minimal reference to the physics of magnetics. Coupled coils are viewed as 2-dim extensions of linear inductors. Some important properties of coupled coils are discussed, and the ideal transformer model is derived as a special case of coupled coils having infinitely large inductance and perfect coupling. A practical transformer model is developed around the ideal transformer with additional inductances accounting for the finite inductance and imperfect coupling.

Chapter 6 introduces the periodic function and the various ways to quantify a periodic function. The concepts of average and rms values as well as some representation methods are covered. Specifically, the representation of sine functions using complex numbers and phasor diagrams is described in detail. This chapter also provides an in-depth treatment of Fourier series representations of periodic functions. This topic may, however, be studied, or postponed to a later time, depending upon the level of students' mathematical skills. Although the inclusion of Fourier series representations in this chapter provides a complete introductory treatment of the concept of periodic functions, omitting it during first reading should present no serious impairment to the study of subsequent topics.

Chapter 7 studies the steady-state analysis of circuits that contain inductors and capacitors. Since no transient response is required, the problem can be treated using techniques described in Chapters 2 and 3. The only difference is that the solution of the resulting set of linear equations now involves complex numbers. An alternative approach based on the use of phasor diagrams is also covered in this chapter. The concepts of active, reactive and apparent powers are introduced here for a two-fold purpose. First, these concepts are essential in electrical engineering. Second, the use of power concepts can sometimes yield very rapid solutions to circuit analysis problems. For students who have little background in complex numbers, Appendix E provides a quick reference to the essential techniques in the manipulation of complex numbers that are necessary for understanding the material covered in this chapter.

Chapter 8 covers the basics of stability, complex frequency, transfer functions and frequency response. While details of the mathematics are left to a mathematics course, essential techniques of analyzing circuits in the frequency domain are covered in this chapter. Construction of Bode diagrams is described using the conventional approach as well as a new approach based on "inverted poles and zeros" which can provide better physical insight into the contributions of the various parameters in a function transfer.

Chapter 9 begins the development of a different approach to the analysis of circuits. Specifically, the aim is to introduce the basic elements of circuit topologies such as trees, co-trees, cutsets and loops, and their use in the setting up of independent Kirchhoff's law equations, which lays the foundation for efficient circuit analysis to be covered in Chapter 10. The matrix representations of the topology of a given circuit are described, which will be helpful in the systematic formulation of state equations to be discussed in Chapter 11. Moreover, the emphasis is on physical inspection since the ultimate aim is to allow rapid and efficient analysis of electric circuits.

Chapter 10 exploits the concepts described in Chapter 9 to formulate independent Kirchhoff's law equations resulting in a minimum number of equations to be solved. The essential methods covered in this chapter are the cutset-voltage method and the loop-current method, both of which require simple inspection of the circuit topology. With this approach, problems that appear to be complicated can even be solved by hand calculations.

Chapter 11 exploits circuit topologies again to set up state equations for general dynamic circuits for which complete solutions are sought. The emphasis is on fast inspection of the topology and a systematic procedure for deriving state equations for any given circuit, however complicated. Using the topological approach, minimal efforts are required in the subsequent manipulation of the equations. To keep the description concise, methods of solving state equations using Laplace transformation are briefly outlined, leaving the details to an engineering mathematics course. Moreover, the essential techniques of Laplace transform are summarized in a self-contained manner in Appendix C. Also, Appendix D summarizes the procedures for partial fraction expansion, which are essential in performing inverse Laplace transformation.

Chapter 12 covers the basics of two-port characterization of linear passive circuits. Although two-port representations are very specific representations of circuits, they are overwhelmingly used in the analysis of electronic circuits. For this reason, this chapter covers some common configurations of connecting electronic circuits such as shunt-series, series-shunt, shunt-shunt, etc. Practical methods of finding two-port parameters are also covered. The treatment of two-port circuits is sometimes more conveniently included in an analogue electronics course. If such is the case, the delivery of the material of Chapter 12 may be omitted or incorporated in an appropriate analogue circuits course.

In writing this book, a conscientious effort has been made to maintain a concise and to-the-point style. Topics in circuit theory are selectively included in this book to cover a minimal but adequate amount of material for grasping the essential techniques. Extra material addressing specific problems and redundant viewpoints that do not improve understanding drastically are eliminated. Wherever appropriate, examples are used to clarify certain concepts and to illustrate analysis methods that may appear complicated from the general statements of procedures. Moreover, since details of methods of solutions of linear and differential equations are deliberately omitted in the main text (but summarized in the appendices), the use of this book in an introductory circuit analysis course assumes a co-requisite of an engineering mathematics course covering the essential mathematical methods.

Practice is an important process of gaining mastery of circuit analysis techniques. To maintain a concise style, each chapter is followed by a few drill problems which provide immediate reinforcement of the salient concepts and techniques discussed in the chapter. There are also specific problems for students to practise using the PSPICE analysis programme. Since PSPICE is a general analysis programme, students are encouraged to use it for solving other circuit problems that may be encountered in their studies. The "SCH" files that describe the schematics of the circuits in the end-of-chapter problems are available to instructors and students from the publishers' homepage "http://www.awl-he.com". Instructors may also contact their local Addison Wesley Longman representatives for the complete solutions manual.

It was very fortunate for me to have studed under Prof. Keith Adams who has given me much inspiration in the course of my study of electric circuits at Melbourne University. I also acknowledge with gratitude the many helpful discussions with my colleagues at Hong Kong Polytechnic University, especially Martin Chow who has generously imparted on me his invaluable experience in teaching circuits at the first-year level. Special thanks are due to Prof. Yim-shu Lee and Prof. Wan-chi Siu, whose encouragement and support made the writing of this book an enjoyable undertaking. To Dr. Geoffrey Cross of the University of Ulster and other referees, I wish to express my sincere appreciation for their advice and criticism which were invaluable in the revision of the preliminary versions of the manuscript. Last, but not least, Ms. Anna Faherty of Addison Wesley Longman deserves special thanks for her professional and enthusiastic support of this project.

Finally, I must thank my wife Belinda for her patience and understanding during the preparation of this book. I owe her a great debt of gratitude.

Hong Kong   C. K. Tse
January 1998

Contents

Chapter 1

Introduction

An electric circuit is formed by interconnecting components having different electric properties. It is therefore important, in the analysis of electric circuits, to know the properties of the involved components as well as the way the components are connected to form the circuit. In this introductory chapter some ideal electric components and simple connection styles are introduced. Without resort to advanced analysis techniques, we will attempt to solve simple problems involving circuits that contain a relatively small number of components connected in some relatively simple fashions. In particular we will derive a set of useful formulae for dealing with circuits that involve such simple connections as "series", "parallel", "ladder", "star" and "delta". This chapter serves as a review of the basic properties of electric circuits. In addition we will briefly introduce the PSPICE analysis programme and how it can be used to help analyze electric circuits.

1.1 Fundamental Quantities

In the study of electric circuits, we deal with the fundamental phenomenon of the movement of electrically charged particles or simply *charged particles*. The fundamental quantities that are used to describe how rapidly charged particles move in a circuit and in what way they do so in the circuit are *current* and *voltage*.

Current is sometimes referred to as the "through" quantity and voltage as the "across" quantity. In the physical context, current is the flow of electric charge through a component or apparatus, whereas voltage is the potential difference between two points in a circuit. Current flows from high potential to low potential. In particular we define the current, I, flowing in a component or apparatus as the amount of charge passing through that component or apparatus per unit of time. Denoting charge by q, we may write current I as

$$I = \lim_{\delta t \to 0} \frac{\delta q}{\delta t} = \frac{dq}{dt}$$

1

Prefix	Multiplier (abbreviation)
Peta	$\times 10^{15}$ (P)
Tera	$\times 10^{12}$ (T)
Giga	$\times 10^{9}$ (G)
Mega	$\times 10^{6}$ (M)
Kilo	$\times 10^{3}$ (k)
Hecto	$\times 10^{2}$ (h)
Deca	$\times 10$ (da)
deci	$\times 10^{-1}$ (d)
centi	$\times 10^{-2}$ (c)
milli	$\times 10^{-3}$ (m)
micro	$\times 10^{-6}$ (μ)
nano	$\times 10^{-9}$ (n)
pico	$\times 10^{-12}$ (p)
femto	$\times 10^{-15}$ (f)

Table 1.1: Prefixes of units

Two other important quantities that are frequently used in describing physical systems are *power* and *energy*. If a small quantity of electric charge δq is displaced from a point A to a point B, then the change in its potential energy (or work done) is equal to $V.\delta q$, where V denotes the voltage between A and B. In order to measure how rapidly work is done, we consider the amount of work done per unit of time. This quantity is called *power*, and is usually denoted by P.

$$P = \lim_{\delta t \to 0} V \frac{\delta q}{\delta t} = V \frac{dq}{dt} = VI$$

The unit of current is the *ampere* (A), that of voltage is the *volt* (V), that of energy is the *joule* (J), and that of power is the *watt* (W). Prefixes are often used to emphasize the significant figures when the magnitudes are too large or too small. Common prefixes and their corresponding multipliers are shown in Table 1.1. For example, 1ns denotes 1×10^{-9}s, and 6kV denotes 6×10^{3}V.

1.2   Direction and Polarity

Current direction indicates the direction of flow of positive charge, and voltage polarity indicates the relative potential between two points. Usually, "+" is assigned to a higher potential point and "−" to a lower potential point. However, during analysis, direction and polarity can be *arbitrarily* assigned on circuit diagrams. Actual direction and polarity will be governed by the sign of the value. Figure 1.1 shows some examples.

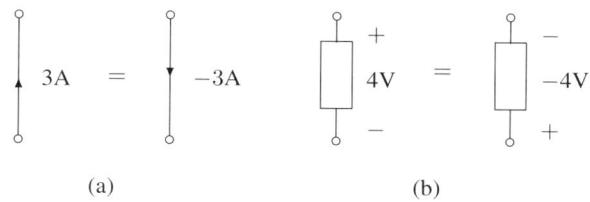

Figure 1.1: Equivalent assignments of (a) current direction; (b) voltage polarity

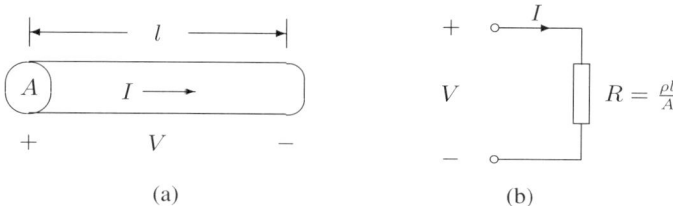

Figure 1.2: Ohm's law. (a) Metal wire; (b) circuit symbol

1.3 Basic Circuit Elements

1.3.1 Resistance

When voltage is applied to a piece of metal wire, as shown in figure 1.2 (a), the current I flowing through the wire is proportional to the voltage V across two points in the wire. This property is known as Ohm's law, which reads

$$V = IR \quad \text{or} \quad I = GV$$

where R is called resistance, and G is called conductance. The resistance R and the conductance G of the same piece of wire is related by $R = 1/G$. Resistance is measured in *ohms* (Ω) and conductance in *siemens* (S or $\mho$).

Any apparatus/device that has this property is called a *resistor*. Study of the physics of resistance shows that it is proportional to the length of the metal wire, l, and inversely proportional to the cross-sectional area, A, i.e.,

$$R = \frac{\rho l}{A}$$

where the proportionality constant ρ is known as the *resistivity* of the metal.

We may calculate the power required to pass current I through a resistor of resistance R using the previously derived formula, i.e.,

$$P = VI$$

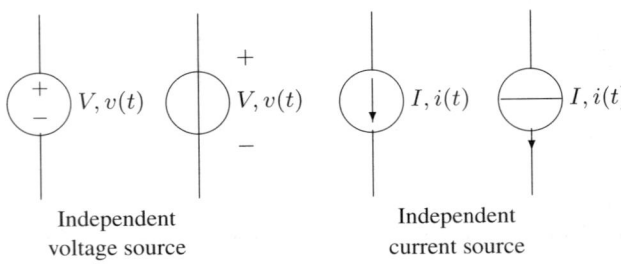

Independent   Independent
voltage source   current source

Figure 1.3: Symbols for independent sources

Using the Ohm's law equation, we get

$$P = \frac{V^2}{R} = GV^2 = I^2 R \geq 0 \qquad \text{for all } V, I$$

The last inequality defines a property called *passivity*.

1.3.2   Independent and Dependent Sources

There are two principal types of source, namely *voltage source* and *current source*. Sources can be either independent or dependent upon some other quantities.

An *independent voltage source* maintains a voltage (fixed or varying with time) which is not affected by any other quantity. Similarly an *independent current source* maintains a current (fixed or time-varying) which is unaffected by any other quantity. The usual symbols are shown in figure 1.3.

Some voltage (current) sources have their voltage (current) values varying with some other variables. They are called *dependent* voltage (current) sources or *controlled* voltage (current) sources, and their usual symbols are shown in figure 1.4.

Remarks — It is not possible to force an independent voltage source to take up a voltage which is different from its defined value. Likewise, it is not possible to force an independent current source to take up a current which is different from its defined value. Two particular examples are short-circuiting an independent voltage source and open-circuiting an independent current source. Both are not permitted.

1.3.3   Circuit

A collection of devices such as resistors and sources in which terminals are connected together by connecting wires is called an *electric circuit*. These wires converge in *nodes*, and the devices are called *branches* of the circuit, as shown in figure 1.5.

The general circuit problem is to find all currents and voltages in the branches of the circuit when the intensities of the sources are known. Such a problem is usually referred to as *circuit analysis*.

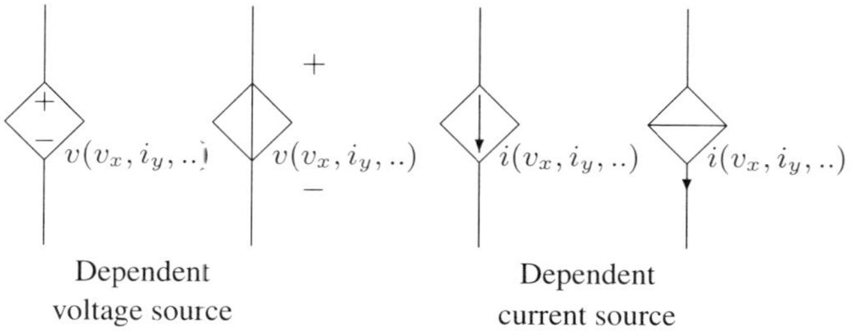

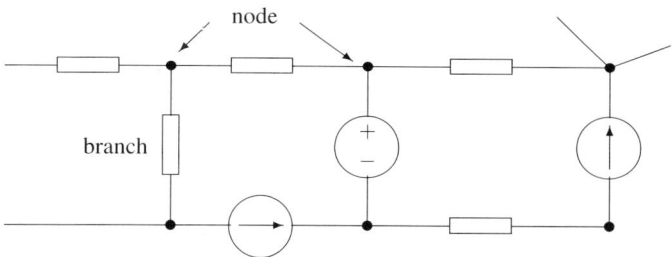

Dependent
voltage source

Dependent
current source

Figure 1.4: Symbols for dependent sources. Variables in brackets are the controlling variables whose values affect the value of the source.

Figure 1.5: Part of an electric circuit

Remarks — While the current in a resistor has a fixed relationship with the voltage across it, the current flowing in a voltage source, or the voltage across a current source, is theoretically unrestricted and can assume whatever value governed by the external circuit.

1.4 Kirchhoff's Laws

The voltage across each element and the current through each element in an electric circuit are governed by two general results which are summarized in Kirchhoff's two laws. Since Kirchhoff's laws are derived from general physical properties of electricity, they are applicable to all kinds of electric circuits.

1.4.1 Kirchhoff's Current Law (KCL)

As a direct consequence of the conservation of charge, namely charge can neither be created nor destroyed, the node, being of negligible physical size, holds no charge. For instance, referring to figure 1.6, the sum of I_1, I_2 and I_3 must equal zero.

Formally, KCL states that *the algebraic sum of the currents in all the branches that converge in a common node is equal to zero.* In mathematical form, for n branches converging into a node, KCL states that

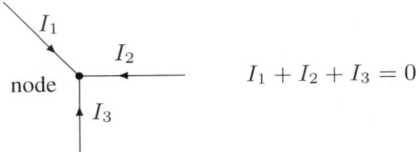

$$I_1 + I_2 + I_3 = 0$$

Figure 1.6: Kirchhoff's current law

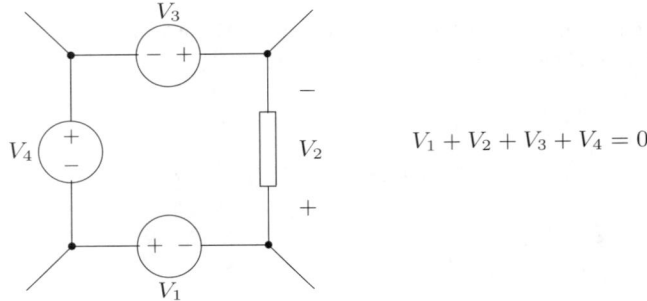

$$V_1 + V_2 + V_3 + V_4 = 0$$

Figure 1.7: Kirchhoff's voltage law

$$I_1 + I_2 + \cdots + I_n = 0$$

where I_k is the current flowing in the kth branch and its direction is assumed to be pointing towards the node.

1.4.2 Kirchhoff's Voltage Law (KVL)

When a charged particle is moved from a point to another, the work done is $V.\delta q$, where V is the voltage across the two points and δq is the amount of charge on the particle. Consider a particular case where the two points are actually the same point in the circuit. In this case, the work done is zero. By the same argument, if a unit charge is moved around a closed path such as the square path shown in figure 1.7, the work done is zero, i.e.,

$$V_1 + V_2 + V_3 + V_4 = 0$$

Formally, KVL states that *the algebraic sum of the voltages between successive nodes in a closed path in a circuit is equal to zero.* In mathematical form, for a closed path with successive nodes $1, 2, \ldots, n$, KVL states that

$$V_{1,2} + V_{2,3} + \cdots + V_{n-1,n} + V_{n,1} = 0$$

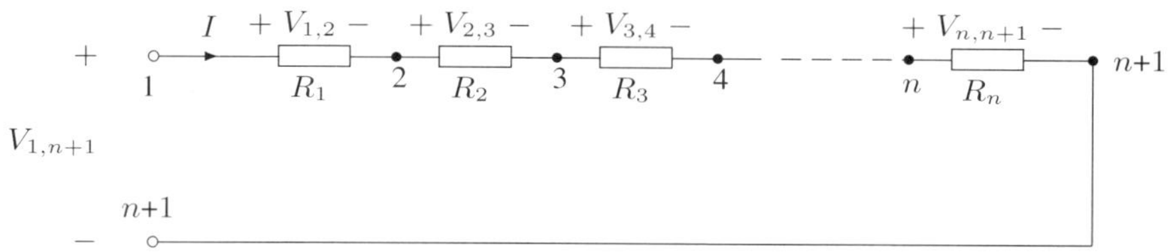

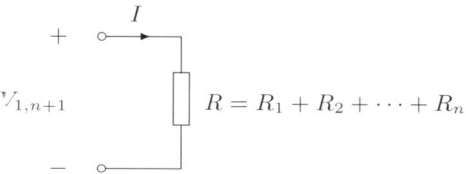

Figure 1.8: Series connection

$$R = R_1 + R_2 + \cdots + R_n$$

Figure 1.9: Equivalent resistance for n resistors in series

where $V_{j,k}$ is the voltage between nodes j and k.

Remarks — Any circuit that has a solution must satisfy Kirchhoff's laws. From the properties of independent sources, we can immediately conclude that a circuit cannot be solved if there exists a loop that is formed exclusively of independent voltage sources. Thus, short-circuiting an independent voltage source, as remarked earlier, is a particular case where KVL is violated. Similarly, a circuit cannot be solved if there exists a node to which only independent current sources are connected. Also, open-circuiting an independent current source is a particular case where KCL is violated.

1.5 Series and Parallel Circuits

1.5.1 Series Circuit

When devices are "chained" up such that each of the nodes is incident to just two devices, the resulting circuit is said to be a series circuit.

A series circuit consisting of n resistors is shown in figure 1.8. Clearly, KVL gives $V_{1,n+1} = V_{12} + V_{23} + \cdots + V_{n,n+1}$. Also, from Ohm's Law, $V_{1,n+1} = (R_1 + R_2 + \cdots + R_n)I$. Therefore, the equivalent circuit seen from nodes 1 and $n+1$ is a resistor of resistance $R_1 + R_2 + \cdots + R_n$, as shown in figure 1.9.

1.5.2 Parallel Circuit

When devices are connected such that one terminal of each device is connected to a node of the circuit while the other terminals of the elements are connected to another node of the circuit, the resulting circuit is said to be a parallel circuit.

A parallel circuit consisting of n resistors is shown in figure 1.10. Note that we use conductance instead of resistance in this case. Clearly, KCL gives $I = I_1 + I_2 + \cdots +$

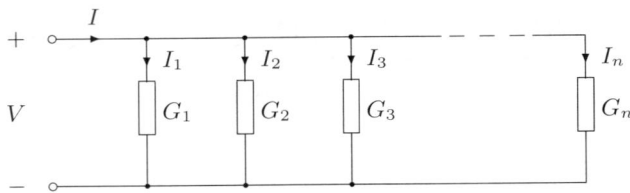

Figure 1.10: Parallel connection

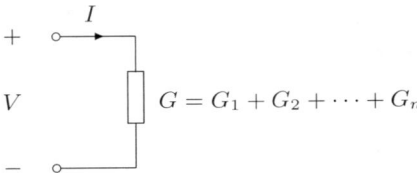

Figure 1.11: Equivalent conductance for n conductors in parallel

I_n. Since all voltages across the resistors are equal to V, we have $I = (G_1 + G_2 + \cdots + G_n)V$, and the equivalent conductance as seen from the left end is $G_1 + G_2 + \cdots + G_n$, as shown in figure 1.11. Note that the expression $G = G_1 + G_2 + \cdots + G_n$ can be written as

$$R = \frac{R_1 R_2 R_3 \cdots R_n}{R_2 R_3 \cdots R_n + R_1 R_3 R_4 \cdots R_n + \cdots + R_1 R_2 \cdots R_{n-1}}$$

For example, if there are only two resistors, the equivalent resistance of the parallel circuit is

$$R = \frac{R_1 R_2}{R_1 + R_2}$$

Remarks — Although the choice between using resistance and conductance in analysis is arbitrary, it is preferable to perform calculations in terms of resistance for the case of series connection, but in terms of conductance in the case of parallel connection. As seen from the above derivation, algebraic brevity is an obvious advantage of making this choice.

Example 1.1: Illustration of series/parallel reduction — It is possible to reduce an assembly of resistors whose configuration is based on series and parallel connections. Referring to the circuit of figure 1.12, we can use an equivalent resistance R_{eq} to replace the circuit such that the input current and voltage are unaffected.

First of all, we observe that R_4 and R_5 are in parallel and can be replaced by an equivalent resistance R' which is given by

$$R' = R_4 \| R_5 = \frac{R_4 R_5}{R_4 + R_5}$$

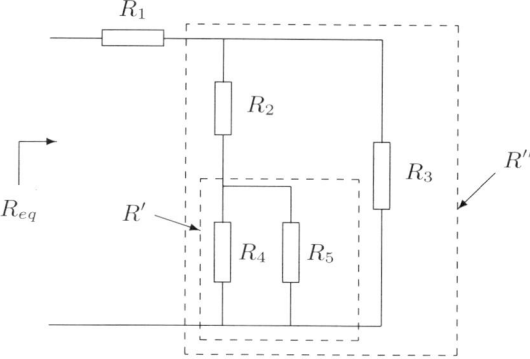

Figure 1.12: Series/parallel reduction process

This R' is connected in series with R_2, and the resulting sub-circuit is connected in parallel with R_3. Thus, using the series formula, followed by the parallel formula, we get an equivalent resistance R'' which represents the part of the circuit covering R_2, R_3, R_4 and R_5, i.e.,

$$R'' = R_3 \| (R_2 + R')$$

Finally, adding R_1 to R'' yields the equivalent resistance R_{eq} as required.

$$R_{eq} = R_1 + \frac{R_3 \left(R_2 + \dfrac{R_4 R_5}{R_4 + R_5} \right)}{R_3 + \left(R_2 + \dfrac{R_4 R_5}{R_4 + R_5} \right)}$$

1.6 Ladder Circuit

The ladder circuit represents a commonly used circuit style that is configured purely on the basis of series and parallel connections. We can derive the equivalent resistance of the ladder circuit by successive applications of the series and parallel reduction formulae introduced in the previous section.

Example 1.2: Simple ladder circuit — Suppose we wish to calculate the equivalent resistance as observed from terminals 0 and 1 for the ladder circuit shown in figure 1.13. First, we observe that R_{12} is in series with the rest of circuit to the right of node 2 including G_{20}. Using the series formula, we can write

$$\frac{V}{I} = R_{12} + \frac{1}{G_2}$$

where G_2 is the equivalent conductance of the sub-circuit beginning at node 2 and including G_{20}. Thus, the ladder circuit can be reduced to the one shown in figure 1.14 (a). Next, we consider G_2 being a parallel connection of G_{20} and another sub-circuit

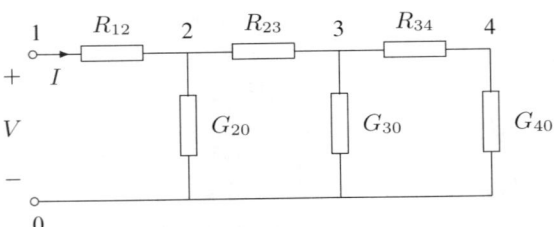

Figure 1.13: Ladder circuit

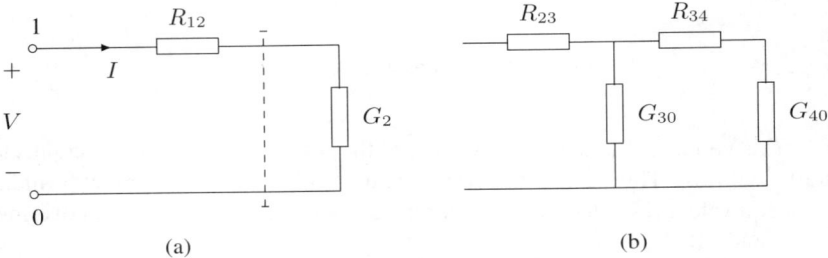

(a)   (b)

Figure 1.14: (a) Equivalent circuit of the ladder circuit; (b) equivalent circuit of R_2'

beginning at node 2 and excluding G_{20}, i.e.,

$$G_2 = G_{20} + \frac{1}{R_2'}$$

where R_2' is the sub-circuit shown in figure 1.14 (b). Thus, we have

$$\frac{V}{I} = R_{12} + \cfrac{1}{G_{20} + \cfrac{1}{R_2'}}$$

Continuing with this process, we get

$$\frac{V}{I} = R_{12} + \cfrac{1}{G_{20} + \cfrac{1}{R_{23} + \cfrac{1}{G_{30} + \cfrac{1}{R_{34} + \cfrac{1}{G_{40}}}}}}$$

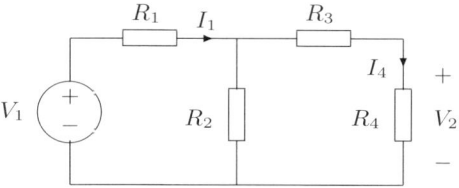

Figure 1.15: A circuit solvable by series and parallel concepts

1.7 Voltage and Current Division Formulae

We have seen how Kirchhoff's laws can be used to generate handy formulae for the equivalent resistance of the simple series and parallel circuits. Sometimes, we may want to know the voltage across a particular resistor in a series circuit, or the current in a particular resistor in a parallel circuit.

In the case of the series circuit shown in figure 1.8, the voltage across resistor R_i is given by Ohm's law as $R_i I$. where I is the current flowing in each resistor. Thus, we can write the *voltage division formula* as

$$V_{i,i+1} = R_i I = \frac{R_i V}{R_1 + R_2 + \cdots + R_n}$$

Likewise, in the case of the parallel circuit shown in figure 1.10, the current in conductor G_i is given by Ohm's law as $G_i V$, where V is the voltage across each conductor. Thus, we can write the *current division formula* as

$$I_i = G_i V = \frac{G_i I}{G_1 + G_2 + \cdots + G_n}$$

Example 1.3: Circuits containing series and parallel connections — Consider the circuit of figure 1.15. Assume that V_1 and all values of resistors are known. The circuit can be solved on the basis of series and parallel connections. In particular we can find V_2 as follows.

The total resistance seen from the source is

$$R = \frac{V_1}{I_1} = R_1 + \cfrac{1}{\cfrac{1}{R_2} + \cfrac{1}{R_3 + R_4}} = R_1 + \frac{R_2(R_3 + R_4)}{R_2 + R_3 + R_4}$$

Therefore,

$$I_1 = \frac{(R_2 + R_3 + R_4)V_1}{(R_3 + R_4)(R_1 + R_2) + R_1 R_2}$$

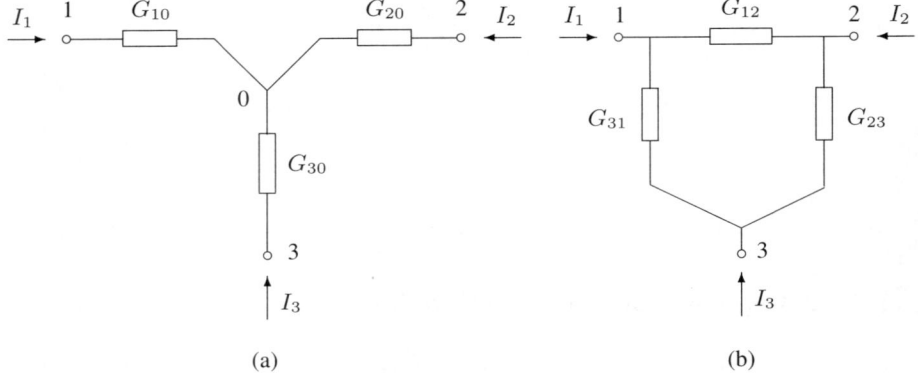

Figure 1.16: Y to Δ transformation

Using the current division formula, we can find

$$
\begin{aligned}
I_4 &= I_1 \times \frac{\left(\dfrac{1}{R_3 + R_4}\right)}{\left(\dfrac{1}{R_3 + R_4}\right) + \dfrac{1}{R_2}} \\
&= \frac{R_2 I_1}{R_2 + R_3 + R_4}
\end{aligned}
$$

Finally, since $V_2 = I_4 R_4$, we have

$$
V_2 = \frac{R_2 R_4 V_1}{(R_1 + R_2)(R_3 + R_4) + R_1 R_2}
$$

1.8   Equivalence of "Star" (Y) and "Delta" (Δ) Circuits

Consider the circuits shown in figure 1.16. The characteristic connection styles of these circuits give them the names Y-circuit and Δ-circuit. The interchange between a Y-circuit and a Δ-circuit is frequently made use of in the analysis of circuits of various types, especially for some circuits which are not reducible by the series/parallel reduction process. For example, the so-called bridge circuit, as shown in figure 1.17, is virtually unsolvable using the series/parallel concept. The star-delta interchange will help solve this problem.

1.8.1   Transformation from Y-Circuit to Δ-Circuit

In the following we will show that a Y-circuit can be replaced by a Δ-circuit without affecting the external voltages and currents. Specifically, given a Y-circuit with three

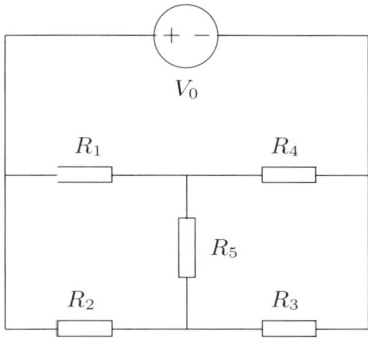

Figure 1.17: Circuit that cannot be solved using the series/parallel concept

conductances G_{10}, G_{20} and G_{30}, converging to a single node 0, as shown in figure 1.16 (a), the problem is to find the values of the conductances in a Δ-circuit such that the two circuits are equivalent. By 'equivalent' we mean that the two circuits can be interchanged without affecting the voltage across any two of the terminals and the current flowing into any one of the terminals. First, consider the Y-circuit.

$$
\begin{aligned}
I_1 &= G_{10}(V_1 - V_0) \\
I_2 &= G_{20}(V_2 - V_0) \\
I_3 &= G_{30}(V_3 - V_0)
\end{aligned}
$$

where V_1, V_2, V_3 and V_0 are voltages of nodes 1, 2, 3 and 0 with reference to a datum. At node 0, $I_1 + I_2 + I_3 = 0$, i.e.,

$$
G_{10}V_1 + G_{20}V_2 + G_{30}V_3 - G_{10}V_0 - G_{20}V_0 - G_{30}V_0 = 0
$$
$$
\Rightarrow \quad V_0 = \frac{G_{10}V_1 - G_{20}V_2 + G_{30}V_3}{G_{10} + G_{20} + G_{30}}
$$

Hence,

$$
\begin{aligned}
I_1 &= \frac{(G_{20} + G_{30})V_1 - G_{20}V_2 - G_{30}V_3}{G_{10} + G_{20} + G_{30}V_3} \times G_{10} \\
I_2 &= \frac{-G_{10}V_1 + (G_{30} + G_{10})V_2 - G_{30}V_3}{G_{10} + G_{20} + G_{30}} \times G_{20} \\
I_3 &= \frac{-G_{10}V_1 - G_{20}V_2 + (G_{10} + G_{20})V_3}{G_{10} + G_{20} + G_{30}} \times G_{30}
\end{aligned}
$$

Now consider the Δ-circuit of figure 1.16 (b).

$$
\begin{aligned}
I_1 &= (G_{12} + G_{31})V_1 - G_{12}V_2 - G_{31}V_3 \\
I_2 &= -G_{12}V_1 + (G_{12} + G_{23})V_2 - G_{23}V_3 \\
I_3 &= -G_{31}V_1 - G_{23}V_2 + (G_{31} + G_{23})V_3
\end{aligned}
$$

By comparing the two sets of equations from the Y-circuit and Δ-circuit, we get

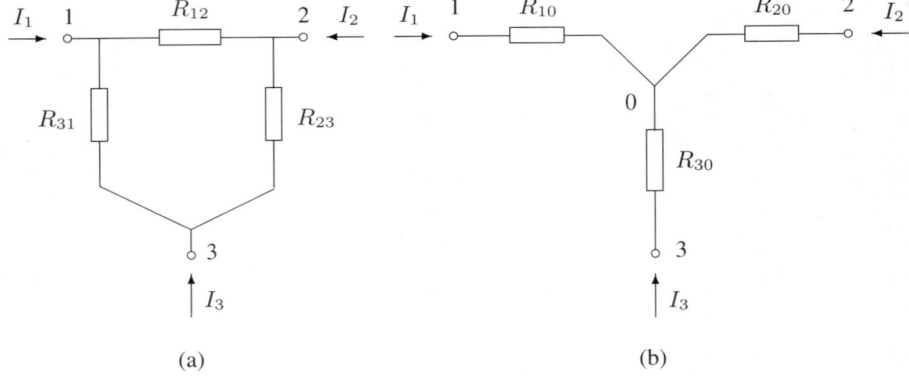

Figure 1.18: Δ to Y transformation

$$
\begin{aligned}
G_{12} &= \frac{G_{10}G_{20}}{G_{10} + G_{20} + G_{30}} \\
G_{23} &= \frac{G_{20}G_{30}}{G_{10} + G_{20} + G_{30}} \\
G_{31} &= \frac{G_{10}G_{30}}{G_{10} + G_{20} + G_{30}}
\end{aligned}
$$

1.8.2 Transformation from Δ-Circuit to Y-Circuit

We may also transform a Δ-circuit into a Y-circuit using a similar procedure. In doing so, however, algebraic brevity can be maintained if the derivation is performed in terms of resistance. Specifically, given a Δ-circuit with resistances R_{12}, R_{23} and R_{31}, as shown in figure 1.18 (a), the problem is to find the values of the resistances in a Y-circuit such that the two circuits are equivalent.

The result is surprisingly similar to that of the Y-to-Δ transformation. Specifically, we observe that the forms of expressions are identical to those of Y-to-Δ transformation, except that resistance is used here in each conversion formula.

$$
\begin{aligned}
R_{10} &= \frac{R_{12}R_{31}}{R_{23} + R_{31} + R_{12}} \\
R_{20} &= \frac{R_{23}R_{12}}{R_{23} + R_{31} + R_{12}} \\
R_{30} &= \frac{R_{31}R_{23}}{R_{23} + R_{31} + R_{12}}
\end{aligned}
$$

Using the above formulae, we can transform any Δ-circuit into a Y-circuit.

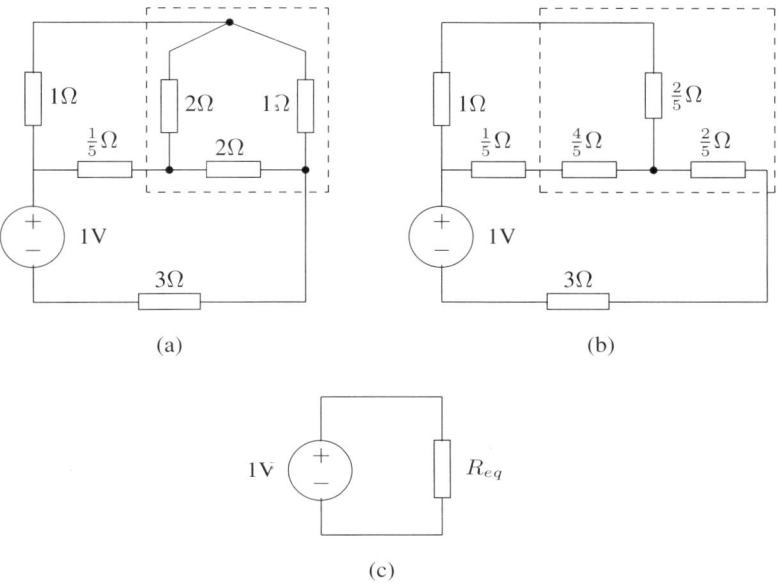

Figure 1.19: Example of Δ-to-Y conversion

Example 1.4: Delta-star conversion — We may apply delta-star converison to solve the particular bridge circuit shown in figure 1.19 (a). Suppose we wish to find the power supplied by the voltage source. The part enclosed in the dashed box is identified as a Δ-circuit. After transforming this Δ-circuit to a Y-circuit, the usual series/parallel reduction technique can be applied. Specifically, the transformed circuit is shown in figure 1.19 (b), from which we can easily write the equivalent resistance R_{eq} connecting the voltage source, as shown in figure 1.19 (c), as

$$R_{eq} = \frac{\left(1 + \frac{2}{5}\right)\left(\frac{1}{5} + \frac{4}{5}\right)}{\left(1 + \frac{2}{5}\right) + \left(\frac{1}{5} + \frac{4}{5}\right)} + \frac{2}{5} + 3 = 3.9833\Omega$$

Hence, the power supplied by the voltage source is $1^2/3.9833 = 0.251$W.

1.9 Computer-Aided Analysis

The use of computer-aided analysis programmes for the study of electric circuits has become increasingly popular. There are many reasons for using computer-aided analysis programmes for analyzing circuits. First, the high computational speed of digital computers offers great saving in time, especially when large circuits are analyzed. Besides, while hand calculation can easily commit errors, the computer always produces accurate answers provided accurate and viable information about a circuit is specified. We should stress, however, that computers do not "think" although they are powerful and efficient workhorses for performing tedious and time-consuming calculations. Without understanding the circuit to be analyzed, we will never use any computer-aided analysis programme effectively and correctly.

In this book we introduce PSPICE which, at the time of writing, is the most widely used programme for analyzing general electric and electronic circuits. In this first chapter we introduce the basic usage of PSPICE, and throughout the book we gradually introduce other features which are relevant to the topics being discussed. The PSPICE programme is part of *The Design Center CAE* (Computer-Aided Engineering) System from MicroSim Corporation.[1] PSPICE Version 7.0 provides an interactive environment for performing circuit analysis. Since PSPICE runs on the PC under the standard Windows environment, it is extremely easy to use.

Usually, we start with drawing the circuit which is to be analyzed. The software component that draws circuits is called *Schematics*. Upon invoking *Schematics*, a window pops up on the screen with the usual menu bar located near the top of the window frame. Under this environment, other software components can be invoked via the pull-down menu. In general, the following processes are involved in analyzing a circuit:

1. Draw the circuit using the interactive drawing utilities and save the schematic diagram in a file.

2. Select the type of analysis to be performed.

3. Invoke the simulation.

4. Examine the results of analysis or display the simulated waveforms.

The flow chart shown in figure 1.20 illustrates the general procedure through which a circuit is analyzed using PSPICE.

Example 1.5: Illustration of circuit analysis using PSPICE — Suppose we wish to use PSPICE to analyze a circuit consisting of a series connection of a 10V voltage source, a 1kΩ resistor and a 9kΩ resistor. The first step is to invoke *Schematics*. All subsequent steps can be invoked from the pull-down menu in the *Schematics* window. The following is a summary of actions used to create and analyze the circuit in PSPICE.

1. We tell PSPICE what circuit to analyze by drawing the circuit and specifying all component values. The drawing procedure is so straightforward that the user manual is hardly needed. From the **Draw** menu, we choose *"Get New Part"* to open up the *"Add Part"* dialogue box in which we can scroll through the *Library* list. The names of the component libraries are self-explanatory. For instance, when we find the library called "source.slb", we can click "OK" and choose VSRC, followed by the standard mouse action of "dragging" the voltage source to any desired location. In a similar fashion, we can add two resistors to the schematic. For the case of resistors, the library is "analog.slb". We may also rotate a component via the **Edit** menu. To complete the circuit, we have to

---

[1] An evaluation CD-ROM or disk set for PSPICE can be obtained from MicroSim Corporation, 20 Fairbanks, Irvine CA 92718, U.S.A.

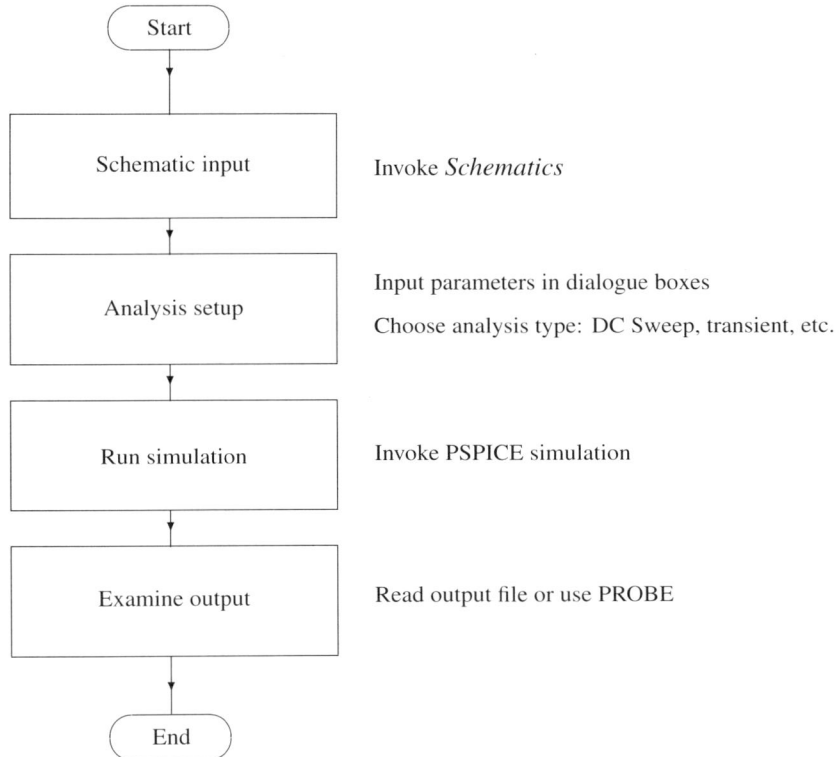

Figure 1.20: Flow chart of using PSPICE

connect the three components together. This can be done by using *"Wire"* from the **Edit** menu, which appears as a pencil cursor. Dragging this pencil cursor to one point of the schematic and clicking the mouse will anchor the wire to that point. Dragging and clicking again at another point will join the two points together. Thus, we can "wire" up the voltage source and the two resistors in series.

2. An arbitrary point in the circuit must be assigned as the ground which will be taken by PSPICE as reference zero voltage during analysis. The process of ground assignment is the same as adding a component to the circuit. This special ground component is a one-terminal component called AGND from the library "port.slb".

3. The next step is to assign values to the components. Double-clicking on the component value will bring up a dialogue box in which we can type in the new value of the component. In the case of the voltage source, we can also type in such specifications as DC value, AC value, etc. We may likewise change the component labels. It is also possible to label a node with any desired name, by double-clicking the node to be labelled. For example, we label the two nodes in

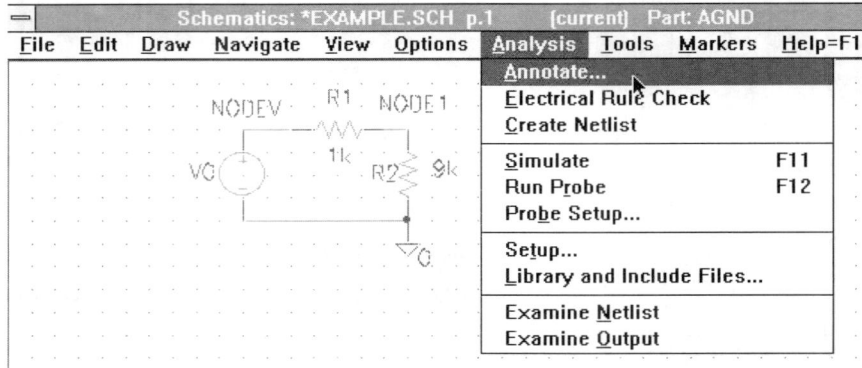

Figure 1.21: Drawing a circuit in *Schematics* for PSPICE simulation

this circuit as NODE1 and NODEV. Figure 1.21 shows the complete schematic
diagram that appears in the *Schematics* window.

4. When the circuit is drawn and values specified, we should store the circuit file
   on disk. From the **File** menu, we choose *"Save As"* to bring up a dialogue box
   in which we can type in the desired file name.

5. We are now ready to analyze the circuit. From the **Analysis** menu, we choose
   "Simulate" to start the analysis. The programme will then perform a series
   of actions, including the creation of some internal files and checking of errors
   in the schematic diagram. If no errors are found, PSPICE will perform the
   analysis. In this case, PSPICE will perform the "bias point calculation".

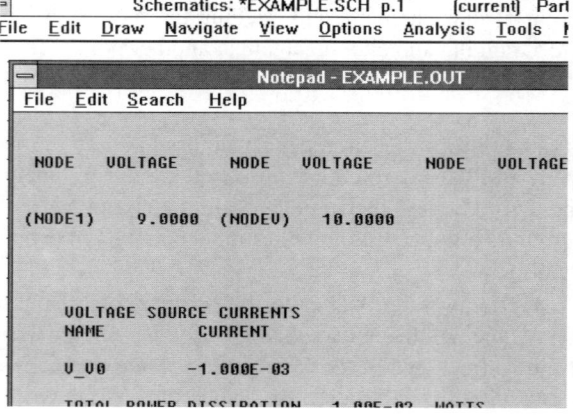

Figure 1.22: Examining an output file from a PSPICE simulation

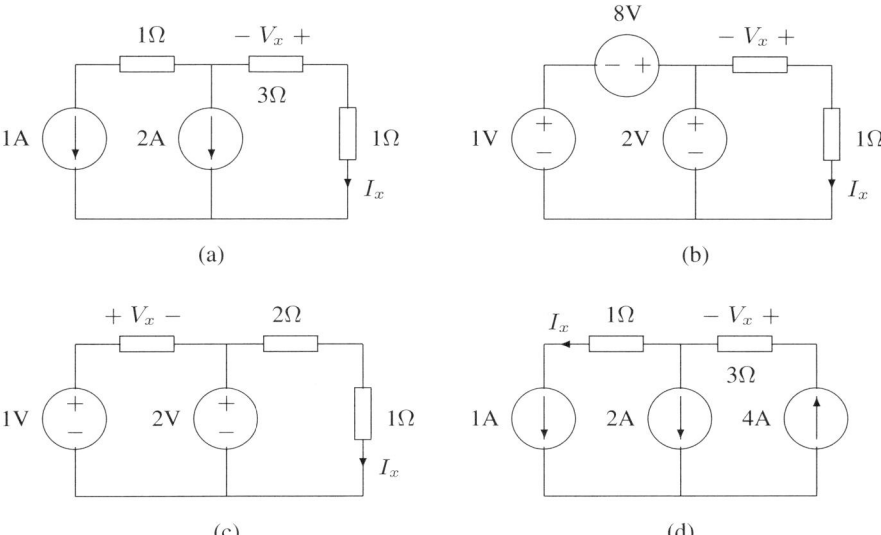

(a)

(b)

(c)

(d)

Figure 1.23: Circuits for Problem 2

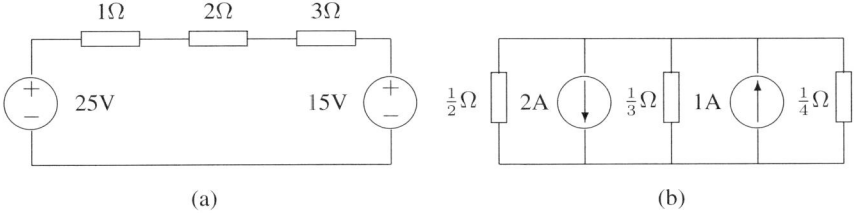

(a)

(b)

Figure 1.24: Circuits for Problem 3

6. Upon completion, PSPICE beeps and saves results in an output file and a data file. We may choose *"Examine Output"* from the **Analysis** menu to read the output file. Figure 1.22 shows the output file generated by PSPICE. Here, we can check the voltages at NODE1 and NODEV. When results have to be inspected graphically, e.g., a transient waveform, we may choose PROBE to display the results. We postpone this feature to Chapters 2 and 4.

Since PSPICE is designed to be user-friendly software, we can easily get acquainted with its usage by simply 'trying it out'. It may be used, for instance, to verify numerical answers to the end-of-chapter problems. Often, results from PSPICE are surprisingly informative and stimulating, especially when something turns out to be contradictory to what we believe it is.

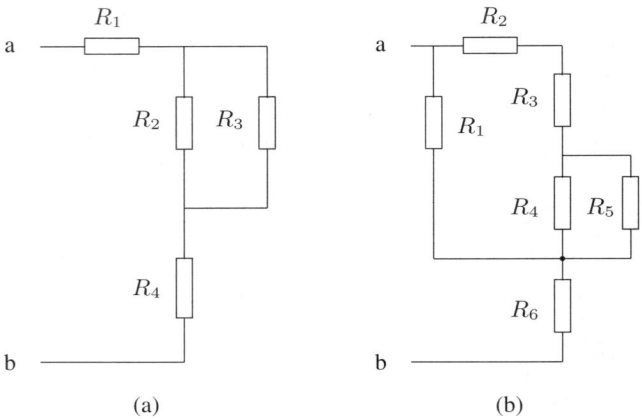

Figure 1.25: Circuits for Problems 4 and 5

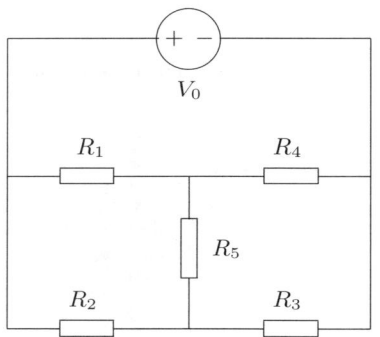

Figure 1.26: Circuit for Problem 6

1.10   Problems

1. A voltage of 3V is applied across a wire of length 10m, cross-sectional area 2mm^2, and resistivity $1.72 \times 10^{-8} \Omega$m. Calculate the resistance of the wire and the current flowing through it.

2. If a circuit violates Kirchhoff's laws, it is not solvable. State which one(s) of the circuits in figure 1.23 is/are not solvable. For each of the solvable circuits, find the unknown current I_x and the unknown voltage V_x.

3. Using the voltage and current division formulae, calculate the voltage across each of the resistors in figure 1.24 (a), and the current across each of the resistors in figure 1.24 (b).

4. For each of the circuits shown in figure 1.25, find the equivalent resistance as seen from the input terminals a and b.

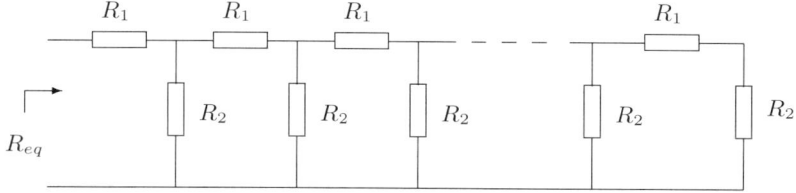

Figure 1.27: Circuit for Problem 7

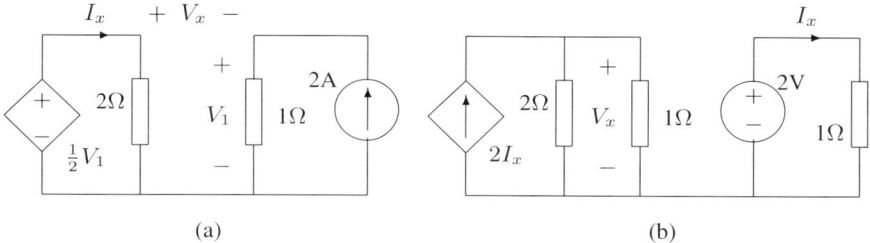

(a)   (b)

Figure 1.28: Circuits for Problem 8

5. For each of the circuits of figure 1.25, find the voltage across R_1 and the current in R_2, assuming that all resistors are 1Ω and the voltage applied across terminals ab is 10V.

6. Using star-delta conversions, find the current in R_5 for the bridge circuit of figure 1.26 in terms of the voltage V_0 and the resistances.

7. Calculate the equivalent resistance of the infinite ladder circuit shown in figure 1.27. Consider two cases separately. (i) $R_1 = R$ and $R_2 = 2R$; (ii) $R_1 = R_2 = R$.

8. For each of the circuits of figure 1.28, calculate the unknown current I_x and voltage V_x.

9. Use PSPICE to verify your answer to Problem 6, assuming $R_1 = R_4 = 10\Omega$,

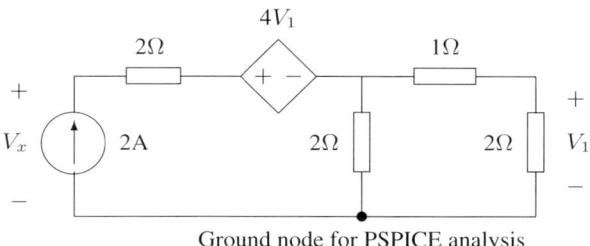

Ground node for PSPICE analysis

Figure 1.29: Circuit for Problem 10

$R_2 = 10\Omega$ and $R_3 = 8\Omega$. Also, verify that the current in R_5 is zero if $R_2 R_4 = R_1 R_3$.

10. PSPICE can be used to analyze circuits containing dependent sources. Explore this feature of PSPICE by analyzing the circuit of figure 1.29. Find in particular the voltage V_x. (Hint: When drawing the circuit in *Schematics*, get the part called voltage-controlled voltage source (E) from the library "analog.slb". This E part can serve as a voltage source whose value is a multiple of another voltage in the circuit. Enter an appropriate multiplier in the dialogue box after double-clicking the E part.)

Chapter 2

Circuit Theorems

Based on Kirchhoff's two laws and some weak assumptions, many useful results can be derived, which drastically simplify the analysis of electric circuits. In this chapter a number of remarkable results, which are now well-known circuit theorems, are discussed, namely Tellegen's theorem, reciprocity theorem, Thévenin theorem, Norton theorem, and the maximum power transfer theorem.

2.1 Tellegen's Theorem

One of the most remarkable results that make use of Kirchhoff's laws is Tellegen's theorem. Since only Kirchhoff's laws are used in the derivation, the result is applicable to all kinds of circuits. Of such great generality, Tellegen's theorem has countless applications.

Consider any linear circuit. Let V_k be the voltage at node k with respect to ground. The voltage across the branch connecting nodes k and l is given by the following KVL equation:

$$V_{kl} = V_k - V_l$$

Suppose the current flowing in this branch is I_{kl}. Multiplying the above KVL equation by I_{kl} gives

$$V_{kl}I_{kl} = (V_k - V_l)I_{kl}$$

Summing over all branches gives

$$\sum_{\text{all branches}} V_{kl}I_{kl} = \sum_{\text{all branches}} (V_k - V_l)I_{kl}.$$

where the symbol $\sum$ denotes summation over all branches.[1] The RHS of the above

---

[1] It is customary in mathematics, for the sake of brevity, to use the symbol $\sum$ to denote a sum of a number of terms. For example, $1 + 2 + 3 + \cdots + n$ can be written as $\sum_{i=1}^{n} i$, and $x_1 + x_2 + \cdots + x_n$ can be written as $\sum_{i=0}^{n} x_i$. In both examples, i is known as the index of summation.

equation can be written as

$$\sum_{\text{all branches}} (V_k - V_l)I_{kl} = \frac{1}{2}\sum_{k=1}^{n}\sum_{l=1}^{n}(V_k - V_l)I_{kl}$$

where the factor 1/2 is due to the fact that the summation on the RHS has taken the product twice, and that $I_{kl} = 0$ if there is no connection (branch) between nodes k and l. Also, since $I_{kl} = -I_{lk}$, we can write

$$(V_k - V_l)I_{kl} = (V_l - V_k)I_{lk}.$$

Hence, we have

$$\sum_{k=1}^{n}\sum_{l=1}^{n}(V_k - V_l)I_{kl} = \sum_{k=1}^{n}\left(V_k\sum_{l=1}^{n}I_{kl}\right) - \sum_{l=1}^{n}\left(V_l\sum_{k=1}^{n}I_{kl}\right)$$

Now, using the KCL equations $\sum_{k=1}^{n}I_{kl} = 0$ and $\sum_{l=1}^{n}I_{kl} = 0$, the RHS of the above equation can be shown to be identically zero. Thus, we have

$$\sum_{k=1}^{n}\sum_{l=1}^{n}V_{kl}I_{kl} = 0$$

This result is known as Tellegen's theorem. A superficial meaning of this theorem is *conservation of instantaneous power,* which means that the sum of the vi products of the branches is zero.

$$\boxed{\sum_{\text{all branches}} vi = 0}$$

This seems to be a trivial conclusion. In a given self-contained circuit, no power is created or destroyed, and hence at any instant the total power generated or absorbed by the branches must be zero.

If we examine the derivation more closely, however, Tellegen's theorem actually means a lot more than just power conservation. In particular, we note that in the derivation of Tellegen's theorem, Kirchhoff's law equations have been used independently of each other. It is thus not necessary that the set of currents and the set of voltages exist simultaneously in the circuit. In other words, the set of currents and the set of voltages do not need to be taken from the same circuit. They can be taken from two different circuits provided the circuits satisfy the same set of Kirchhoff's law equations, i.e., have exactly the same configuration.

Example 2.1: Illustration of Tellegen's theorem — Figure 2.1 shows two different circuits. These two circuits, however, have the same configuration. We let v_i and i_i be the voltage and current in branch i of Circuit A, and let $\hat{v}_i$ and $\hat{i}_i$ be the voltage and current in branch i of Circuit B. From Tellegen's theorem, the following are all true:

$$\sum v_i i_i = 0$$

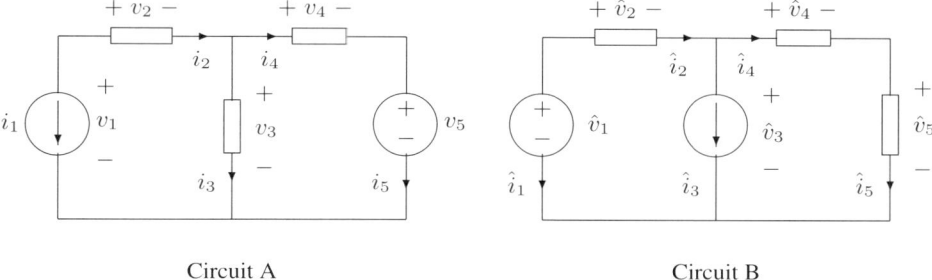

Circuit A   Circuit B

Figure 2.1: Tellegen's theorem

$$\sum v_i \hat{i}_i = 0$$

$$\sum \hat{v}_i i_i = 0$$

$$\sum \hat{v}_i \hat{i}_i = 0$$

$$\sum (v_i - \hat{v}_i)(i_i - \hat{i}_i) = 0$$

$$\sum (v_i + \hat{v}_i)(i_i + \hat{i}_i) = 0$$

$$\sum (v_i - \hat{v}_i)(i_i + \hat{i}_i) = 0$$

$$\sum (v_i + \hat{v}_i)(i_i - \hat{i}_i) = 0$$

The last four equations follow from the fact that if two sets of voltages satisfy the same KVL equation, then the set of sums or differences of the two sets of voltages also satisfies the same KVL equation. Similarly, if two sets of currents satisfy the same KCL equation, then the set of sums or differences of the two sets of currents also satisfies the same KCL equation. Thus, we may imagine that a circuit exists with the same configuration as Circuits A and B, and its corresponding branch voltages and currents are equal to $v_i \pm \hat{v}_i$ and $i_i \pm \hat{i}_i$.

2.2 Linearity and Reciprocity

2.2.1 Linear Elements and Dependent Sources

A circuit element is said to be linear if the voltage across its terminal, v, and the current flowing through it, i, are related by a linear function. For example, a *linear resistor* is characterized by

$$v = Ri \quad \text{or} \quad i = Gv$$

where R and G are constant. This has been referred to as the Ohm's law equation in Chapter 1.

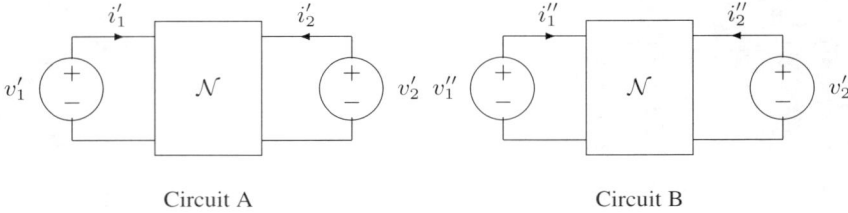

Circuit A   Circuit B

Figure 2.2: General reciprocity

A *linear capacitor* and a *linear inductor* are characterized by, respectively,

$$i = C\frac{dv}{dt} \quad \text{and} \quad v = L\frac{di}{dt}$$

where C and L are constants. If the differentiation with respect to time is treated as a linear operator p, we can write $i = pCv$ for the linear capacitor, and $v = pLi$ for the linear inductor. For all these cases, we have $i = kv$ or $v = k'i$ as the constitutive relation for the linear element concerned.

Furthermore, we define a *linear dependent voltage source* as a voltage source whose value is a linear combination of voltages and currents in other branches of the circuit, i.e.,

$$v = m_1v_1 + m_2v_2 + \cdots + k_1i_1 + k_2i_2 + \cdots$$

where the ms and ks are constants, and, v_n and i_n are voltage and current of branch n. Similarly, a *linear dependent current source* is defined by

$$i = s_1v_1 + s_2v_2 + \cdots + r_1i_1 + r_2i_2 + \cdots$$

where the ss and rs are constants, and, v_n and i_n are voltage and current of branch n. Note that the defining function for a linear dependent source can include as many voltages and currents as desired.

2.2.2   Yet Another Special Application of Tellegen's Theorem

Suppose $\mathcal{N}$ is a circuit containing only linear elements. Thus, for any branch in $\mathcal{N}$, we can write $i_n = k_nv_n$. Now we connect two voltage sources to $\mathcal{N}$ at some arbitrary points, as shown in Circuit A of figure 2.2. In the same way, we connect another set of two voltage sources to $\mathcal{N}$, as shown in Circuit B of figure 2.2.

Now we make two applications of Tellegen's theorem to $\mathcal{N}$. In the first application, we take voltages from Circuit A and currents from Circuit B, giving

$$\sum_{\mathcal{N}} v_n'i_n'' - v_1'i_1'' - v_2'i_2'' = 0$$

$$\Rightarrow \sum_{\mathcal{N}} v_n'k_nv_n'' - v_1'i_1'' - v_2'i_2'' = 0$$

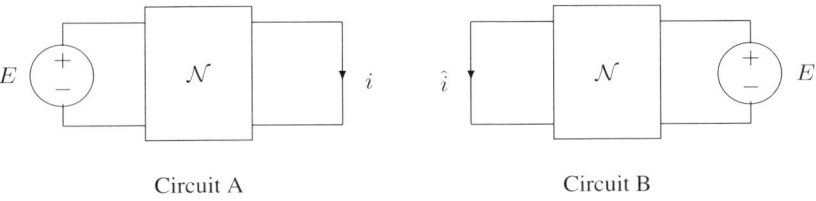

Circuit A   Circuit B

Figure 2.3: Reciprocity theorem

In the second application, we take voltages from Circuit B and currents from Circuit A, giving

$$\sum_{\mathcal{N}} v_n'' i_n' - v_1'' i_1' - v_2'' i_2' = 0$$

$$\Rightarrow \sum_{\mathcal{N}} v_n'' k_n v_n' - v_1'' i_1' - v_2'' i_2' = 0$$

Combining the results from the above two applications of Tellegen's theorem, we obtain

$$\boxed{v_1' i_1'' + v_2' i_2'' = v_1'' i_1' + v_2'' i_2'}$$

We refer to this relation as *general reciprocity*, which is valid for circuits containing exclusively linear elements.[2]

2.2.3 Reciprocity Theorem

Consider the particular way of termination of $\mathcal{N}$ shown in figure 2.3 in which $v_2' = v_1'' = 0$ and $v_1' = v_2'' = E$. Using the general reciprocity relation derived in the previous subsection, we have

$$\hat{i} = i$$

This result is sometimes referred to as the *reciprocity theorem*, although it is clearly a special case of the general reciprocity relation shown earlier. In simple words, this reciprocity theorem states that if we measure the current due to a voltage excitation, then the same current will result when the positions of the current and voltage excitation are interchanged. To illustrate this idea, consider the network $\mathcal{N}$ being excited (driven) by a voltage source E, as shown in figure 2.3. We measure the short-circuit current at the output of Circuit A and note its value i. Then, we remove E from Circuit A and place it to the output of Circuit B. Now, we measure the short-circuit current in the input of Circuit B and note its value $\hat{i}$. The reciprocity theorem tells us that $i = \hat{i}$.

---

[2]It can be shown that the general reciprocity holds for any circuit containing ideal transformers, resistors, capacitors and inductors. We postpone the discussion of ideal transformers to Chapter 5.

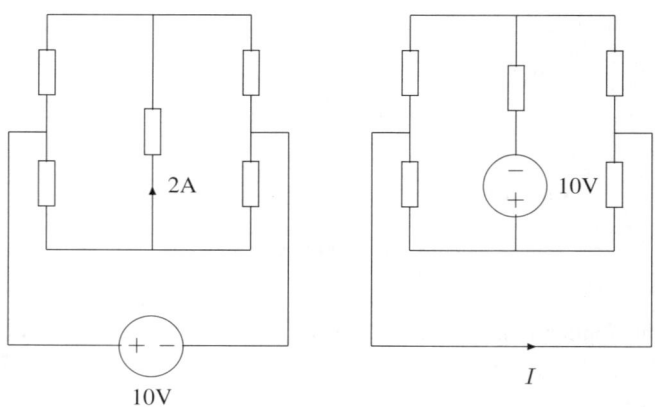

Figure 2.4: Example of application of reciprocity theorem

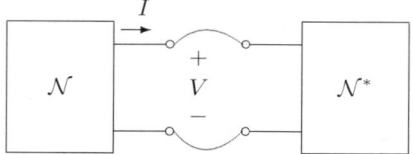

Figure 2.5: Model of connecting circuits

(Note that we may also state another reciprocity theorem in terms of current excitation and voltage measurement. See Problem 3 of this chapter.)

Example 2.2: Application of reciprocity theorem — Suppose we wish to find the current I in the circuit shown on the right of figure 2.4. The situation is exactly the same as in figure 2.3 if we consider the 10V source and the 2A current on the left figure as E and i respectively. From the reciprocity theorem, we have $I = 2A$.

2.3   Thévenin and Norton Theorems

Probably the most frequently used theorems for simplification of circuits are due to Thévenin and Norton. In this section, we will derive the main results and discuss their use. Consider a circuit $\mathcal{N}$ with two terminals connecting to an external apparatus $\mathcal{N}^*$, i.e., another circuit, as shown in figure 2.5. If we analyze this system, we will end up with a set of equations. Moreover, if the circuit contains only linear elements, linear dependent sources and independent sources, a linear set of equations is generated. If we proceed by eliminating all variables except V and I, we will end up with a linear equation of the following form:

$$aV + bI - c = 0$$

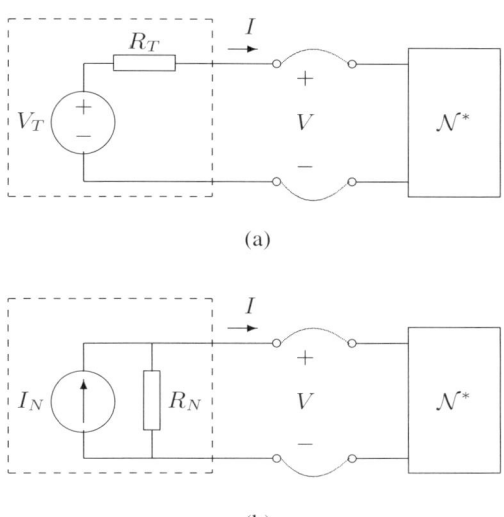

(a)

(b)

Figure 2.6: (a) Thévenin equivalent circuit; (b) Norton equivalent circuit

where a, b and c are independent of V and I. There are two cases to consider.

Case 1 — If a is non-zero, we may express the voltage V in terms of I by dividing the above linear equation by a, i.e.,

$$V = \frac{-b}{a}I + \frac{c}{a} = -R_T I + V_T$$

where b/a and c/a are denoted by R_T and V_T, respectively, in the last equality.

Case 2 — If b is non-zero, likewise, we may express the current I by dividing the above linear equation by b, i.e.,

$$I = \frac{-a}{b}V + \frac{c}{b} = -\frac{V}{R_N} + I_N$$

where b/a and c/b are denoted by R_N and I_N, respectively, in the last equality.

For Case 1, we can actually find a circuit that gives the linear equation $V = -R_T I + V_T$. Obviously, such a circuit is simply a series connection of resistor R_T and voltage source V_T, as represented in figure 2.6 (a). Since this circuit has the same property as the original circuit $\mathcal{N}$, from the point of view of $\mathcal{N}^*$, it is usually referred to as an "equivalent circuit" for circuit $\mathcal{N}$. This result is known as Thévenin's theorem which states that *any two-terminal circuit comprising linear resistors, linear dependent sources and independent sources can be represented by a series combination of a resistance and an independent voltage source.*

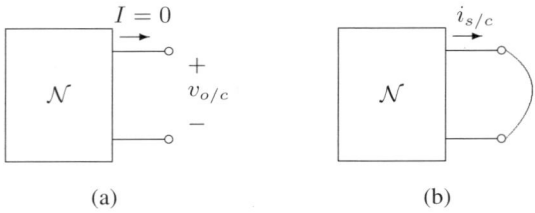

Figure 2.7: Determining Thévenin and Norton circuits. (a) Open-circuit measurement; (b) short-circuit measurement

Likewise, for Case 2, we can find a circuit that satisfies the linear equation $I = -V/R_N + I_N$. Here, an obvious realization for this equation is a parallel connection of resistor R_N and current source I_N, as represented in figure 2.6 (b). Such a circuit is also an equivalent circuit for $\mathcal{N}$. This result is known as Norton's theorem which states that *any two-terminal circuit comprising linear resistors, linear dependent sources and independent sources can be represented by a parallel combination of a resistance and an independent current source.*

Remarks — We will see later that both the Thévenin and Norton theorems extend to circuits with capacitors and inductors as long as the element relations are linear, i.e., circuits for which a linear set of equations can be written. Furthermore, there will be no Thévenin equivalence if $a = 0$ which corresponds to the case where $\mathcal{N}$ is an independent current source. Likewise, there will be no Norton equivalence if $b = 0$ which corresponds to the case where $\mathcal{N}$ is an independent voltage source.

2.4   Determination of Thévenin and Norton Equivalent Circuits

2.4.1   Formal Procedure

Suppose we wish to determine the Thévenin and Norton equivalent circuits for a given circuit $\mathcal{N}$. The problem is to find the values of V_T, R_T, I_N and R_N. Obviously, since $V = -R_T I + V_T$, we can determine V_T very easily by measuring the terminal voltage V with I set to zero. This corresponds to an open-circuit measurement of voltage V, as shown in figure 2.7 (a). Similarly, since $I = -V/R_N + I_N$, we can determine I_N by measuring I with V set to zero. This corresponds to a short-circuit measurement of current I, as shown in figure 2.7 (b). Now, let $v_{o/c}$ and $i_{s/c}$ be the open-circuit voltage and short-circuit current, respectively. We may write

$$
\begin{aligned}
V_T &= v_{o/c} \\
I_N &= i_{s/c}
\end{aligned}
$$

Also, we may re-write the connecting linear equations for the above cases of open-circuit measurement and short-circuit measurement as, respectively,

$$
-\frac{v_{o/c}}{R_N} + I_N = 0 \quad \text{and} \quad -R_T i_{s/c} + V_T = 0
$$

But since $V_T = v_{o/c}$ and $I_N = i_{s/c}$, the above equations give

$$R_N = R_T = \frac{v_{o/c}}{i_{s/c}}$$

In summary, the general procedure for determining the Thévenin and Norton equivalent representations for a given circuit is as follows:

1. Find the open-circuit voltage $v_{o/c}$, as in figure 2.7 (a).
2. Find the short-circuit current $i_{s/c}$, as in figure 2.7 (b).
3. The values of R_T and V_T are given by

$$\boxed{R_T = \frac{v_{o/c}}{i_{s/c}} \quad \text{and} \quad V_T = v_{o/c}}$$

4. The values of R_N and I_N are given by

$$\boxed{R_N = \frac{v_{o/c}}{i_{s/c}} \quad \text{and} \quad I_N = i_{s/c}}$$

Clearly, when the Thévenin equivalent circuit has been found, we can easily work out the Norton equivalent circuit, and vice versa, by using $I_N = V_T/R_T$, $V_T = I_N R_N$ and $R_N = R_T$.

Example 2.3: Finding Thévenin and Norton equivalent circuits — In this example we attempt to determine the Thévenin and Norton equivalent circuits for the circuit of figure 2.8 (a). By inspection, the circuit has an open-circuit voltage given by the voltage division formula as

$$v_{o/c} = 2 \times \frac{1}{1+1} = 1\mathrm{V}$$

and a short-circuit current given by the current division formula as

$$i_{s/c} = \frac{2}{1+0.5} \times \frac{1}{2} = \frac{2}{3}\mathrm{A}$$

Thus, we have

$$V_T = 1\mathrm{V} \quad \text{and} \quad I_N = \frac{2}{3}\mathrm{A}$$

Also, the values of R_T and R_N are given by

$$R_T = R_N = \frac{v_{o/c}}{i_{s/c}} = \frac{1}{2/3} = \frac{3}{2}\Omega$$

The equivalent circuits are shown in figures 2.8 (c) and (d).

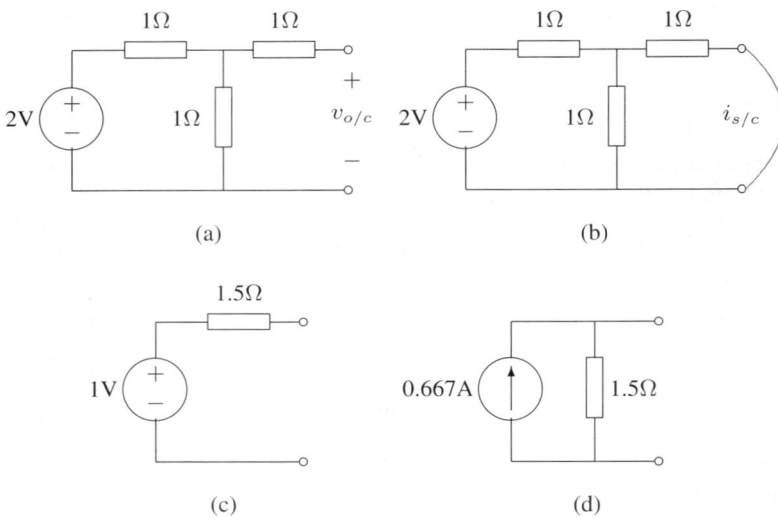

Figure 2.8: Example of finding Thévenin and Norton circuits. (a) Circuit with open-circuit terminals; (b) circuit with short-circuit terminals; (c) Thévenin equivalent circuit; (d) Norton equivalent circuit

2.4.2 Successive Interchange Between Thévenin and Norton Circuits

One of the most important applications of the Thévenin and Norton theorems is in the simplification of circuits. Essentially, when a suitably isolated part of the circuit is replaced by a Thévenin equivalent circuit, one node can be eliminated, and likewise, when a suitably isolated part of the circuit is replaced by a Norton equivalent circuit, one loop can be eliminated. Thus, a circuit containing a number of resistors, current sources and voltage sources may be systematically solved by applying the Thévenin and Norton theorems appropriately and continuously.

Example 2.4: Circuit simplification — The circuit shown in figure 2.9 (a) is to be simplified to a Thévenin equivalent circuit. We will demonstrate here a simple technique based on successive interchange of Thévenin and Norton representations. First of all, we isolate the left side of the circuit and replace it with a Thévenin equivalent circuit, as shown in figure 2.9 (b). Next we combine this Thévenin equivalent circuit with the 2Ω resistor to form another Thévenin equivalent circuit representing a bigger part of the original circuit. Next we convert this Thévenin equivalent circuit into a Norton representation, as in figure 2.9 (c), and combine it with the 1A current source. The result is another Norton equivalent circuit representing a yet bigger part of the original circuit, as shown in figure 2.9 (d). Finally, converting this back to a Thévenin circuit and combining with the 3V at the very right end, the final Thévenin equivalent circuit is obtained, as shown in figure 2.9 (e).

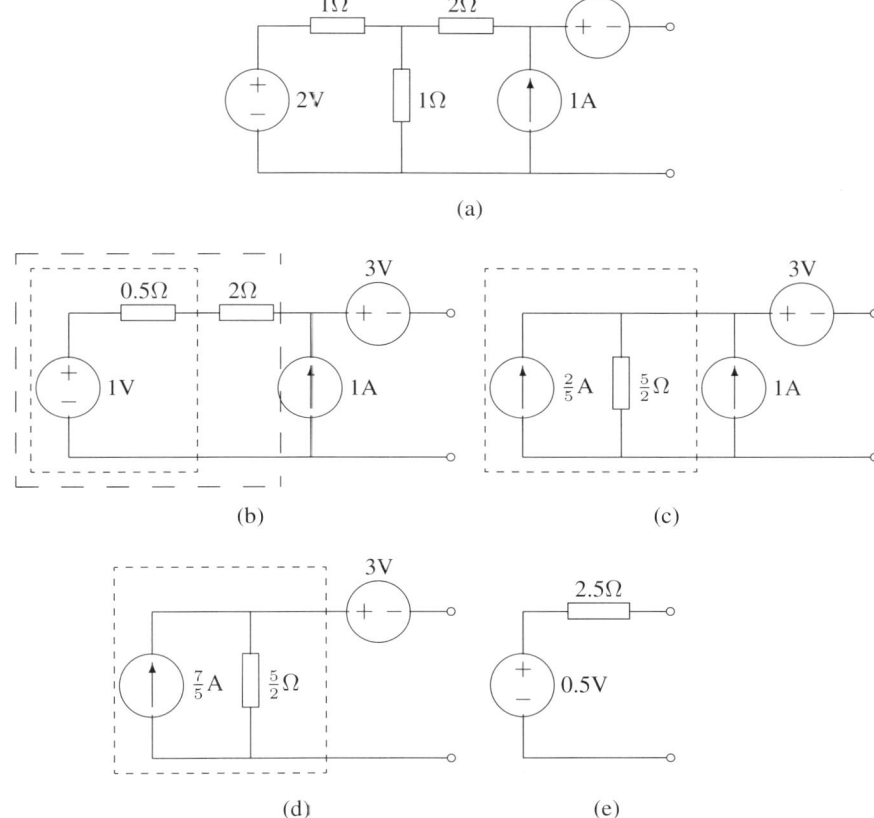

Figure 2.9: Successive interchange of Thévenin and Norton circuits

2.4.3   Equivalent Resistance by Inspection

Sometimes, we are interested only to find the equivalent resistance while the Thévenin equivalent voltage and Norton equivalent current are either readily obtained or less important. In such cases, we may use the following short-cut to obtain the equivalent resistance R_T or R_N.

We first short-circuit all voltage sources and open-circuit all current sources, leaving behind a resistive circuit. Then, R_T and R_N is simply equal to the equivalent resistance seen from the terminals.

Example 2.5: Finding equivalent resistance by inspection — Consider the circuit shown in figure 2.10 (a). After we short out all voltage sources and open up all current sources, we obtain the circuit shown in figure 2.9 (b). The resistance seen from the terminals is equal to 1.5Ω. Thus, $R_T = R_N = 1.5\Omega$.

Moreover, if we wish to find the complete equivalent circuit, we need also to find

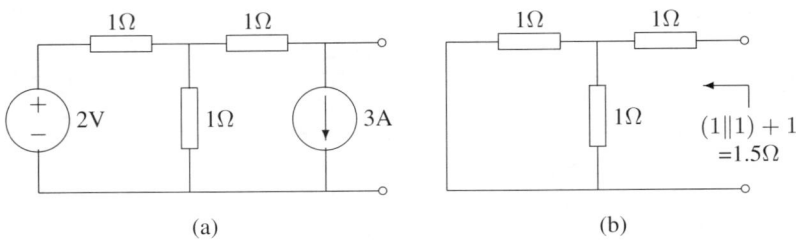

(a)   (b)

Figure 2.10: Finding equivalent resistance by inspection

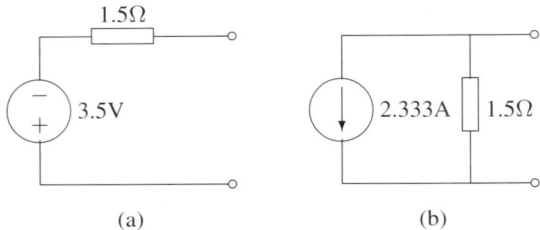

(a)   (b)

Figure 2.11: (a) Thévenin equivalent circuit; (b) Norton equivalent circuit for the circuit of Example 2.5

either $v_{o/c}$ or $i_{s/c}$. In this example, we note that the circuit is almost the same as the one in Example 2.3, with an additional current source of 3A connected right next to the terminals. Hence, the short-circuit current would differ from the current calculated in Example 2.3 by 3A. Taking into account the opposite current direction, we have

$$i_{s/c} = \frac{2}{3} - 3 = -2.333\text{A}$$

This gives $I_N = -2.333$A. We may, if required, calculate the Thévenin voltage using the relation $V_T = R_N I_N$, giving $V_T = -3.5$V. Hence, we may draw the equivalent circuits as in figure 2.11.

Remarks — In the above example, the method of calculating $i_{s/c}$ is somewhat *ad hoc*. But that is all we can do for the time being. In later chapters, we will develop more techniques to find $v_{o/c}$ and $i_{s/c}$ for circuits of arbitrary configurations.

Example 2.6: The bridge circuit — The bridge circuit shown in figure 2.12 (a) could be a difficult circuit to solve, since the series/parallel reduction technique and the voltage/current division principle are of little help. A possible way to get around this problem is to use the delta-star transformation introduced in Chapter 1, so that the usual series/parallel reduction procedure can be applied to the transformed circuit. However, if we wish to find the current through the middle resistor R_5, a much more efficient way is to work out the Thévenin equivalent circuit seen by R_5.

Taking the usual procedure, we first find the open-circuit voltage across A and B

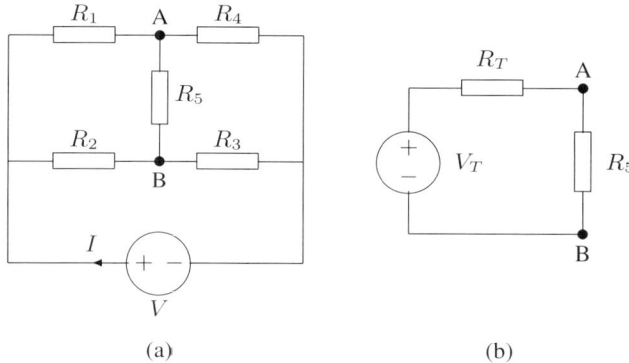

Figure 2.12: (a) Bridge circuit; (b) equivalent circuit for calculating current in R_5

by removing R_5.

$$v_{o/c} = V \times \left(\frac{R_4}{R_1 + R_4} - \frac{R_3}{R_2 + R_3} \right)$$

Next, we find the short-circuit current at AB by shorting out R_5.

$$
\begin{aligned}
i_{s/c} &= \text{(current in } R_1) - \text{(current in } R_4) \\
&= I \times \left(\frac{G_1}{G_1 + G_2} - \frac{G_4}{G_3 + G_4} \right)
\end{aligned}
$$

where I is the current supplied by the voltage source. The equivalent resistance R_T is given by

$$
\begin{aligned}
R_T &= \frac{v_{o/c}}{i_{s/c}} \\
&= \frac{G_1 + G_2 + G_3 + G_4}{(G_1 + G_4)(G_2 + G_3)} \\
&= (R_1 \| R_4) + (R_2 \| R_3)
\end{aligned}
$$

Hence, the equivalent circuit can be drawn as in figure 2.12 (b), from which we can determine the current flowing in R_5 as

$$\text{Current in } R_5 = \frac{V_T}{R_T + R_5}$$

2.5 Maximum Power Transfer Theorem

Consider a circuit represented by its Thévenin equivalent circuit and terminated by a load resistance R_L, as shown in figure 2.13 (a). The current delivered to the load is given by

$$I = \frac{V_T}{R_T + R_L}$$

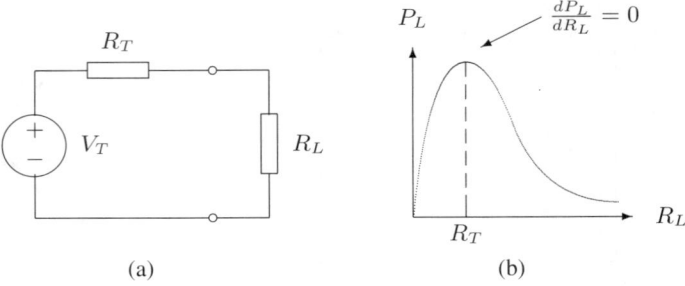

Figure 2.13: (a) Model for studying maximum power transfer; (b) power transfer versus R_L

and the power delivered to the load is

$$\begin{aligned} P_L &= I^2 R_L \\ &= \frac{V_T^2 R_L}{(R_T + R_L)^2} \end{aligned}$$

From the above equation, we see that P_L is a function of R_L, a graphical representation of which is shown in figure 2.13 (b). An interesting question arises here. What is the maximum power delivered to the load if R_L is allowed to vary? To answer this question, we differentiate the expression of P_L with respect to R_L, i.e.,

$$\frac{dP_L}{dR_L} = V_T^2 \frac{R_T - R_L}{(R_T + R_L)^3}$$

Setting $dP_L/dR_L = 0$ gives

$$\boxed{R_L = R_T}$$

This result is known as the *maximum power transfer theorem,* which states that *given a fixed source with a fixed internal resistance R_T, maximum power transfer takes place when R_L is equal to the given R_T.*

   Remarks — It is often misunderstood that maximum power transfer occurs when R_T equals the load resistance. This is obviously wrong because if we are allowed to vary R_T, maximum power transfer should always occur when $R_T = 0$. Indeed, we have to be cautious when attempting to apply the maximum power transfer theorem, and specifically we are dealing with the problem of choosing R_L, not R_T, to transfer maximum power from the source to the load.

   Example 2.7: Maximum power transfer — For the circuit of figure 2.13 (a), we vary R_L from 0Ω to 200Ω and tabulate the power dissipated in R_L. Assume that $V_T = 10$V and $R_T = 100\Omega$.

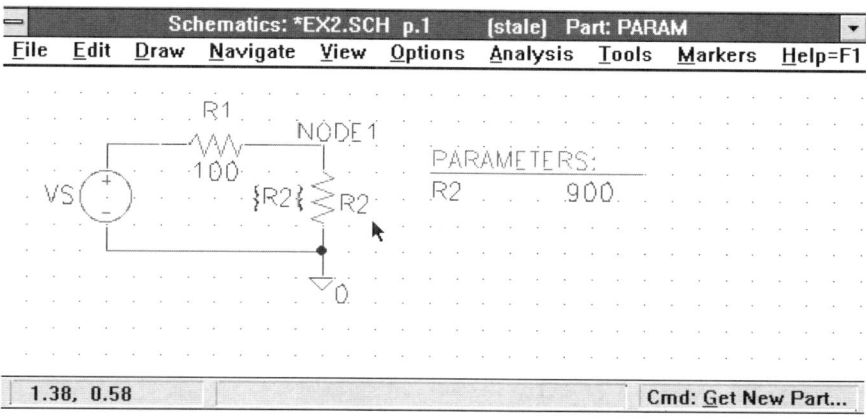

Figure 2.14: Circuit drawn in *Schematics* for PSPICE analysis of maximum power transfer

R_L (Ω)	0	20	40	60	80	100
P_L (W)	0	0.139	0.204	0.234	0.247	0.250

R_L (Ω)	120	140	160	180	200
P_L (W)	0.248	0.243	0.237	0.230	0.222

From the above table we clearly see that the power dissipation in R_L reaches a maximum at $R_L = 100\Omega$. We may also conclude this result from the maximum power transfer theorem.

2.6 Using Sweep Mode and PROBE in PSPICE

PSPICE can be used to determine the value of load resistance that gives maximum power transfer. Essentially we make use of the sweep feature of PSPICE to "sweep" the load resistor through a range of values and tell PSPICE to display the power dissipated in R2 for the range of resistance values swept. We use the software component PROBE to display graphically the power versus the resistance value. The following example illustrates how a component parameter can be chosen as a sweep variable for PSPICE analysis and how the software component PROBE can be used to display variables that are functions of some voltages and currents in the circuit.

Example 2.8: Using sweep mode in PSPICE for determining the maximum power transfer condition — As outlined previously in Chapter 1, the first step is to draw the circuit by invoking *Schematics*. Figure 2.14 shows the *Schematics* window in which the circuit to be analyzed is drawn. Our aim is to find the value of R2 for which maximum power is transferred to it from the voltage source. To perform the required analysis, we need to select the sweep mode, invoke simulation, and display the power curve graphically. The following summarizes the procedure.

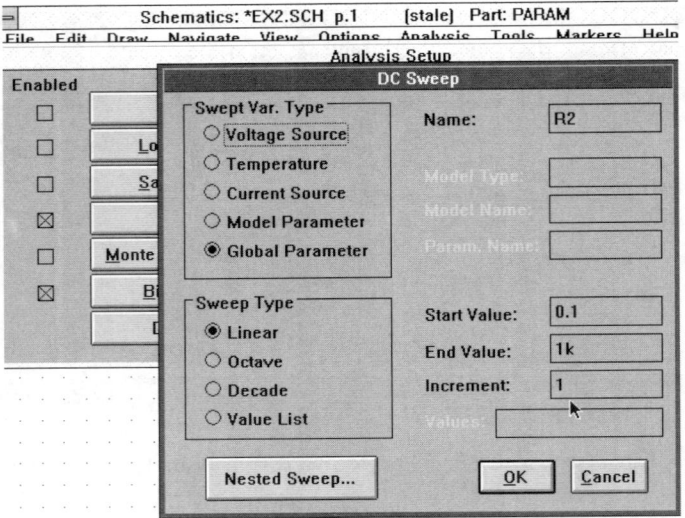

Figure 2.15: Dialogue box for specifying sweeping parameters

1. To select sweep mode for resistor R2, we double-click on its value to bring up a *"Set Attribute Value"* dialogue box in which we enter "{R2}". Here, the curly brackets are important. Then, from the **Draw** menu, we choose *"Get New Part"* to get the part PARAM from the library "special.slb" and drag it to anywhere in the schematic. Clicking on PARAM brings up a dialogue box in which we enter R2 (without curly brackets) for NAME1 and also 9k for VALUE1. This 9k value is needed for PSPICE's bias-point calculation and is not relevant to our analysis here. Finally, from the **Analysis** menu, we choose *"Setup"* and click *"DC Sweep"* to bring up a dialogue box, as shown in figure 2.15, in which we enter the range of values to be swept and specify the sweep variable and type.

2. We are now ready to analyze the circuit. From the **Analysis** menu, we choose *"Simulate"* to start the analysis as usual (see Section 1.9). PSPICE beeps when it has finished analyzing the circuit.

3. Since we wish to examine the result graphically, we may use PROBE to display the power curve against variation of R2. Upon choosing *"Probe"* from the **Analysis** menu, the PROBE window pops up. There we choose *"Add"* from the **Trace** menu to bring up the *"Add Trace"* dialogue box in which we can specify any function we wish to display. For this example, we enter "V(NODE1)*I(R2)" which is the power dissipation in R2. Upon clicking "OK", PROBE will display the required power curve, as shown in figure 2.16.

Remarks — If the sweep variable is a voltage source or current source, we do not need to get PARAM. All we have to do is to bring up the *"DC Sweep"* dialogue

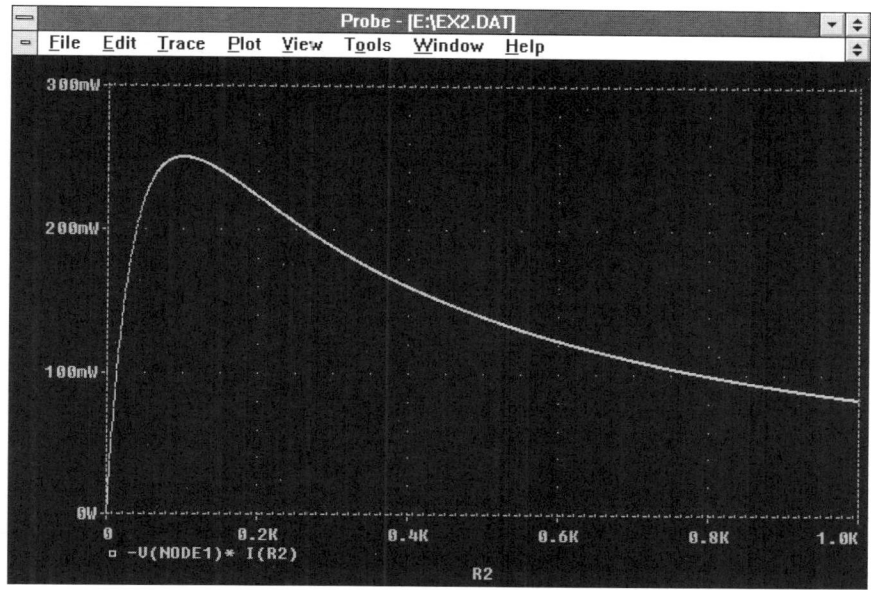

Figure 2.16: Power dissipation in R2 displayed by PROBE

box by choosing *"Setup"* in the **Analysis** menu, select *"Voltage Source"* or *"Current Source"* as the sweep variable, and enter the appropriate name and range.

2.7 Problems

1. For the circuits shown in figure 2.1, verify that the following relations are true in addition to the ones given in Example 2.1. (i) $\sum v_i(i_i + \hat{i}_i)$; (ii) $\sum i_i(2v_i + 3\hat{v}_i)$; and (iii) $\sum(4v_i + 8\hat{v}_i)(5i_i - 2\hat{i}_i)$.

2. For the circuits of figure 2.17, the following measurements are taken: $i_s = -1$A, $\hat{v}_s = 3$V, $v_L = 2$V. By applying Tellegen's theorem and making suitable

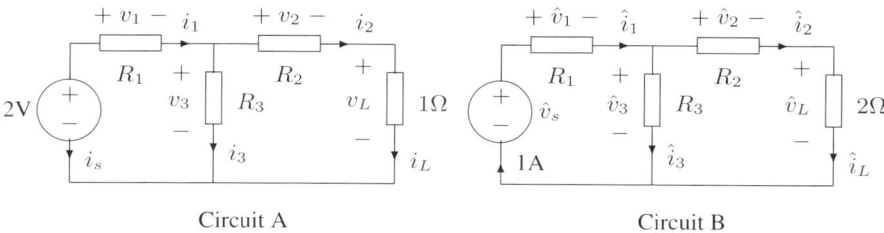

Circuit A   Circuit B

Figure 2.17: Circuits for Problem 2

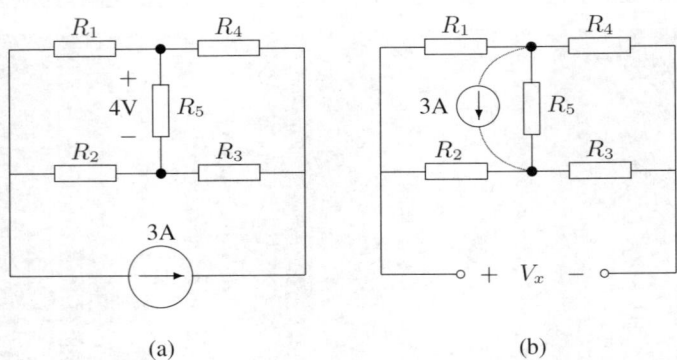

(a)   (b)

Figure 2.18: Circuits for Problem 4

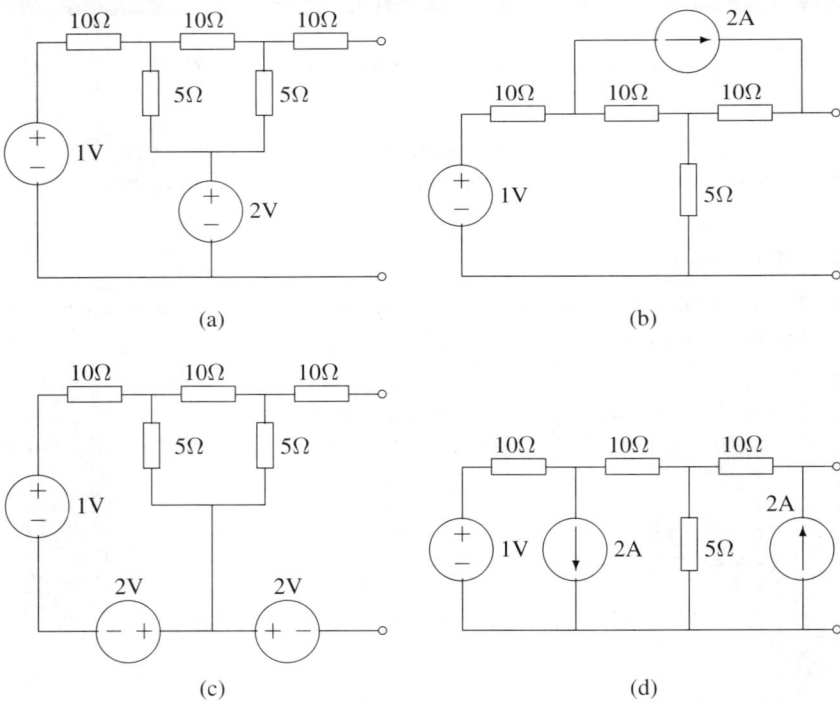

(a)   (b)

(c)   (d)

Figure 2.19: Circuits for Problem 6

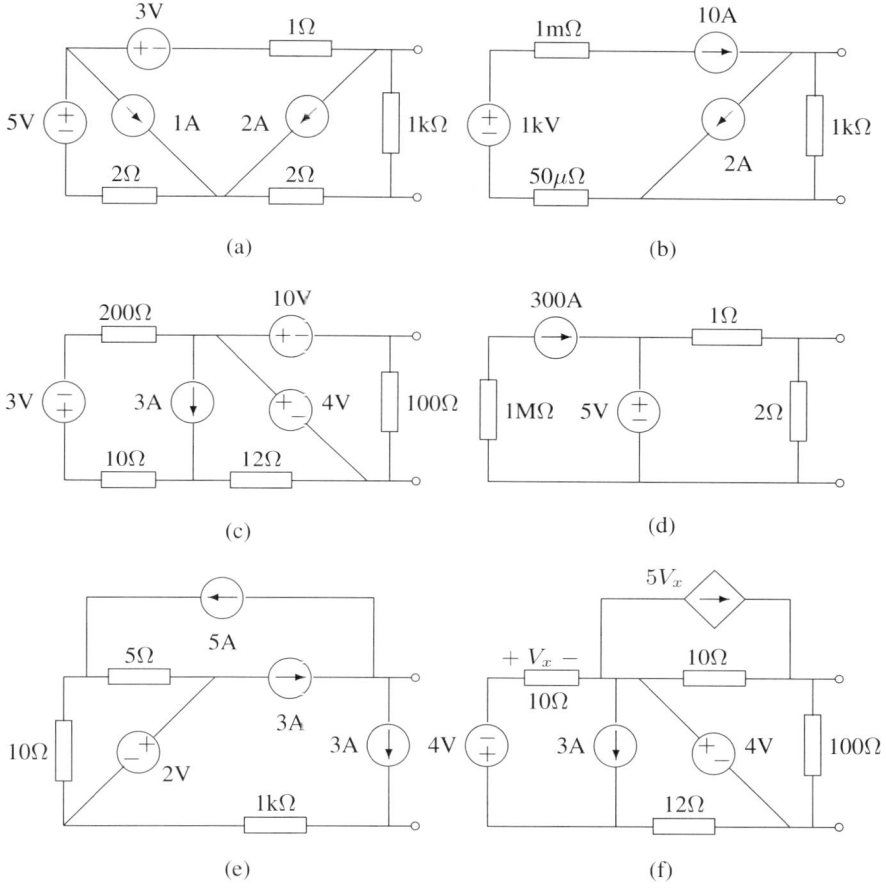

Figure 2.20: Circuits for Problems 5 and 10

assumption, determine $\hat{v}_L$ and $\hat{i}_L$. Calculate also the value of i_3. Explain why, in Circuit A, current i_3 is not flowing in the downward direction as would be expected for such a simple ladder configuration.

3. In Section 2.2.3, the reciprocity theorem has been stated in terms of voltage excitation and current measurement. Re-state the reciprocity theorem in terms of current excitation and voltage measurement. (Hint: Examine the derivation of the reciprocity theorem carefully and see what happens if current and voltage are interchanged throughout the derivation.)

4. Suppose measurement is taken for the voltage across R_5 in the circuit of figure 2.18 (a). Using the reciprocity theorem, find V_x in the circuit shown in figure 2.18 (b).

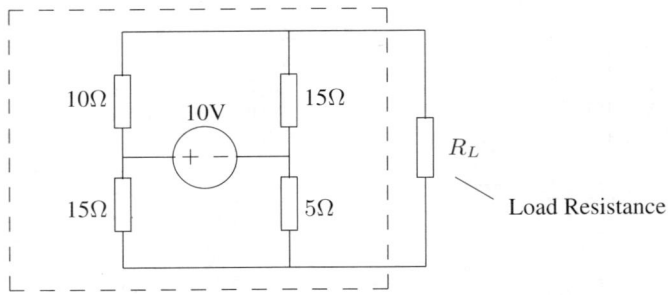

Figure 2.21: Circuit for Problem 8

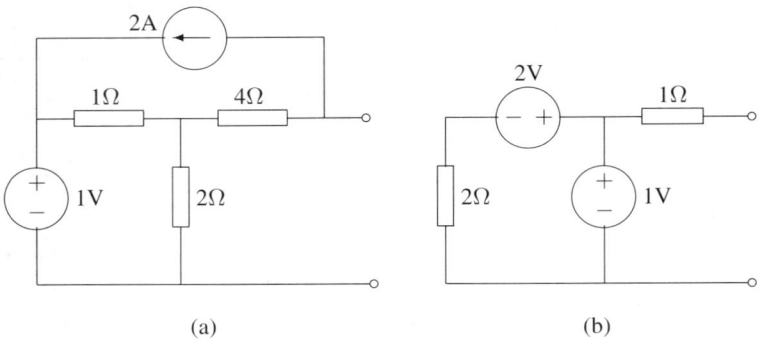

(a)   (b)

Figure 2.22: Circuits for Problem 9

5. Simplify the circuits of figures 2.20 (a)–(e) to their Norton and Thévenin equivalent circuits, as far as possible. State for which of the following circuit(s) Norton and/or Thévenin equivalent circuits cannot be found.

6. The method of successively interchanging Norton and Thévenin equivalent representations will fail for the circuits of figures 2.19 (a) and (b). Verify that circuits (a) and (b) are equivalent to circuits (c) and (d), respectively. (This problem illustrates a useful technique of relocating sources such that the circuit can be solved by the usual successive interchange technique.)

7. Apply the successive interchange technique to find the Thévenin and Norton equivalent circuits for the circuits of figure 2.19.

8. For the circuit of figure 2.21, find the value of the load resistance that gives the maximum power transfer from the voltage source to the load. Hence, find the maximum possible power transfer from the source to the load. Verify your answer by performing a DC sweep of R_L using PSPICE. Follow the procedure outlined in Example 2.8.

9. Use PSPICE to find the Thévenin and Norton resistance for the circuits of figure 2.22. Use the DC sweep function of PSPICE to get the answers. (Hint: Apply a sweeping DC voltage at the terminals and determine the slope of the I-V curve from PROBE.)

10. Use PSPICE to find the Thévenin and Norton equivalent circuits for the circuits of figure 2.20 (f). Use the part called voltage-controlled-current-source from the library "analog.slb" to simulate the dependent source. Double-click it to enter an appropriate multiplier after a proper connection is made.

Chapter 3

Mesh and Nodal Analyses

The mesh and nodal methods are introduced in this chapter as general analysis methods that can be used to analyze electric circuits of arbitrary configuration. While the mesh analysis method is applicable to planar circuits only, the nodal analysis method is applicable to any circuit. However, these methods do not in general yield the most efficient solutions, and with the help of graph theory which is to be introduced in Chapter 9, we can derive more efficient solution approaches, of which the mesh and nodal methods are particular cases. Nonetheless, the mesh and nodal methods suffice for most practical purposes, and, because of their simplicity, have become the most widely used techniques for solving general circuit problems.

3.1 Mesh and Planarity

A *mesh* is a loop which does not contain any other loops within it. For example, in the circuit shown in figure 3.1, there are altogether four meshes. A *planar circuit*

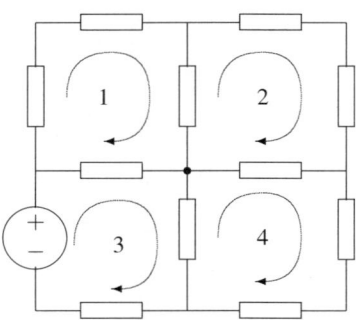

Figure 3.1: Mesh

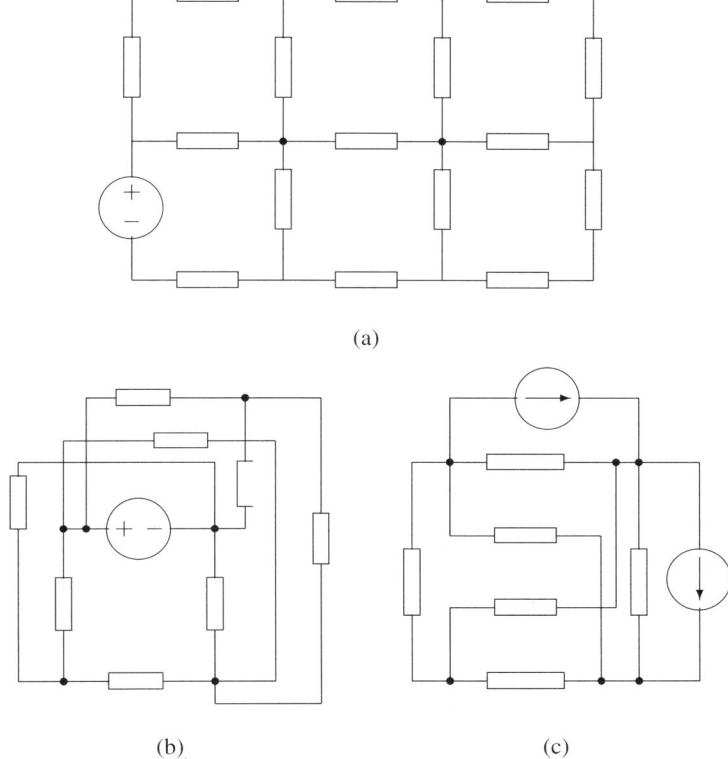

(a)

(b)   (c)

Figure 3.2: (a) Planar circuit; (b) non-planar circuit; (c) planar or non-planar?

is a circuit which contains no branches that pass over or under any other branches. For example, the circuit in figure 3.2 (a) is planar, and the one in figure 3.2 (b) is non-planar. It should be noted that planarity is a property of the circuit configuration which is independent of the way it is drawn. Thus, even though a circuit may look deceptively non-planar in the way it is drawn, it could in fact be planar (see the circuit in figure 3.2 (c) for example).[1]

3.2 Mesh Analysis

Basically, mesh analysis initially assumes that the currents flowing around the meshes are independent. For each mesh, a KVL equation is written, and the resulting system of linear equations is solved to obtain the unknown mesh currents. The number of equations involved is equal to the number of meshes in the circuit. Also, the current through a particular branch is generally a linear combination of the mesh currents.

---

[1] The mesh is defined in this way to distinguish it from a loop which is any closed path in a circuit.

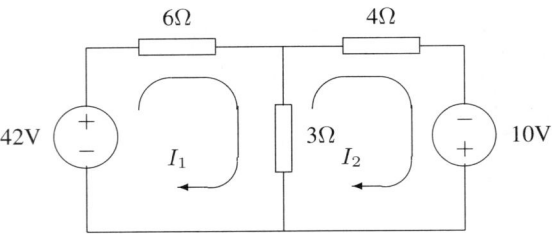

Figure 3.3: Illustration of the mesh method

3.2.1 Application of the Mesh Method to Circuits Containing No Current Sources

Before embarking on the mesh analysis, we must check that the circuit is planar since the mesh method is only valid for planar circuits. The following example illustrates the simplest application of the mesh method to circuits containing no independent current sources.

Example 3.1: Illustration of mesh method — The circuit shown in figure 3.3 is planar and has two meshes. We will take the following steps to solve this circuit:

Step 1: Assign mesh currents. As shown in the figure, two mesh currents are defined, namely I_1 and I_2. They are assumed independent of each other. Thus, the current through the 3Ω resistor is $I_1 - I_2$.

Step 2: Write down the KVL equation for each mesh.

$$\text{Mesh 1:} \quad -42 + 6I_1 + 3(I_1 - I_2) = 0$$
$$\text{Mesh 2:} \quad 3(I_2 - I_1) + 4I_2 - 10 = 0$$

Step 3: Rearrange the set of KVL equations to a system of linear equations.

$$9I_1 - 3I_2 = 42$$
$$-3I_1 + 7I_2 = 10$$

Step 4: Solve the system of equations.[2]

$$I_1 = \frac{\begin{vmatrix} 42 & -3 \\ 10 & 7 \end{vmatrix}}{\begin{vmatrix} 9 & -3 \\ -3 & 7 \end{vmatrix}} = 6\text{A}$$

$$I_2 = \frac{\begin{vmatrix} 9 & 42 \\ -3 & 10 \end{vmatrix}}{\begin{vmatrix} 9 & -3 \\ -3 & 7 \end{vmatrix}} = 4\text{A}$$

---

[2]Appendix B discusses the methods for solving systems of simultaneous equations.

3.2.1.1   General Solution and Superposition

The mesh analysis method in general leads to a linear system of equations from which the unknown mesh currents are solved. Denoting the unknown mesh currents by the vector $\bar{I}$, the system of equations in matrix form can be written as

$$\mathcal{R}\bar{I} = \mathcal{U}$$

where $\mathcal{R}$ is called the *resistance matrix* and $\mathcal{U}$ is the *source vector* constructed from the voltage sources, $V_1, V_2, \ldots, V_n$, i.e.,

$$\mathcal{R}\bar{I} = a_1 V_1 + a_2 V_2 + \cdots + a_n V_n$$

Upon solving this matrix equation we get $\bar{I}$ as a sum of $V_1, V_2, \ldots, V_n$.

$$\begin{aligned} \bar{I} &= \mathcal{R}^{-1} a_1 V_1 + \mathcal{R}^{-1} a_2 V_2 + \cdots + \mathcal{R}^{-1} a_n V_n \\ &= b_1 V_1 + b_2 V_2 + \cdots + b_n V_n \end{aligned}$$

In other words, $\bar{I}$ is a linear combination of the voltage sources. This result illustrates an important property of linear circuits called *superposition*. It should be noted that superposition is not a law, but is simply a consequence of linearity.

Example 3.2: Demonstration of superposition from mesh analysis — For the circuit of Example 3.1, the equation to be solved can be re-written in matrix form as

$$\begin{pmatrix} 9 & -3 \\ -3 & 7 \end{pmatrix} \begin{pmatrix} I_1 \\ I_2 \end{pmatrix} = \begin{pmatrix} 42 \\ 10 \end{pmatrix}$$

The 2×2 matrix in the LHS of the above equation is the resistance matrix, and the 2-dim vector in the RHS is the source vector. The unknown currents can be expressed as

$$\begin{pmatrix} I_1 \\ I_2 \end{pmatrix} = \begin{pmatrix} 9 & -3 \\ -3 & 7 \end{pmatrix}^{-1} \begin{pmatrix} 42 \\ 10 \end{pmatrix}$$

which can be put as

$$\begin{pmatrix} I_1 \\ I_2 \end{pmatrix} = \begin{pmatrix} 9 & -3 \\ -3 & 7 \end{pmatrix}^{-1} \begin{pmatrix} 42 \\ 0 \end{pmatrix} + \begin{pmatrix} 9 & -3 \\ -3 & 7 \end{pmatrix}^{-1} \begin{pmatrix} 0 \\ 10 \end{pmatrix}$$

Clearly, the individual contributions of the 42V and the 10V source combine in a linear fashion. If we denote the resistance matrix by $\mathcal{R}$, the 42V source by V_1 and the 10V source by V_2, we can write the unknown currents as a linear combination of V_1 and V_2:

$$\begin{aligned} \begin{pmatrix} I_1 \\ I_2 \end{pmatrix} &= \mathcal{R}^{-1} \begin{pmatrix} V_1 \\ 0 \end{pmatrix} + \mathcal{R}^{-1} \begin{pmatrix} 0 \\ V_2 \end{pmatrix} \\ &= \mathcal{R}^{-1} \begin{pmatrix} 1 \\ 0 \end{pmatrix} V_1 + \mathcal{R}^{-1} \begin{pmatrix} 0 \\ 1 \end{pmatrix} V_2 \\ &= A V_1 + B V_2 \end{aligned}$$

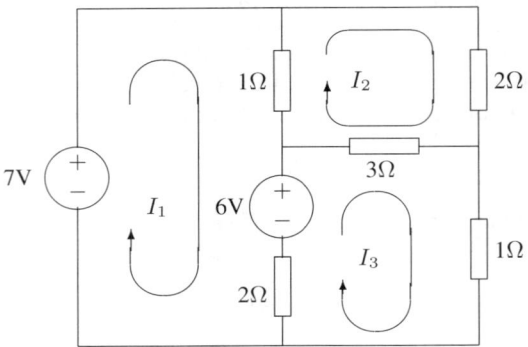

Figure 3.4: Application of the mesh method to a circuit containing no current sources

3.2.1.2   The Resistance Matrix $\mathcal{R}$ and Source Vector $\mathcal{U}$

The following properties of $\mathcal{R}$ and $\mathcal{U}$, mainly from inspection of the general result (not from a rigorous proof though), will help establish the system of equations very quickly for any circuit that does not contain current sources.

1. The resistance matrix $\mathcal{R}$ is *symmetric.*[3] Thus, we need only find the upper or lower triangle of $\mathcal{R}$.

2. The diagonal element $\mathcal{R}_{jj}$ is found as the sum of all the "associated" resistances for mesh j. For the circuit of figure 3.3, $\mathcal{R}_{11} = 6+3 = 9\Omega$, and $\mathcal{R}_{22} = 4+3 = 7\Omega$.

3. Other elements in $\mathcal{R}$ are found as signed values of the associated resistances. That is, $\mathcal{R}_{ij}$ is the resistance which lies along both meshes i and j. A positive sign is assigned if both mesh currents go in the same direction through that resistance, and a negative sign is assigned otherwise. In the circuit of figure 3.3, I_1 and I_2 go in opposite directions through the common branch. Thus, $\mathcal{R}_{12} = \mathcal{R}_{21} = -3\Omega$.

4. The source vector $\mathcal{U}$ is derivable from simple inspection of the circuit. To find the jth element of $\mathcal{U}$ corresponding to mesh j, we simply add up the values of all voltage sources that appear in mesh j with an appropriate sign attached to each voltage such that if the mesh direction goes into the "−" terminal of the voltage source, the voltage value is positive, and vice versa. Thus, for the circuit of figure 3.3, the first element of $\mathcal{U}$ is +42V and the second element is +10V.

Example 3.3: Writing mesh equations by inspection — Consider the three-mesh circuit shown in figure 3.4. The standard procedure involves writing the KVL equations for the three meshes and hence solving these equations for the unknown mesh currents.

$$\text{Mesh 1:}\qquad (I_1 - I_2) \times 1 + 6 + (I_1 - I_3) \times 2 - 7 = 0$$

---

[3]The symmetry in $\mathcal{R}$ can also be established formally from the general reciprocity theorem.

Mesh 2: $(I_2 - I_1) \times 1 + I_2 \times 2 + (I_2 - I_3) \times 3 = 0$

Mesh 3: $-6 + (I_3 - I_2) \times 3 + I_3 \times 1 + (I_3 - I_1) \times 2 = 0$

Alternatively, since the circuit has no current sources, we may readily write down the resistance matrix making use of the above-mentioned properties.

$$\mathcal{R} = \begin{bmatrix} 1+2 & -1 & -2 \\ -1 & 1+3+2 & -3 \\ -2 & -3 & 2+3+1 \end{bmatrix}$$

The source vector can be constructed by inspecting the sources associated with each mesh. Specifically, for mesh 1, there are two voltage sources. The mesh current goes into the "−" terminal of the 7V source and into the "+" terminal of the 6V source. Thus, the first element of the source vector is 7−6V. Also, since mesh 2 has no voltage source, the second element of the source vector is 0. Likewise, the third element of the source vector is 6V. This leads to the following matrix equation:

$$\begin{bmatrix} 1+2 & -1 & -2 \\ -1 & 1+3+2 & -3 \\ -2 & -3 & 2+3+1 \end{bmatrix} \begin{bmatrix} I_1 \\ I_2 \\ I_3 \end{bmatrix} = \begin{bmatrix} 7-6 \\ 0 \\ 6 \end{bmatrix}$$

Upon solving this matrix equation, we get $I_1 = 3$A, $I_2 = 2$A and $I_3 = 3$A.

3.2.2 Application of the Mesh Method to Circuits With Current Sources

Circuits that contain independent current sources require special attention when using the mesh method. Essentially, for the branch that contains a current source, its current is already known. It is thus meaningless to assign an unknown current to it as the sum or difference of two mesh currents. However, the mesh method initially assumes that the number of unknowns is equal to the number of meshes. We must therefore eliminate the redundancy by ignoring one or more KVL equation(s). The procedure is illustrated by the following example.

Example 3.4: Circuit containing a current source — Consider the circuit shown in figure 3.5 (a) which contains a current source. We will take the following steps to solve this circuit.

Step 1: Assign mesh currents as usual to the circuit, i.e., I_1, I_2 and I_3.

Step 2: Open-circuit the current source temporarily. This gives rise to a "super-mesh", as drawn using a dotted line in figure 3.5 (b).

Step 3: Write down the KVL equations in terms of I_1, I_2 and I_3.

Mesh 2: $(I_2 - I_1) \times 1 + I_2 \times 2 + (I_2 - I_3) \times 3 = 0$

Super-mesh: $-7 + (I_1 - I_2) \times 1 + (I_3 - I_2) \times 3 + I_3 \times 1 = 0$

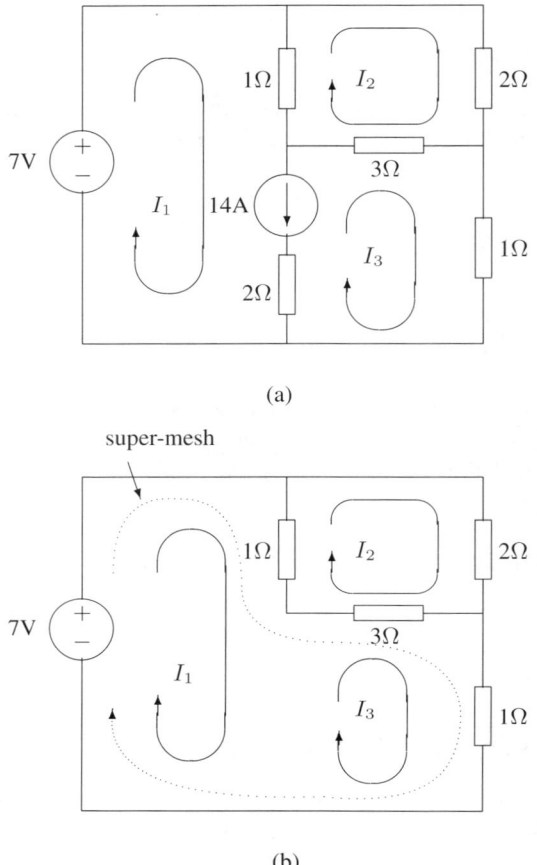

(a)

(b)

Figure 3.5: Application of the mesh method to a circuit containing a current source

Step 4: Write down the relation between the current source and the mesh currents.

$$14 = I_1 - I_3$$

Step 5: Solve the system of equations.

$$-I_1 + 6I_2 - 3I_3 = 0$$
$$I_1 - 4I_2 + 4I_3 = 7$$
$$I_1 + 0 - I_3 = 14$$

$$\Rightarrow \quad I_1 = 15A, \quad I_2 = 3A, \quad I_3 = 1A$$

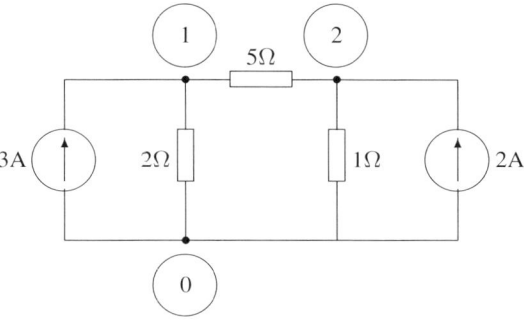

Figure 3.6: Illustration of the nodal method

3.3 Nodal Analysis

With validity extended to non-planar circuits, the nodal method is a more general method compared with the mesh method. For this reason, and also for its simplicity, the nodal method is a popular approach, on which many computer analysis programs are developed. For instance, the PSPICE analysis programme uses a variation of the nodal technique in the formulation of Kirchhoff's law equations.

3.3.1 Application of the Nodal Method to Circuits Containing No Voltage Sources

Basically the nodal method involves setting up KCL equations for $(n-1)$ nodes, where n is the total number of nodes in a given circuit. Note that every electric component represents a branch and each branch is terminated by two nodes, one at each end. Let V_k be the voltage at node k with respect to a chosen ground node. For each node (except the ground node), we write down one KCL equation. Therefore, we have $(n-1)$ equations and $(n-1)$ unknown voltages. We will use a simple example to illustrate the nodal method.

Example 3.5: Illustration of nodal method — The circuit in figure 3.6 has three nodes, i.e., $n = 3$. We choose node 0 as the ground node. Thus we expect two equations to be set up and solved for the following two unknowns.

$$
\begin{aligned}
V_1 &= \text{Voltage at node 1 w.r.t. node 0} \\
&\quad \text{(i.e. Voltage across nodes 1 and 0)} \\
V_2 &= \text{Voltage at node 2 w.r.t. node 0} \\
&\quad \text{(i.e. Voltage across nodes 2 and 0)}
\end{aligned}
$$

The KCL equations for nodes 1 and 2 are

$$
\text{Node 1:} \qquad -3 + \frac{V_1}{2} + \frac{V_1 - V_2}{5} = 0
$$

$$\text{Node 2:} \qquad \frac{V_2 - V_1}{5} + \frac{V_2}{1} - 2 = 0$$

$$\Rightarrow \quad \begin{bmatrix} \frac{1}{2} + \frac{1}{5} & -\frac{1}{5} \\ -\frac{1}{5} & \frac{1}{5} + 1 \end{bmatrix} \begin{bmatrix} V_1 \\ V_2 \end{bmatrix} = \begin{bmatrix} 3 \\ 2 \end{bmatrix}$$

Solving this matrix equation gives

$$V_1 = 5\text{V} \quad \text{and} \quad V_2 = 2.5\text{V}$$

3.3.1.1   General Solution and Superposition

As in the mesh method, we observe that the node voltages are linear combination of the current sources. The algebra goes as follows. Denoting the unknown node voltages by the vector $\bar{V}$, the the system of equations in matrix form can be written as

$$\mathcal{G}\bar{V} = \mathcal{U}$$

where $\mathcal{G}$ is called the *conductance matrix*, and $\mathcal{U}$ is the *source vector* constructed from the current sources, $I_1, I_2, \ldots, I_n$, i.e.,

$$\mathcal{G}\bar{V} = a_1 I_1 + a_2 I_2 + \cdots + a_n I_n$$

Upon solving this matrix equation we get $\bar{V}$ as a sum of $I_1, I_2, \ldots, I_n$.

$$\begin{aligned} \bar{V} &= \mathcal{G}^{-1} a_1 I_1 + \mathcal{G}^{-1} a_2 I_2 + \cdots + \mathcal{G}^{-1} a_n I_n \\ &= b_1 I_1 + b_2 I_2 + \cdots + b_n I_n \end{aligned}$$

Thus, $\bar{V}$ is a linear combination of the current sources.

Example 3.6: Demonstration of superposition from nodal analysis — Consider the circuit of Example 3.5. The solution can be put in the following matrix form.

$$\begin{pmatrix} V_1 \\ V_2 \end{pmatrix} = \begin{pmatrix} \frac{7}{10} & -\frac{1}{5} \\ -\frac{1}{5} & \frac{6}{5} \end{pmatrix}^{-1} \begin{pmatrix} 3 \\ 2 \end{pmatrix}$$

which can be expanded to

$$\begin{pmatrix} V_1 \\ V_2 \end{pmatrix} = \begin{pmatrix} \frac{7}{10} & -\frac{1}{5} \\ -\frac{1}{5} & \frac{6}{5} \end{pmatrix}^{-1} \begin{pmatrix} 3 \\ 0 \end{pmatrix} + \begin{pmatrix} \frac{7}{10} & -\frac{1}{5} \\ -\frac{1}{5} & \frac{6}{5} \end{pmatrix}^{-1} \begin{pmatrix} 0 \\ 2 \end{pmatrix}$$

Again, the individual contributions of the 3A and the 2A source combine in a linear fashion. If we denote the conductance matrix by $\mathcal{G}$, the 3A source by I_1 and the 2A source by I_2, we can write the unknown currents as a linear combination of I_1 and I_2, i.e.,

$$\begin{aligned} \begin{pmatrix} V_1 \\ V_2 \end{pmatrix} &= \mathcal{G}^{-1} \begin{pmatrix} I_1 \\ 0 \end{pmatrix} + \mathcal{G}^{-1} \begin{pmatrix} 0 \\ I_2 \end{pmatrix} \\ &= \mathcal{G}^{-1} \begin{pmatrix} 1 \\ 0 \end{pmatrix} I_1 + \mathcal{G}^{-1} \begin{pmatrix} 0 \\ 1 \end{pmatrix} I_2 \\ &= A I_1 + B I_2 \end{aligned}$$

The above result demonstrates again the principle of superposition.

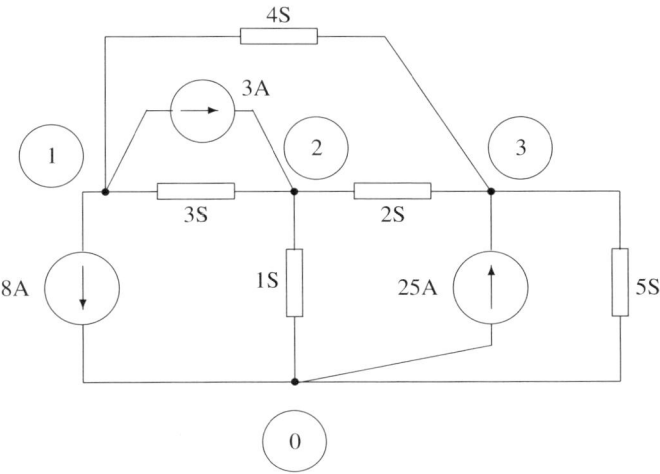

Figure 3.7: Application of the nodal method to a circuit containing no voltage sources

3.3.1.2 *The Conductance Matrix $\mathcal{G}$ and Source Vector $\mathcal{U}$*

The following properties of $\mathcal{G}$ and $\mathcal{U}$, similar to those in the case of the mesh method, will help establish the system of equations very quickly for any circuit that does not contain voltage sources.

1. The conductance matrix $\mathcal{G}$ is *symmetric*. Thus, we need only find the upper or lower triangle of $\mathcal{G}$.

2. The diagonal element $\mathcal{G}_{jj}$ is found as the sum of all the "associated" conductances for node j. For the circuit of figure 3.6, $\mathcal{G}_{11} = \frac{1}{2} + \frac{1}{5} = \frac{7}{10}$S, and $\mathcal{G}_{22} = 1 + \frac{1}{5} = \frac{6}{5}$S.

3. Other elements in $\mathcal{G}$ are found as negative values of associated conductances. That is, $\mathcal{G}_{ij}$ is *minus* the conductance between nodes i and j. (Note: the minus sign is always attached to the conductance value in the case of off-diagonal elements.) Thus, $\mathcal{G}_{12} = \mathcal{G}_{21} = -\frac{1}{5}$S for the circuit of figure 3.6.

4. The source vector $\mathcal{U}$ is constructed such that the jth element of $\mathcal{U}$ is the sum of currents entering node j. For the circuit of figure 3.6, $\mathcal{U}$ is simply $[3 \; 2]^T$.

Example 3.7: Writing nodal equations by inspection — Consider the four-node circuit shown in figure 3.7. We assign node 0 as the reference node. The unknowns are V_1, V_2 and V_3, and the conductance matrix $\mathcal{G}$ is

$$\mathcal{G} = \begin{bmatrix} 4+3 & -3 & -4 \\ -3 & 3+2+1 & -2 \\ -4 & -2 & 4+2+5 \end{bmatrix}$$

Also, the source vector can be constructed by inspection. The complete nodal equation in matrix form is

$$
\begin{bmatrix}
4+3 & -3 & -4 \\
-3 & 3+2+1 & -2 \\
-4 & -2 & 4+2+5
\end{bmatrix}
\begin{bmatrix}
V_1 \\
V_2 \\
V_3
\end{bmatrix}
=
\begin{bmatrix}
-11 \\
3 \\
25
\end{bmatrix}
$$

Solving this matrix equation gives

$$
\begin{bmatrix}
V_1 \\
V_2 \\
V_3
\end{bmatrix}
=
\begin{bmatrix}
1\text{V} \\
2\text{V} \\
3\text{V}
\end{bmatrix}
$$

3.3.2   Application of the Nodal Method to Circuits With Voltage Sources

For circuits that contain voltage sources, some special treatment is needed to eliminate the redundancy of KVL equations, since additional information is available on the node voltages. For example, if nodes i and j are connected by a voltage source, then $V_j - V_i$ is already given. We shall illustrate the application of nodal method for such circuits by an example.

Example 3.8: Circuit containing a voltage source — Consider the circuit shown in figure 3.8 (a) which contains a voltage source between nodes 2 and 3. The nodal method aims to find the three unknowns V_1, V_2 and V_3, which are voltages at nodes 1, 2 and 3, respectively, all with respect to node 0.

The special procedure here is to treat nodes 2 and 3 as being merged together to form one *super-node,* as illustrated in figure 3.8 (b). Then, we write down KCL equations for node 1 and the super-node:

Node 1: $8 + (V_1 - V_2) \times 3 + 3 + (V_1 - V_3) \times 4 = 0$

Super-node: $(V_2 - V_1) \times 3 - 3 + V_2 \times 1 + (V_3 - V_1) \times 4 + V_3 \times 5 + 25 = 0$

Apparently we have only two equations, but with three unknowns. However, the voltage source dictates that

$$V_3 - V_2 = 2$$

Hence, we actually have a system of three equations with three unknowns:

$$
\left\{
\begin{array}{rcl}
7V_1 - 3V_2 - 4V_3 & = & -11 \\
-7V_1 + 4V_2 + 9V_3 & = & -22 \\
-V_2 + V_3 & = & 2
\end{array}
\right.
$$

from which the unknown voltages can be found.

3.4   Use of Superposition in Solving Circuit Problems

Let us review the meaning of superposition by considering again the application of the mesh method to a circuit containing more than one independent source. In particular,

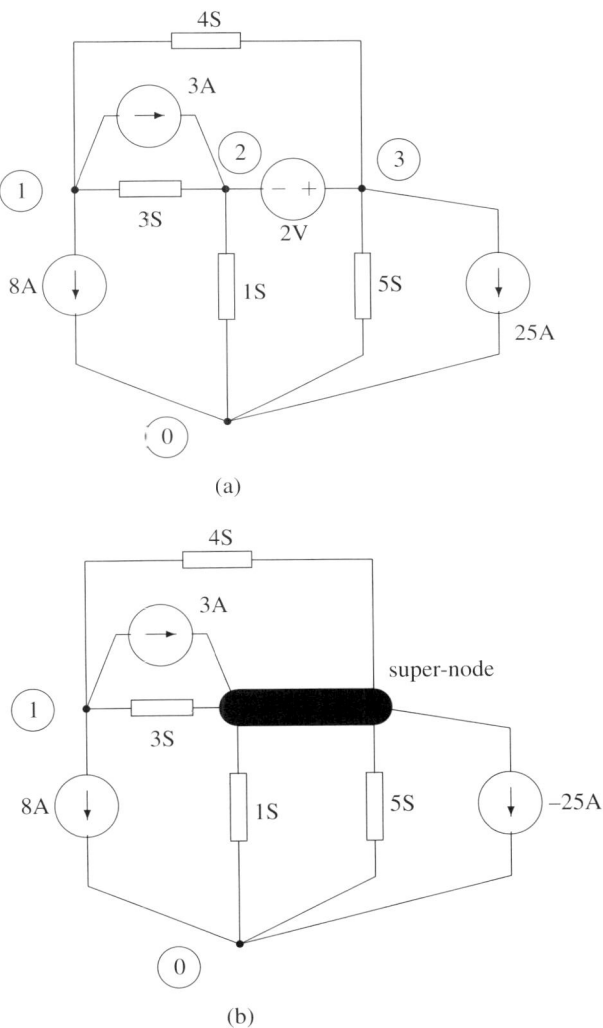

Figure 3.8: Application of the nodal method to a circuit with a voltage source

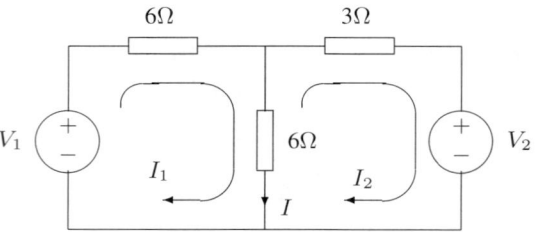

Figure 3.9: Superposition

the mesh equation for the circuit of figure 3.9 is given by

$$\begin{pmatrix} I_1 \\ I_2 \end{pmatrix} = \begin{pmatrix} 12 & -6 \\ -6 & 9 \end{pmatrix}^{-1} \begin{pmatrix} V_1 \\ -V_2 \end{pmatrix}$$
$$= \begin{pmatrix} \frac{1}{8} & \frac{1}{12} \\ \frac{1}{12} & \frac{1}{6} \end{pmatrix} \begin{pmatrix} V_1 \\ -V_2 \end{pmatrix}$$

The current I is given by

$$I = I_1 - I_2 = \frac{1}{24} V_1 + \frac{1}{12} V_2$$

which clearly shows how V_1 and V_2 algebraically superpose to give I. Plain knowledge of this property, however, does not mean we can solve the circuit in a better way. The real question is whether we can turn the phenomenon into a useful procedure for solving circuits that contain more than one source.

Indeed the importance of superposition lies in its practical use for solving linear circuit problems. From the above example, we can see that if we let $V_2 = 0$ and calculate I, we would expect to get $V_1/24$. Alternatively if we let $V_1 = 0$ and calculate I, we would expect to get $V_2/12$. Hence, a useful trick to solve this circuit is to consider only one source at a time, in which case the circuit reduces to a very simple single-source circuit. The computation of I when either $V_1 = 0$ or $V_2 = 0$ is extremely easy. The principle of superposition tells us that the actual value of I is simply equal to the sum of the values computed in the two cases.

Example 3.9: Application of superposition — Consider the circuit shown at the top of figure 3.10, which contains three independent sources. We can exploit the principle of superposition to obtain the unknown voltage V_x. Basically we consider three separate cases, in each of which only one source is present.

Case 1: As shown on the lower left of figure 3.10, only the 1V source is present in the circuit while the 2V source is shorted and the 1A source is opened. We get by inspection

$$V_x' = \frac{1}{2}\text{V}$$

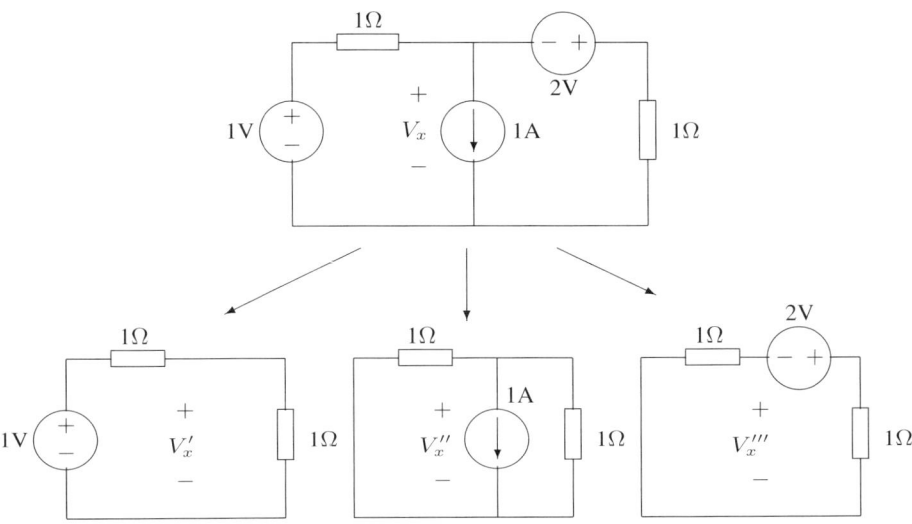

Figure 3.10: Application of superposition

where the prime indicates that the value is the contribution from the 1V source.

Case 2: As shown in the lower middle of figure 3.10, only the 1A source is present in the circuit while the other two voltage sources are shorted out. Here, we have

$$V_x'' = -\frac{1}{2}\text{V}$$

where the double-prime indicates that the value is the contribution from the 1A source.

Case 3: As shown on the lower right of figure 3.10, only the 2V source is present in the circuit while the 1V source is shorted and the 1A source is opened. Here, we have

$$V_x''' = -1\text{V}$$

where the triple-prime indicates that the value is the contribution from the 2V source.

Finally we apply the principle of superposition to obtain V_x. This simply involves adding up the contributions from the above three cases, i.e.,

$$V_x = V_x' + V_x'' + V_x''' = -1\text{V}$$

Remarks — To conclude this chapter, we point out an important observation regarding the number of simultaneous Kirchhoff's law equations which are to be solved. For the mesh method, the number of equations is equal to the number of meshes in the circuit. This is true for both cases with and without current sources. For the nodal method, the number of equations is equal to the number of nodes minus one. This is again true for both cases with and without voltage sources. In Chapters 9 and 10, graph theory will be gainfully exploited to formulate independent Kirchhoff's law equations, resulting in minimum number of equations to be solved.

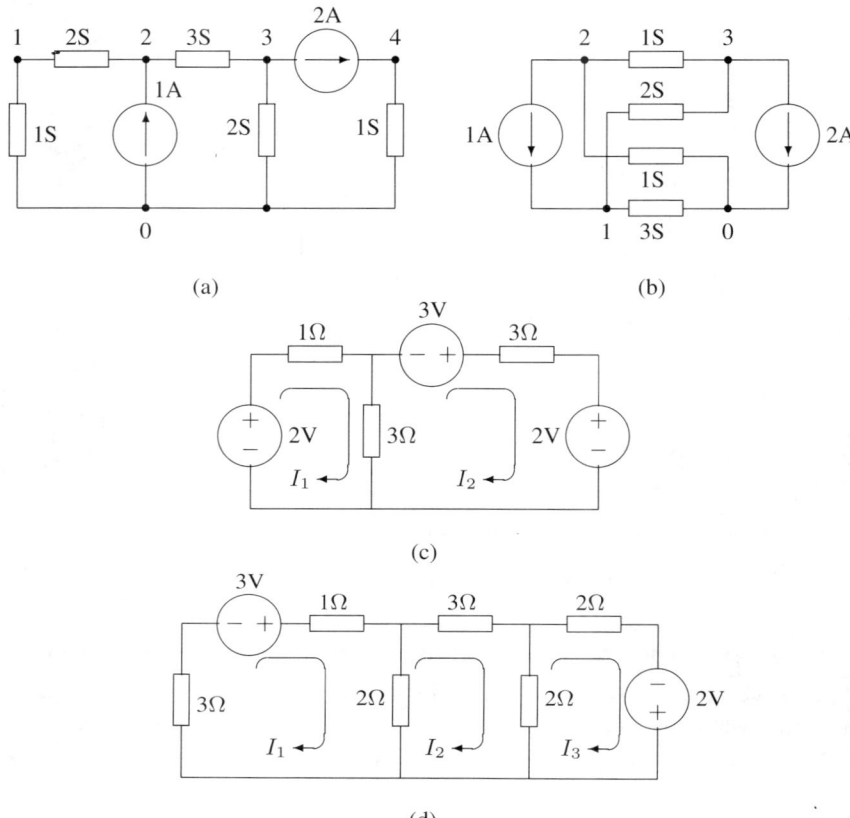

Figure 3.11: Circuits for Problem 1

3.5 Problems

1. State whether circuit (b) of figure 3.11 is planar or non-planar. For circuits (a) and (b) of figure 3.11, derive the nodal equations, and for circuits (c) and (d), derive the mesh equations. Put the equations in standard matrix form.

2. By using the nodal method, find V_x in the circuit of figure 3.12 (a). Derive the conductance matrix and source vector. Hence, write down the system of equations in matrix form. Demonstrate the principle of superposition.

3. For the circuit of figure 3.12 (b), set up the nodal equations using the concept of a super-node or otherwise. Find I_y.

4. By using the mesh method, find I_x in the circuit of figure 3.13 (a). Derive the resistance matrix and write the system of equations in matrix form. Demonstrate the principle of superposition.

5. For the circuit of figure 3.13 (b), set up the mesh equations using the concept of a super-mesh or otherwise. Find V_y.

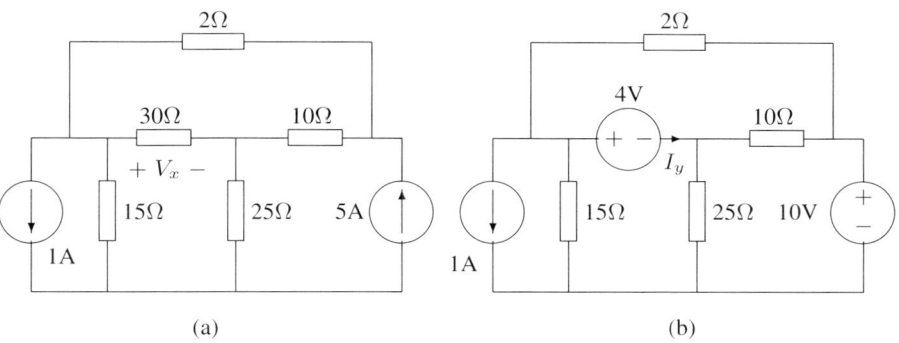

(a)   (b)

Figure 3.12: Circuits for Problems 2 and 3

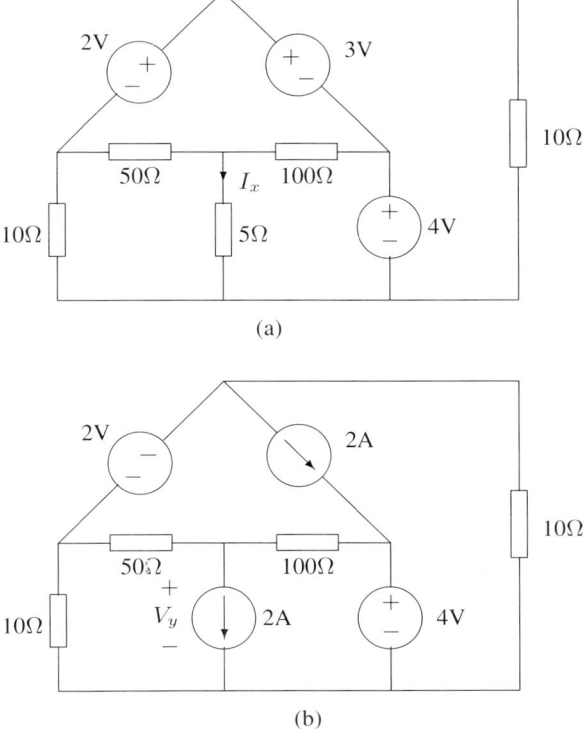

(a)

(b)

Figure 3.13: Circuits for Problems 4 and 5

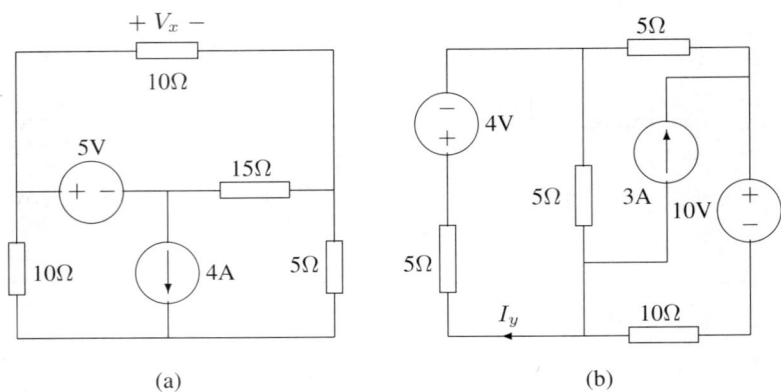

(a)   (b)

Figure 3.14: Circuits for Problem 6

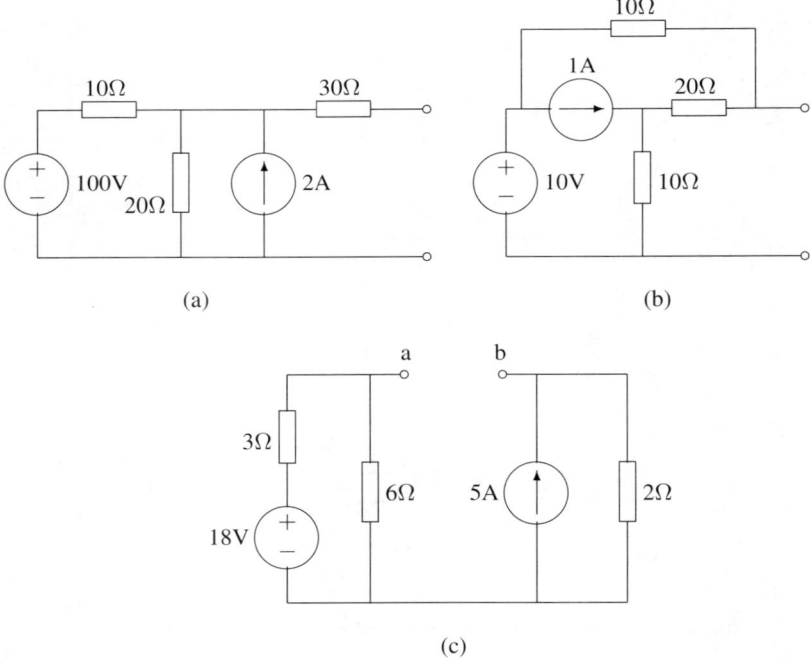

(a)   (b)

(c)

Figure 3.15: Circuits for Problems 9 and 10

6. Explain the principle of superposition, and discuss its validity. Find V_y. Use superposition to determine V_x and I_y in the circuits of figure 3.14.

7. Using the nodal method, find V_x for circuit (a) of figure 3.14, with the 5V source replaced by a dependent voltage source of intensity $3V_x$. Using the mesh method, find I_y for circuit (b) of figure 3.14, with the 3A source replaced by a dependent current source of intensity $4I_y$.

8. Consider a circuit containing m meshes and n nodes. Explain how you would choose the solution method among the mesh and nodal approaches. How would the number of current sources and voltage sources affect your choice of solution method?

9. For each of the circuits shown in figure 3.15, calculate the open-circuit voltage by using superposition, and hence derive the Thévenin equivalent circuit. For circuit (c), calculate the value of resistor that must be connected to a and b to develop maximum power in the resistor.

10. Use PSPICE to verify superposition. In particular, for each of the circuits in figure 3.15, find the open-circuit voltages and short-circuit currents for the cases when only one source is present, others being zero. Demonstrate how individual results are combined linearly to give the answers found previously in Problem 9.

Chapter 4

First-Order Transients

So far we have been discussing circuits that contain resistances and sources. For such circuits, the currents and voltages acquire their values instantaneously whenever there is a sudden change in circuit configuration. For example, in the circuit of figure 4.1, v and i are zero before the switch is closed, and become V and V/R instantly as the switch is closed. In this chapter we will study a different class of circuits in which currents and voltages vary as exponential functions of time after the circuit experiences a sudden change in its configuration. The main constituent elements that cause such behaviour are the capacitor, the inductor and the resistor. We will develop formal methods of analysis based on solutions of the describing first-order differential equations, and will also develop fast techniques, mainly by inspection, for obtaining transient expressions and waveforms. This chapter concludes with an illustrative example of using PSPICE for transient analysis of electric circuits.

4.1 Linear Capacitors and Inductors

The symbols for the capacitor and the inductor are shown in figure 4.2. For the linear capacitor, the charge stored in it, q_C, is proportional to the terminal voltage, v_C, i.e.,

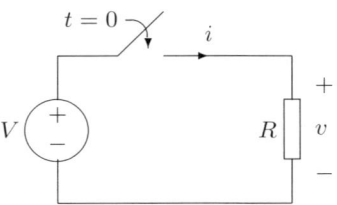

Figure 4.1: Resistive circuit. $i(t) = V/R$ and $v(t) = V$ for all $t \geq 0$

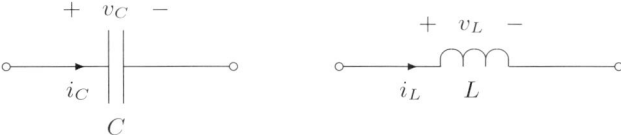

Figure 4.2: Capacitor and inductor

$q_C = C \times v_C$, where C is the proportionality constant known as *capacitance*. Also, since the current flowing through it is dq_C/dt, we may write

$$i_C = C \frac{dv_C}{dt}$$

For the linear inductor, the magnetic flux induced in it, ψ, is proportional to the current through it, i_L, i.e., $\psi = L \times i_L$, where L is the proportionality constant known as *inductance*. Also, since the terminal voltage is $d\psi/dt$, we may write

$$v_L = L \frac{di_L}{dt}$$

The unit of capacitance is the *farad* (F), and that of inductance is the *henry* (H).

Remarks — (i) The sign convention follows the one defined in figure 4.2. With this convention, i.e., current going into the "+" terminal of the element, the capacitor and inductor equations *do not* contain a minus sign in the RHS; this may appear in some physics texts due to a different convention. (ii) When voltage and current are interchanged, the capacitor behaves exactly as an inductor. This is known as *duality*, and the capacitor is said to be the *dual* of the inductor, and vice versa. We will take a closer look at this property in a later section. (iii) Capacitance and inductance sometimes appear as properties of many physical devices, and can be fabricated as circuit components. For the time being, we treat them as circuit elements whose voltage and current are related by the above equations.

4.2 Series and Parallel Connections of Capacitors and Inductors

The capacitance of a capacitor is the measure of its charge storage capacity. When a voltage is applied to a capacitor, a certain amount of charge is stored in it. Intuitively, when a voltage is applied to a parallel combination of n capacitors, as shown in figure 4.3 (a), the total amount of charge stored in the entire assembly is simply the cummulative sum of the charge stored in the n capacitors. Thus, we may write the equivalent capacitance of n capacitors in parallel as $C_1 + C_2 + \cdots + C_n$.

We can formally derive the equivalent capacitance of n capacitors connected in parallel as follows. First, the KCL equation in this case reads

$$i = C_1 \frac{dv_{C1}}{dt} + C_2 \frac{dv_{C2}}{dt} + \cdots + C_n \frac{v_{Cn}}{dt}$$

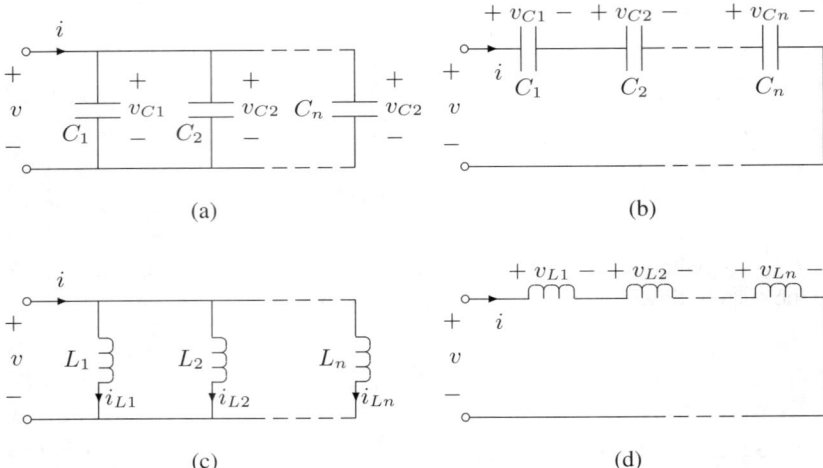

Figure 4.3: (a) Parallel capacitors; (b) series capacitors; (c) parallel inductors; (d) series inductors

Since the n capacitors are connected in parallel, their terminal voltages are identical, i.e.,

$$v_{C1} = v_{C2} = \cdots = v_{Cn} = v$$

Thus, we have

$$i = (C_1 + C_2 + \cdots + C_n)\frac{dv}{dt}$$

Hence, the equivalent capacitance, C_{eq}, of n capacitors in parallel is

$$\boxed{C_{eq} = C_1 + C_2 + \cdots + C_n}$$

When a voltage is applied to n capacitors connected in series, each capacitor will take a portion of the total applied voltage. Referring to figure 4.3 (b), the KVL equation reads

$$v = v_{C1} + v_{C2} + \cdots + v_{Cn}$$

Upon differentiation with respect to time, the above KVL equation becomes

$$\frac{dv}{dt} = \frac{dv_{C1}}{dt} + \frac{dv_{C2}}{dt} + \cdots + \frac{dv_{Cn}}{dt}$$

Putting $dv_{Cj}/dt = i/C_j$ in the above equation, for $j = 1$ to n, yields

$$\frac{dv}{dt} = \frac{i}{C_1} + \frac{i}{C_2} + \cdots + \frac{i}{C_n}$$

Thus, the equivalent capacitance, C_{eq}, of n capacitors in series is given by

$$\frac{1}{C_{eq}} = \frac{1}{C_1} + \frac{1}{C_2} + \cdots + \frac{1}{C_n}$$

Now consider n inductors connected in parallel. Referring to figure 4.3 (c), the KCL equation is

$$i = i_{L1} + i_{L2} + \cdots + i_{Ln}$$

which gives

$$\frac{di}{dt} = \frac{di_{L1}}{dt} + \frac{di_{L2}}{dt} + \cdots + \frac{di_{Ln}}{dt}$$

Putting $di_{Lj}/dt = v/L_j$ in the above equation, for $j = 1$ to n, yields

$$\frac{di}{dt} = \frac{v}{L_1} + \frac{v}{L_2} + \cdots + \frac{v}{L_n}$$

Thus, the equivalent inductance, L_{eq}, of n inductors in parallel is given by

$$\frac{1}{L_{eq}} = \frac{1}{L_1} + \frac{1}{L_2} + \cdots + \frac{1}{L_n}$$

Finally, for n inductors connected in series, as shown in figure 4.3 (d), the KVL equation reads

$$v = v_{L1} + v_{L2} + \cdots + v_{Ln}$$

Putting $v_{Lj} = L_j(di/dt)$ in the above equation, for $j = 1$ to n, yields

$$v = (L_1 + L_2 + \cdots + L_n)\frac{di}{dt}$$

Thus, the equivalent inductance, L_{eq}, of n inductors in series is given by

$$L_{eq} = L_1 + L_2 + \cdots + L_n$$

Example 4.1: Capacitor network — In figure 4.4, the parallel combination of C_1 and C_2 is connected in series with C_3. The equivalent capacitance observed from terminals a and b is given by

$$\frac{1}{C_{eq}} = \frac{1}{C_3} + \frac{1}{C_1 + C_2} \quad \text{or} \quad C_{eq} = \frac{C_3(C_1 + C_2)}{C_1 + C_2 + C_3}$$

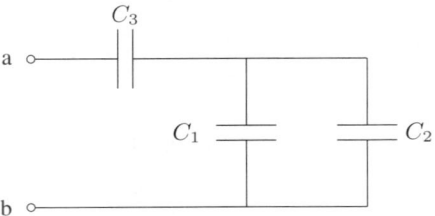

Figure 4.4: Series/parallel connection of capacitors

4.3   Steady State and Transient

When a circuit contains capacitors and/or inductors, it is called a dynamic circuit in the sense that voltages and currents in the circuit are varying as functions of time. To keep things simple, we consider a dynamic linear circuit being driven by some DC driving force. If the circuit is stable, all voltages and currents in the circuit will converge to a set of *unique fixed values,* regardless of what values they assume initially. The scenario may be described in terms of two distinct phases of development, namely *transient period* and *steady state.*

Suppose the circuit starts with some arbitrary initial condition at $t = 0$. For $t > 0$, the values of voltages and currents in this circuit move away from the initial condition, and continue to move according to certain laws of motion which depend on the initial condition as well as the way the circuit is connected. Such motion is called *transient motion.* If the circuit is stable, the motion will eventually settle to a *steady state* in which all voltages and currents are in a stable equilibrium state, i.e., fixed values in the case of DC driven circuits. While the transient motion depends on initial condition, the set of values of the voltages and currents in the steady state is unique.

4.4   Simple First-Order RC Circuit

Figure 4.5 shows one of the simplest linear dynamic circuits. The circuit contains one resistor and one capacitor, and at $t = 0$ the circuit is closed resulting in a constant voltage source driving a series combination of resistance and capacitance. We begin our analysis by considering the voltage at node 1:

$$v_1(t) = \begin{cases} 0 & t < 0 \\ V_o & t \geq 0 \end{cases}$$

Also, the current is

$$i(t) = \frac{v_R}{R} = C\frac{dv_C}{dt}$$

and the voltage across the resistor is

$$v_R(t) = v_1(t) - v_C(t)$$

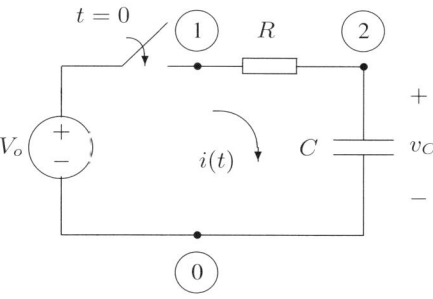

Figure 4.5: Simple RC circuit

Hence,

$$\frac{v_1(t) - v_C(t)}{CR} = \frac{dv_C}{dt}$$

or

$$\frac{dv_C}{dt} + \frac{v_C}{CR} = \frac{v_1(t)}{CR}$$

The above equation is the *first-order differential equation* that governs the transient motion of the circuit. Suppose the capacitor initially has zero charge (voltage), i.e.,

$$v_C(0^+) = 0$$

The general solution is

$$v_c(t) = Ae^{-\frac{t}{CR}} + V_o \quad \text{for } t \geq 0$$

where A is determined from the initial condition as

$$v_C(0^+) = 0 = A + V_o$$

$$\Rightarrow \quad A = -V_o$$

Hence, for $t \geq 0$, we obtain

$$v_c(t) = V_o \left(1 - e^{-\frac{t}{CR}}\right)$$

$$i(t) = C\frac{dv_C}{dt} = \frac{V_o}{R}e^{-\frac{t}{CR}}$$

4.4.1 Time Constant

The waveforms of $v_C(t)$ and $i(t)$ are shown in figure 4.6. These waveforms are typical waveforms of first-order transients. It is interesting to note that the product CR has

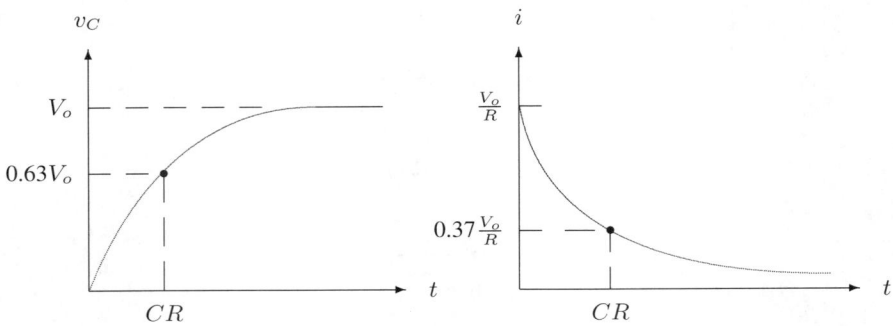

Figure 4.6: Waveforms of v_C and i in simple RC circuit

the unit of time,[1] and the motion approaches the steady state at a rate dictated by the value of CR. If CR is large, the circuit is slow and has a prolonged transient period. However, if CR is small, the circuit reaches the steady state very quickly. From the expression of $v_c(t)$, we can conclude that at $t = CR$ the capacitor voltage has reached 63% of the final steady state value (since $1 - e^{-1} \approx 0.63$). Similarly, the current $i(t)$ has fallen to 37% of its initial value by $t = CR$. Thus, the product CR is an important quantity which describes the characteristic of the transient motion. It is often referred to by the engineers as the *time-constant* of the first-order circuit.[2]

4.4.2   Energy Loss in Charging a Capacitor

Consider, again, the simple RC circuit shown in figure 4.5. The instantaneous power "dissipated" in the capacitor, by definition, is given by

$$p_C(t) = v_C(t)i(t) = \frac{V_o^2}{R}e^{-\frac{t}{CR}}\left(1 - e^{-\frac{t}{CR}}\right)$$

The energy transferred to the capacitor from $t = 0$ to $t = \infty$ is

$$
\begin{aligned}
E_C &= \int_0^\infty v_C(t)i(t)\,dt \\
&= \frac{V_o^2}{R}\int_0^\infty e^{-\frac{t}{CR}}\left(1 - e^{-\frac{t}{CR}}\right)\,dt \\
&= \frac{V_o^2}{R}\left[CR - \frac{1}{2}CR\right] \\
&= \frac{1}{2}CV_o^2
\end{aligned}
$$

---

[1] From the equation $i = C(dv_C/dt)$, we know that $C(\delta v_C/i)$ has the unit of time, and so does CR.

[2] In mathematics, the reciprocal of the time constant is known as the eigenvalue of the first-order system.

The energy dissipated in R from $t = 0$ to $t = \infty$ is

$$
\begin{aligned}
E_R &= \int_0^\infty v_R(t)i(t)\,dt \\
&= \int_0^\infty i(t)^2 R\,dt \\
&= \frac{V_o^2}{R} \int_0^\infty e^{\frac{-2t}{CR}}\,dt \\
&= \frac{1}{2}CV_o^2
\end{aligned}
$$

The energy supplied by the voltage source is

$$
\begin{aligned}
E_V &= \int_0^\infty V_o i(t)\,dt \\
&= \frac{V_o^2}{R} \int_0^\infty e^{\frac{-t}{CR}}\,dt \\
&= CV_o^2
\end{aligned}
$$

We can immediately verify that

$$
E_V = E_R + E_C
$$

which is simply a result of energy conservation. Now, let us increase the value of the resistance ten-fold. Repeating the derivation, we will end up with the same energy expressions, i.e.,

$$
\begin{aligned}
E_C &= \frac{1}{2}CV_o^2 \\
E_R &= \frac{1}{2}CV_o^2 \\
E_V &= CV_o^2
\end{aligned}
$$

Thus, the above energy expressions remain true, regardless of what value R takes. In other words, although the resistor is the physical device through which energy is dissipated in the charging process, its value is immaterial to the amount of energy loss. A vanishingly small resistance will still dissipate the same amount of energy, which is equal to E_R as given above.

Example 4.2: Charging a capacitor — Suppose the same capacitor shown in figure 4.5 is given an initial voltage V_i at $t = 0$. Assuming that $V_o > V_i$, we can calculate the energy loss when it is charged to V_o as follows. First, the differential equation describing the circuit for $t > 0$ is

$$
\frac{dv_C}{dt} + \frac{v_C}{CR} = \frac{V_o}{CR}
$$

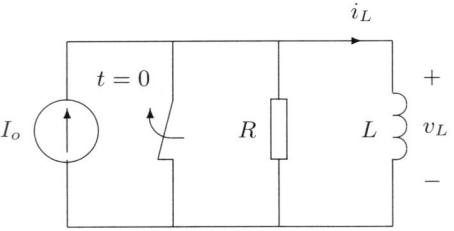

Figure 4.7: Simple RL circuit

The general solution, as given before, is

$$v_C(t) = Ae^{\frac{-t}{CR}} + V_o$$

where A can be found from the initial condition, $v_C(0^+) = V_i$, as $A = V_i - V_o$. Thus, the solution is

$$v_C(t) = V_i + (V_o - V_i)(1 - e^{\frac{-t}{CR}})$$

The current is therefore given by

$$i(t) = C\frac{dv_C}{dt} = \frac{V_o - V_i}{R}e^{\frac{-t}{CR}}$$

Thus, the energy dissipated in R from $t = 0$ to ∞ is

$$E_R = \int_0^\infty i(t)^2 R \, dt = \frac{(V_o - V_i)^2}{R^2}\int_0^\infty e^{\frac{-2t}{CR}} \, dt = \frac{1}{2}C(V_o - V_i)^2$$

4.5   Simple First-Order RL Circuit

Another simple first-order circuit is the RL circuit shown in figure 4.7. In this circuit, a constant current source is injected to a parallel combination of resistance and inductance at $t = 0$. For $t \geq 0$, we have

$$v_L(t) = L\frac{di_L}{dt}$$

and

$$v_R(t) = (I_o - i_L(t))R$$

Hence,

$$(I_o - i_L(t))R = L\frac{di_L}{dt}$$

or

$$\frac{di_L}{dt} + \frac{R}{L}i_L(t) = \frac{RI_o}{L}$$

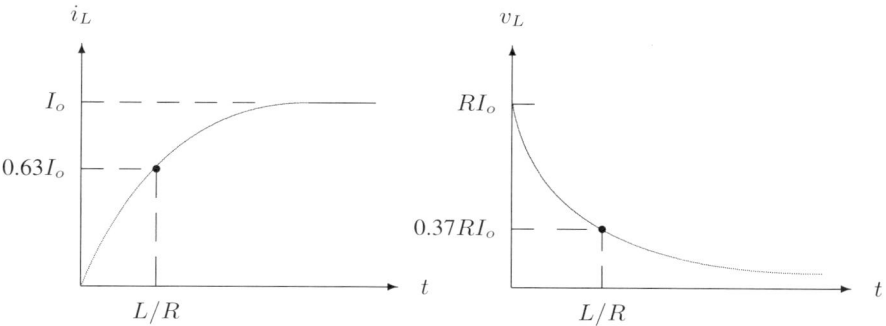

Figure 4.8: Waveforms of i_L and v_L in simple RL circuit

Suppose initially the inductor current is zero, i.e., $i(0^+) = 0$. The general solution is given by

$$i_L(t) = Ae^{\frac{-Rt}{L}} + I_o$$

where A is determined from the initial condition as

$$i_L(0^+) = 0 = A + I_o$$

$$\Rightarrow \quad A = -I_o$$

Hence, for $t \geq 0$, we obtain

$$i_L(t) = I_o\left(1 - e^{\frac{-Rt}{L}}\right)$$
$$v_L(t) = RI_o e^{\frac{-Rt}{L}}$$

Remarks — The waveforms of i_L and v_L, as shown in figure 4.8, demonstrate an interesting phenomenon called *duality*. Essentially we observe that i_L in the RL circuit has exactly the same response as v_C in the RC circuit. We say that the dual of the RC circuit is the RL circuit, with $C = x$ F replaced by $L = x$ H, and $R = y$ Ω replaced by $G = y$ S. Also, it can be shown that the time constants for this pair of circuits have the same magnitude. For the RC circuit the time constant is $CR = xy$ s. For the LR circuit the time constant according to the above derivation is L/R which is also equal to xy s.

4.6 Expressions for First-Order Transients by Inspection

In determining the transient expressions for voltages and currents in a first-order dynamic circuit, we always start with finding the transient expressions for the capacitor voltage or the inductor current. This is because the value of the capacitor voltage or inductor current is *almost always* continuous at the time the circuit configuration is

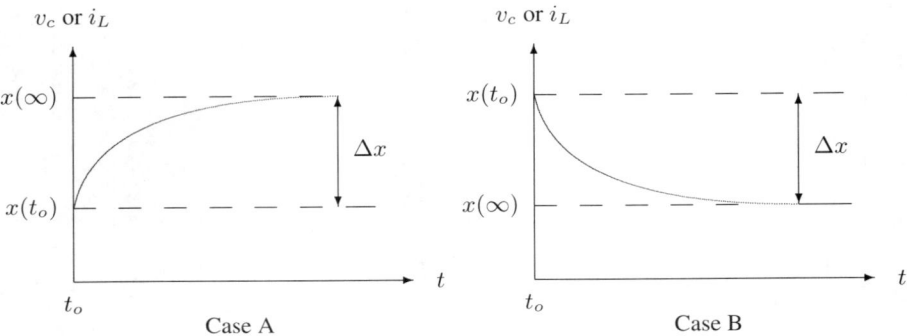

Figure 4.9: Universal waveforms of first-order transients

changed (the switch is turned on/off), and hence the initial condition is the same as the steady state condition prior to the change of circuit configuration. We will discuss in a later section the rare situation where continuity of the capacitor voltage or inductor current is not guaranteed.

The formal procedure, as outlined previously, involves the derivation of the first-order differential equations and the use of the initial condition to determine the general expressions. However, for first-order transients, we may determine the expressions very quickly by inspection.

Essentially, the capacitor voltage or inductor current in a given first-order dynamic circuit takes either of the waveforms shown in figure 4.9. Let x denote either the capacitor voltage or inductor current. Referring to figure 4.9, the expressions for the two possible waveforms can be easily derived. In Case A, $x(t)$ starts from $x(t_o)$ and rises exponentially towards $x(t_o) + \Delta x$. The expression for $x(t)$ is thus given by

$$\text{Case A:} \quad x(t) = x(t_o) + \Delta x \left(1 - e^{-(t-t_o)/\tau}\right)$$

where τ is the time constant. In Case B, $x(t)$ starts from $x(t_o)$ and falls exponentially towards $x(t_o) - \Delta x$. The expression is

$$\text{Case B:} \quad x(t) = x(\infty) + \Delta x \, e^{-(t-t_o)/\tau}$$

It should be noted that once we know the initial value $x(t_o)$, the final value $x(\infty)$ and the time constant τ, $x(t)$ will be determined for all t. Thus, the problem is reduced to

1. finding the time constant;

2. finding initial and final values;

3. hence determining whether the Case A or Case B expression should be used.

Example 4.3: Deriving transient expressions by inspection — Suppose an inductor in a first-order RL circuit has an initial current of 1A at $t = 0$. The circuit is let go until a steady state is reached. The current of the inductor is then found to be 3A. This corresponds to a transient motion of Case A. Thus, we can immediately write the current expression as

$$i_L(t) = 1 + (3-1)(1 - e^{-t/\tau}) = 3 - 2e^{-t/\tau}$$

Suppose the steady state has been reached by $t = 3$s. At $t = 3$s, some switching action causes the inductor current to move towards -2A. Clearly, the transient motion from 3A to -2A belongs to Case B, the starting time being $t = 3$s. Using the Case B expression, we can write the current expression for $t > 3$s as

$$i_L(t) = -2 + 5e^{-(t-3)/\tau}$$

4.7 Determination of the Time Constant

Basically, for any given RC circuit,

$$\tau = C_{eq}R_{eq}$$

and any RL circuit,

$$\tau = \frac{L_{eq}}{R_{eq}}$$

where C_{eq}, L_{eq} and R_{eq} are the equivalent capacitance, inductance and resistance of the circuit. Sometimes the circuit may look more complicated than just an RC or RL circuit, though it can still be first-order. In that case, we should attempt to simplify the circuit to the equivalent simple RC or RL form so that τ can be easily determined. The following procedure may be used for this purpose:

1. Short-circuit all voltage sources and open-circuit all current sources.
2. Place switches in their final positions.
3. Reduce resistances to one equivalent resistance, if possible.
4. Reduce capacitances to one equivalent capacitance, if possible.
5. Reduce inductances to one equivalent inductance, if possible.

Note that if the circuit is first-order, it would not contain both inductors and capacitors, and either step 4 or 5 is always possible. We will illustrate the procedure with two examples.

Example 4.4: Transient expressions for switched RC circuit — In the circuit shown in figure 4.10 (a), the switch originally connects the 50V source to an RC load.

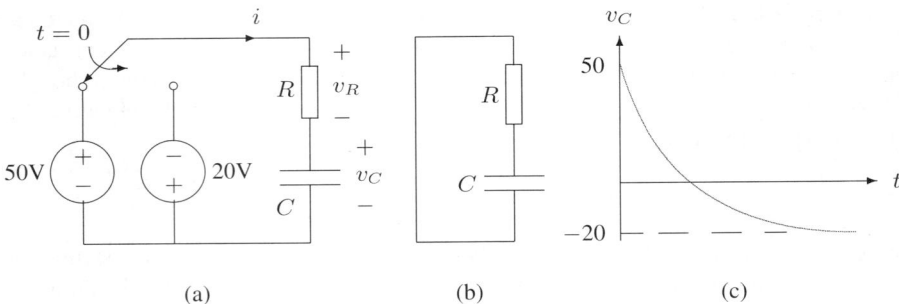

Figure 4.10: RC circuit with readily determined initial and final conditions

At $t = 0$, the switch is thrown to the right, connecting the -20V source to the RC load. Obviously the initial and final values of the capacitor voltage are

$$v_C(0^+) = 50\text{V} \quad \text{and} \quad v_C(\infty) = -20\text{V}$$

Thus, it starts from 50V and decays down to -20V exponentially, as shown in figure 4.10 (c). This corresponds to Case B of figure 4.9. The equivalent RC circuit, using the above reduction procedure, is shown in figure 4.10 (b), from which the time constant is found as

$$\tau = CR$$

Using the Case B expression, we have

$$v_C(t) = -20 + 70e^{-t/CR}$$

Furthermore, we may determine $v_R(t)$ and $i(t)$, respectively, as

$$v_R(t) = -20 - v_C(t) = -70e^{-t/CR}$$

and

$$i(t) = \frac{v_R(t)}{R} = \frac{-70}{R}e^{-t/CR}$$

Example 4.5: RC circuit with non-trivial final condition — In the circuit of figure 4.11 (a), the initial condition is easily determined by inspection, but the final condition is non-trivial. First of all, let us calculate its time constant by reducing it to an equivalent RC model. The circuit after the switch is thrown to the right is shown in figure 4.11 (b). This circuit is reducible to the one shown in figure 4.11 (c). The equivalent capacitance and resistance are found by straightforward inspection as

$$
\begin{aligned}
C_{eq} &= \frac{C_3(C_1 + C_2)}{C_1 + C_2 + C_3} \\
R_{eq} &= R_1 + R_2
\end{aligned}
$$

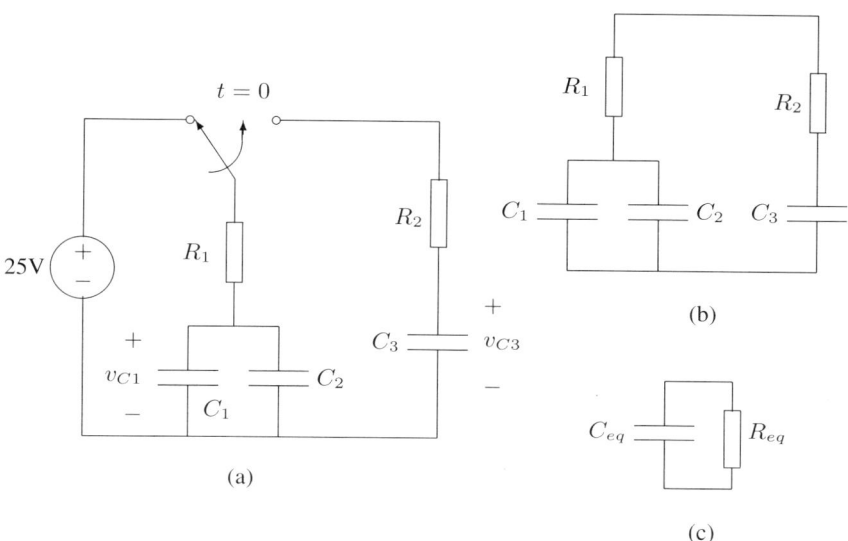

Figure 4.11: RC circuit with non-trivial final condition

Hence, the time constant is

$$\tau = \frac{C_3(C_1 + C_2)(R_1 + R_2)}{C_1 + C_2 + C_3}$$

The initial values are

$$v_{C1}(0^+) = 25, \quad v_{C2}(0^+) = 25, \quad v_{C3}(0^+) = 0.$$

To find the final values of the capacitor voltages, we note from figure 4.11 (b) that

$$(C_1 + C_2)\frac{dv_{C1}}{dt} + C_3\frac{dv_{C3}}{dt} = 0$$

which gives

$$(C_1 + C_2)v_{C1} + C_3 v_{C3} = \text{constant}$$

for all $t \geq 0$. The initial condition dictates that the constant be $25(C_1 + C_2)$. Thus, we have, for all $t \geq 0$,

$$(C_1 + C_2)v_{C1} + C_3 v_{C3} = 25(C_1 + C_2)$$

Since the final values must satisfy $V_{C1}(\infty) = V_{C3}(\infty)$, we have

$$v_{C1}(\infty) = v_{C2}(\infty) = v_{C3}(\infty) = \frac{25(C_1 + C_2)}{C_1 + C_2 + C_3}$$

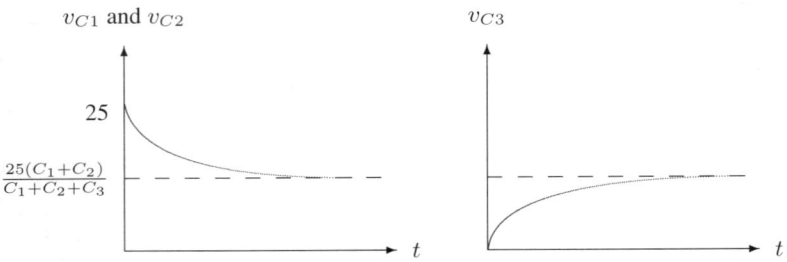

Figure 4.12: Waveforms for the circuit of Example 4.4

Thus, v_{C1} and v_{C2} belong to Case B, whereas v_{C3} belongs to Case A. The waveforms are shown in figure 4.12, from which we have

$$v_{C1}(t) = v_{C2}(t) = \frac{25(C_1 + C_2)}{C_1 + C_2 + C_3} + \frac{25C_3}{C_1 + C_2 + C_3} e^{-t/C_{eq}R_{eq}}$$

and

$$v_{C3}(t) = \frac{25(C_1 + C_2)}{C_1 + C_2 + C_3} \left(1 - e^{-t/C_{eq}R_{eq}}\right)$$

4.8   Initial Conditions and Continuity

Usually before the switch is turned on/off, the circuit already assumes a steady state in which the values of all voltages and currents are supposedly known. Now the switch is turned on/off at $t = 0$. Immediately after this, we are not certain about the voltage and current values. Consider the variable x. The question is whether x is continuous, i.e., whether or not the following is true.

$$x(0^-) = x(0^+)$$

In fact, if x is capacitor voltage or inductor current, the above equation is *always* true for practical circuits. For this reason, *we always derive the expression for v_C or i_L in the first place* since we can always find $v_C(0^+)$ or $i_L(0^+)$.[3] Voltages and currents across resistances, on the other hand, are not always continuous. Therefore we are not always certain about the value of $v_R(0^+)$, even though we know $v_R(0^-)$.

The fact that capacitor voltages and inductor currents cannot be discontinuous can also be appreciated from their constitutive relations. Since $i = C(dv/dt)$ for a capacitor, any jump in capacitor voltage gives rise to infinite current. Likewise, since $v = L(di/dt)$ for an inductor, any jump in inductor current gives rise to infinite voltage. Infinite current and voltage are both not permitted in physical circuits.

---

[3]We will see in Chapter 11 that capacitor voltages and inductor currents are state variables, in terms of which the dynamics of a circuit is described.

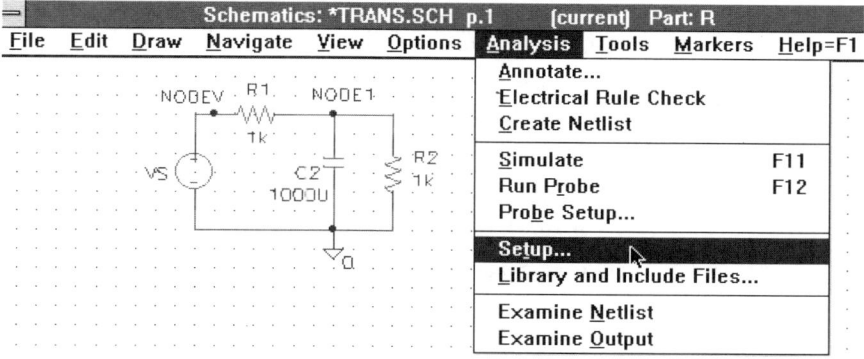

Figure 4.13: Circuit drawn in *Schematics* for PSPICE transient analysis

Of course, theoretically we may tolerate infinite current or voltage, and we may expect capacitor voltage or inductor current to be discontinuous under certain hypothetical conditions. For example, if a capacitor is switched abruptly to a voltage source, then its voltage must be forced to jump from one value to another, and hence $v_C(0^-) \neq v_C(0^+)$. Likewise, if an inductor is switched abruptly to a current source, then its current must be forced to jump from one value to another, and hence $i_L(0^-) \neq i_L(0^+)$.[4] In theory, we can claim continuity of a capacitor voltage (inductor current) if the capacitor (inductor) is not abruptly forced to take up a certain value imposed by independent voltage (current) sources when a switch is closed (opened). However, we should bear in mind that physical circuits always have resistance to prevent abrupt switching and hence continuity is always guaranteed for capacitor voltages and inductor currents in real circuits.

4.9   Transient Analysis with PSPICE

When the circuit to be analyzed contains capacitors and inductors, PSPICE can be used to generate the transient waveforms of the desired voltages and currents. The following example demonstrates how transient analysis can be invoked in PSPICE and how several waveforms can be displayed simultaneously using PROBE.

Example 4.6: Using PSPICE for transient analysis — As usual we start with specifying the circuit to be analyzed by drawing it in *Schematics*, as shown in figure 4.13. The analysis in general involves assigning initial values, selecting transient analysis, invoking simulation, and displaying the required waveforms. A summary of the procedure for analyzing the transient response of the circuit is as follows.

1. To specify initial values for the capacitor voltage, we double-click the capacitor to bring up a dialogue box in which we can enter the desired initial value. In

---

[4]A formal statement of the continuity condition is given in the Continuity theorem, which involves the graph theoretic concepts of loops and cutsets.

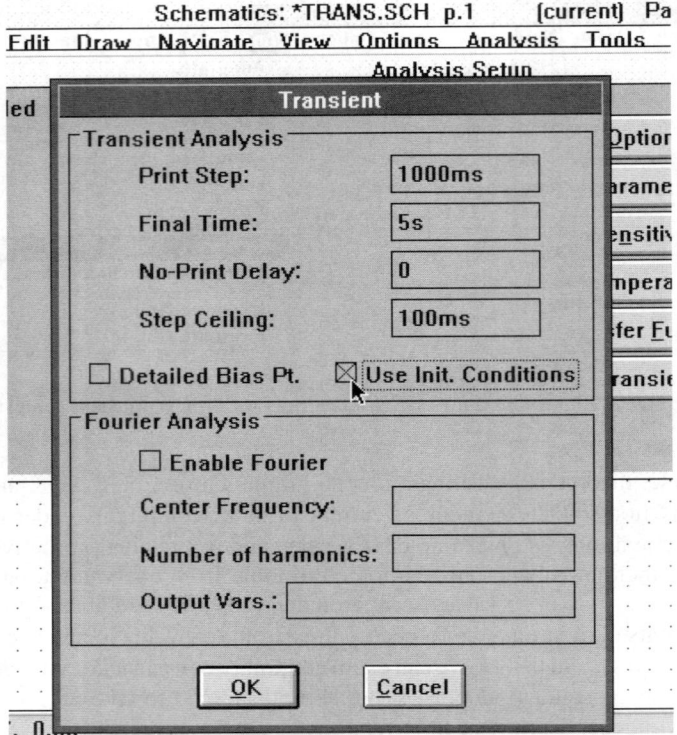

Figure 4.14: Dialogue box for specifying parameters for transient analysis

this example, we set the initial value to 3V. In PSPICE, a zero initial value is assigned by default to all capacitor voltages and inductor currents whose initial values are unspecified.

2. To select transient analysis as the type of analysis to be performed, we choose *"Setup"* from the **Analysis** menu and click *"Transient"* to bring up a dialogue box, as shown in figure 4.14, for entering "print step", "final time", "no-print delay" and "step ceiling". Here, "print step" refers to the step size at which the waveform is to be plotted, "no-print delay" is the amount of time from $t=0$ that is not plotted, and "step ceiling" is the minimum calculation interval. The "print step" must not be smaller than the "step ceiling". Furthermore, we may specify whether initial values are to be used for the analysis by clicking the appropriate check box.

3. We can now invoke the simulation process by choosing *"Simulate"* from the **Analysis** menu.

4. After PSPICE has finished the analysis, we may invoke PROBE to display the desired waveforms. (See also Example 2.8.) In the PROBE window, we choose

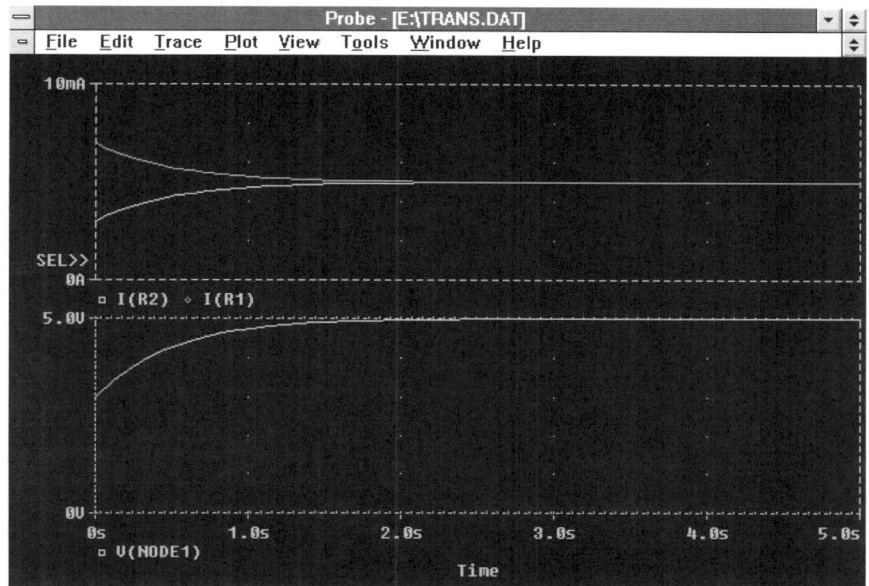

Figure 4.15: Transient waveforms from PSPICE. Upper trace in upper graph: current in R1; lower trace in upper graph: current in R2; Trace in lower graph: voltage across C2

"Add" from the **Trace** menu to select the voltage or current to be plotted. Moreover, we may either plot several waveforms in the same graph by choosing *"Add"* repeatedly, or plot them on separate graphs by choosing *"Add plot"* from the **Plot** menu. As shown in figure 4.15, the two resistor currents are plotted on one graph and the capacitor voltage is plotted on a separate graph. It is also possible to modify the scales of the axes using the **Plot** menu.

Remarks — PSPICE does not allow the circuit to be analyzed to contain a node that has no DC path to the ground node. If two capacitors are connected in series, the node between the capacitors will cause PSPICE to report an error in the output file. In this case, it may read "node X is floating". (PSPICE's output file may be examined by choosing *"Examine Output"* from the **Analysis** menu.) To get PSPICE back to work, a resistor may be inserted between this "floating" node and the ground node. The value of this extra resistor should be large enough to remain essentially as an open-circuit.

4.10 Problems

1. A 400μF capacitor is charged by a voltage source of 10V. Assume that the capacitor has an initial voltage of 3V. Calculate the energy loss when the capacitor is charged to 10V.

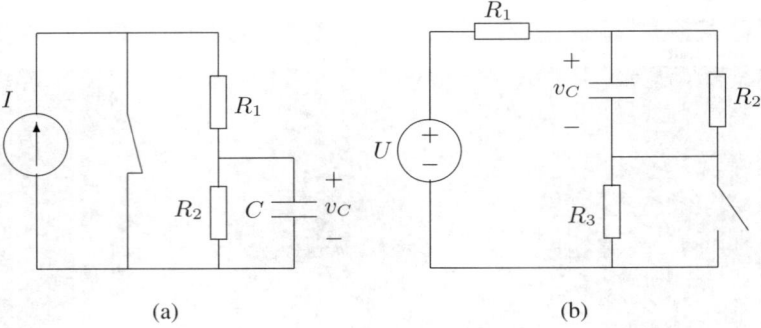

Figure 4.16: Circuits for Problems 3 and 4

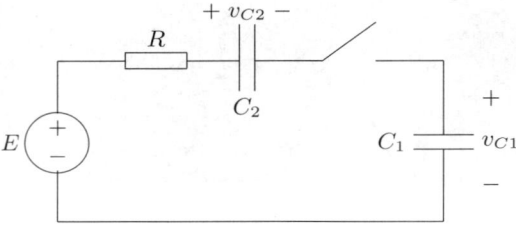

Figure 4.17: Circuit for Problem 5

2. The same capacitor in Problem 1 is charged up by the same 10V voltage source. However, the charging process is stopped when the capacitor voltage reaches 8V. What is the energy loss during the charging process? Do you need to know any other parameters in order to calculate the loss in this case?

3. For the circuit shown in figure 4.16 (a), find the value of the capacitor voltage at $t = 0^+$ and $t = \infty$, assuming that the circuit has been in a steady state for $t < 0$, and that the switch is opened at $t = 0$. Calculate the time constant, and write down the expression for the capacitor voltage.

4. For the circuit shown in figure 4.16 (b), find the value of the capacitor voltage at $t = 0^+$ and $t = \infty$, assuming that the circuit has been in a steady state for $t < 0$, and that the switch is closed at $t = 0$. Calculate the time constant, and write down the expression for the capacitor voltage.

5. For the circuit shown in figure 4.17, assume that $v_{c1}(0^-) = E/2$, $v_{c2}(0^-) = 0$, $C_1 = C_2$, and the switch is closed at $t = 0$. Find the expressions for $v_{c1}(t)$ and $v_{c2}(t)$ for $t > 0$. What is the energy dissipated in R from $t = 0$ to $t = \infty$?

6. A first-order dynamic circuit is composed of a capacitor, a voltage source, a set of switches and a number of resistors. Steady-state measurements of the capacitor voltage are taken for three particular sets of switch states. The measurement for state A is 5V, for state B is 8V, and for state C is 6V. The time constant of

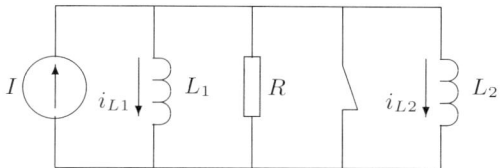

Figure 4.18: Circuit for Problem 7

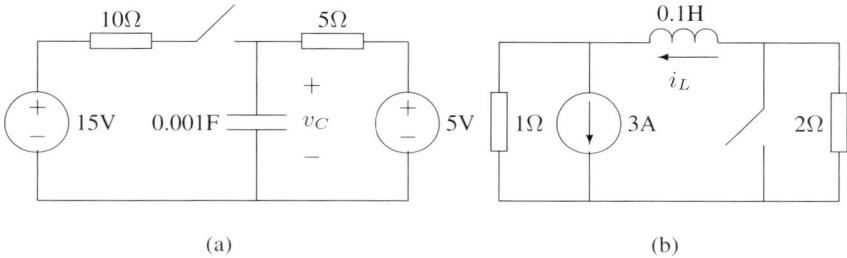

(a)   (b)

Figure 4.19: Circuit for Problem 9

the circuit is 2s when it is in state A, 4s in state B, and 8s in state C. Derive the transient expression for the capacitor voltage when the circuit is switched (a) from state A to state C; (b) from state B to state A; and (c) from state C to state B.

7. Verify that the circuit of figure 4.18 is first-order after the switch is opened. Find the expressions for $i_{L1}(t)$ and $i_{L2}(t)$, assuming the switch is opened at $t = 0$, $i_{L1}(0^-) = I/2$, $i_{L2}(0^-) = 2I$ and $L_1 = L_2$.

8. Assuming continuous capacitor voltage and inductor current, state which of the following concerning a first-order circuit is/are correct:

   (a) If $v_C(0^-) = 10V$ and $v_C(\infty) = 16V$, then $v_C(t) = 10 + 6(1 - e^{-t/\tau})$.
   (b) If $v_C(0^-) = 1V$ and $v_C(\infty) = 5V$, then $v_C(t) = 1 + 5e^{-t/\tau}$.
   (c) If $v_R(0^+) = 10V$ and $v_R(\infty) = 0V$, then $v_R(t) = 10e^{-t/\tau}$.
   (d) If $v_R(0^-) = 10V$ and $v_R(\infty) = 0V$, then $v_R(t) = 10e^{-t/\tau}$.
   (e) If $i_L(0^-) = 5A$ and $i_L(\infty) = 9A$, then $i_L(t) = 5 + 4(1 - e^{-t/\tau})$.
   (f) If $i_L(0^+) = 5A$ and $i_L(\infty) = 1A$, then $i_L(t) = 1 + 4e^{-t/\tau}$.

9. For the circuits in figure 4.19, assume that the switch is closed at $t = 0$. Find the voltage across the capacitor v_C and the current through the 10Ω resistor in circuit (a), for $t > 0$. Find also the current in the inductor i_L and the voltage across the 1Ω resistor in circuit (b), for $t > 0$.

10. Use PSPICE to analyze the circuits shown in figure 4.20. In particular, obtain the transient waveforms for all capacitor voltages. Initial values are given in the circuit diagrams. (Note that in circuit (a), the $100M\Omega$ resistor is not supposed

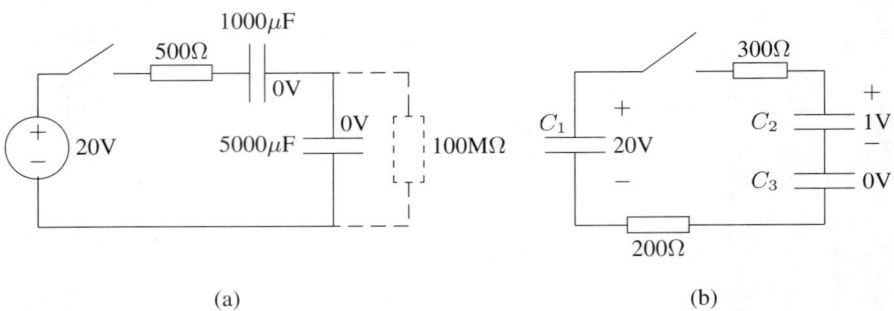

(a)   (b)

Figure 4.20: Circuit for Problem 10

to be part of the circuit to be analyzed (dashed box), and is needed only for PSPICE analysis since PSPICE cannot allow any node in a circuit without a DC path to ground. The presence of the huge resistance is mainly to provide a DC path for the node between the two capacitors in circuit (a). It has no effect on the transient response. When using PSPICE to analyze circuit (b), you should also insert two extra resistors appropriately.)

Chapter 5

Mutual Inductance and Transformers

In Chapter 4 we define a linear inductor as a circuit element whose voltage is proportional to the rate of change of its current, i.e., $v_L = L(di_L/dt)$. The validity of this defining relation relies on the assumption that the flux linking the inductor's coil is proportional to the current flowing through the coil, i.e., $\Phi = Li_L/N$, where Φ is the flux linkage per turn and N is the number of turns of the coil. If we have two inductors, we would expect their voltages to be proportional to the rate of change of their respective currents. If this is true, the two inductors are independent of each other, and as shown in figure 5.1 (a), the flux due to the current in one coil links only to the coil itself. The two coils, in this case, are said to be *uncoupled*. In reality, since magnetic flux can penetrate into the air and extend to a distance, the flux linking a coil may have been due partly to currents in other inductances in the vicinity. In this chapter we will study the properties of the coupled inductor pair and its practical use in voltage and current transformation. In particular we will introduce the *transformer*, in both idealized and practical forms, as a special four-terminal circuit element having galvanic isolation between two pairs of terminals.

5.1 Equations of Coupled Inductors

In the case of two inductors in proximity, as illustrated in figure 5.1 (b), assuming a linear relation between flux linkage and current, the voltage in one inductor would be *partially proportional* to the rate of change of its own current and *partially proportional* to the rate of change of the current in the other inductor, i.e.,

$$
\begin{aligned}
v_1 &= L_{11}\frac{di_1}{dt} + L_{12}\frac{di_2}{dt} \\
v_2 &= L_{21}\frac{di_1}{dt} + L_{22}\frac{di_2}{dt}
\end{aligned}
$$

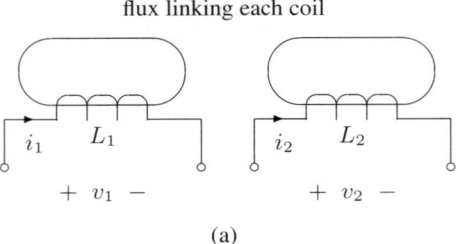

flux linking each coil

(a)

flux linking both coils

(b)

Figure 5.1: (a) Uncoupled inductors and (b) coupled inductors

where

$$
\begin{aligned}
L_{11} &= \quad \text{self-inductance of coil 1} \\
L_{22} &= \quad \text{self-inductance of coil 2} \\
L_{12} &= \quad \text{trans-inductance of coil 1 from coil 2} \\
L_{21} &= \quad \text{trans-inductance of coil 2 from coil 1}
\end{aligned}
$$

Note that uncoupled inductors may be viewed as a particular case of the above general situation, with $L_{12} = L_{21} = 0$.

5.2 Mutual Inductance

It is interesting to consider the energy stored in the pair of coupled inductors shown in figure 5.1 (b) from $t = t_1$ to t_2. In general, this stored energy, W, is given by

$$
W = \int_{t_1}^{t_2} (v_1 i_1 + v_2 i_2) dt
$$

To find W, we put v_1 and v_2 directly into the RHS of the above expression, giving

$$
W = \frac{1}{2} \left[L_{11} i_1^2 + L_{22} i_2^2 \right]_{t_1}^{t_2} + \int_{t_1}^{t_2} \left(L_{12} i_1 \frac{di_2}{dt} + L_{21} i_2 \frac{di_1}{dt} \right) dt
$$

There are two possible ways to expand this expression. Eliminating di_1/dt gives

$$W = \frac{1}{2}\left[L_{11}i_1^2 + L_{22}i_2^2 + 2L_{21}i_1i_2\right]_{t_1}^{t_2} + \int_{t_1}^{t_2}(L_{12} - L_{21})i_1\frac{di_2}{dt}\,dt$$

Alternatively, eliminating di_2/dt gives

$$W = \frac{1}{2}\left[L_{11}i_1^2 + L_{22}i_2^2 + 2L_{12}i_1i_2\right]_{t_1}^{t_2} + \int_{t_1}^{t_2}(L_{21} - L_{12})i_2\frac{di_1}{dt}\,dt$$

If we let t_2 tend to infinity, $i_1 = \cos t$ and $i_2 = \sin t$, then the first expansion implies $L_{12} \leq L_{21}$ since W must be finite. But the second expansion implies $L_{21} \leq L_{12}$ for the same reason. Therefore, we must have

$$L_{12} = L_{21}$$

For brevity, we denote L_{12} and L_{21} by M, and refer to it as *mutual inductance*. Thus, in general, the energy stored in the coupled inductors is

$$W = \left[\frac{1}{2}L_{11}i_1^2 + \frac{1}{2}L_{22}i_2^2 + Mi_1i_2\right]_{t_1}^{t_2}$$

Suppose, at $t = t_1$, the energy stored is zero, and $i_1 = i_2 = 0$. Then the energy stored at any time t is

$$W(t) = \begin{cases} \frac{1}{2}\left[\left(\sqrt{L_{11}}i_1 + \frac{Mi_2}{\sqrt{L_{11}}}\right)^2 + \left(L_{22} - \frac{M^2}{L_{11}}\right)i_2^2\right] & \text{if } L_{11} \neq 0 \\[3mm] \frac{1}{2}\left[\left(\sqrt{L_{22}}i_2 + \frac{Mi_1}{\sqrt{L_{22}}}\right)^2 + \left(L_{11} - \frac{M^2}{L_{22}}\right)i_1^2\right] & \text{if } L_{22} \neq 0 \end{cases}$$

Since $W(t) \geq 0$ for all i_1 and i_2, we must have

$$L_{11}L_{22} \geq M^2$$

In summary, the following results concerning L_{11}, L_{12}, L_{21} and L_{22} are true:

(i)	$L_{12} = L_{21} = M$
(ii)	$L_{11}L_{22} \geq M^2$

Without ambiguity, from now on, we will simply use L_1 to denote the self inductance L_{11}, and L_2 to denote the self inductance L_{22}.

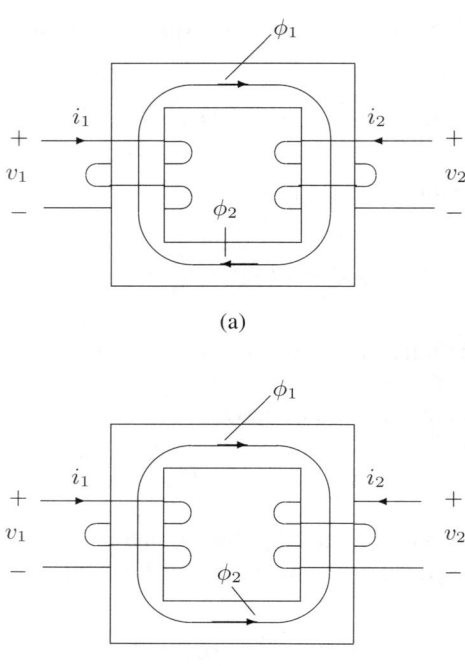

(a)

(b)

Figure 5.2: Polarity and direction. Note the difference in the winding direction of the right-hand coil in the two cases.

5.3 Direction and Polarity — The Dot Convention

The physical orientation of the coils and current directions can affect the way the coils are coupled. In the case of two coils wound on the same magnetic core, the effect of coupling would be completely different for the two cases shown in figure 5.2. In figure 5.2 (a), the two coils produce flux flowing in the same direction through the core,[1] whereas in figure 5.2 (b), they produce flux flowing in opposite directions. Suppose the equations for the coupled coils of figure 5.2 (a) are

$$v_1 \;=\; L_1 \frac{di_1}{dt} + M \frac{di_2}{dt}$$
$$v_2 \;=\; M \frac{di_1}{dt} + L_2 \frac{di_2}{dt}$$

Then, the equations for the coupled coils of figure 5.2 (b) should be adjusted to

$$v_1 \;=\; L_1 \frac{di_1}{dt} - M \frac{di_2}{dt}$$

---

[1] The right-hand screw rule should be used to determine the direction of magnetic flux flowing through the core.

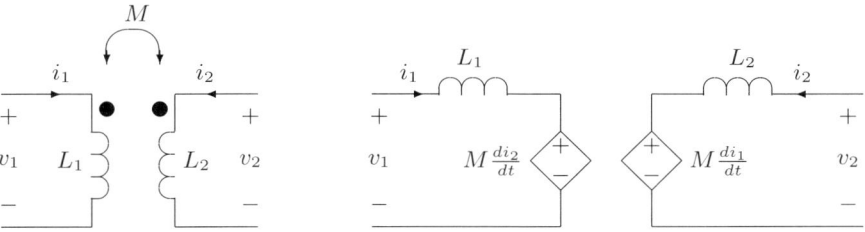

Figure 5.3: Dot convention and circuit model

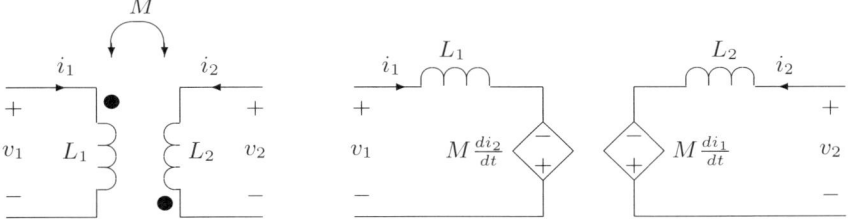

Figure 5.4: Dot convention. (Note the current direction)

$$v_2 = -M\frac{di_1}{dt} + L_2\frac{di_2}{dt}$$

Thus, to avoid confusion, we must specify the current directions as well as winding directions for the coupled coils. This is done using the *dot convention*. Basically, coupled inductors are marked with "•" on one side of the terminals. For each inductor, we assume that current enters the side marked with a "•", and that voltages on all coupled coils have positive polarity on the "•" side. The circuit symbol and the corresponding model are shown in figure 5.3. Hence, using this convention, the coupled coils of figure 5.2 (b) are represented by the circuit symbol in figure 5.4.

Example 5.1: Coupled equations — Figure 5.5 shows a circuit containing a coupled inductor pair. The switch is closed at $t = 0$, with all currents and voltages being zero initially. Suppose we wish to find the current in the resistor, i_R. Using the equa-

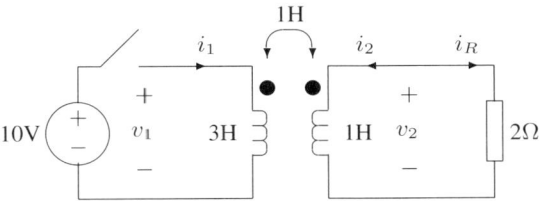

Figure 5.5: Circuit with coupled inductors

tions of coupled coils and the Ohm's law equation for the resistor, we can write down the describing equations, for $t > 0$, as

$$10 = 3\frac{di_1}{dt} + \frac{di_2}{dt}$$

$$v_2 = \frac{di_1}{dt} + \frac{di_2}{dt}$$

$$v_2 = -2i_2$$

Eliminating i_1 from the first and second equations gives

$$2\frac{di_2}{dt} = 3v_2 - 10$$

Hence, putting $v_2 = -2i_2$, we have

$$\frac{di_2}{dt} + 3i_2 = -5$$

Solving this differential equation yields

$$i_2(t) = Ae^{-3t} - \frac{5}{3}$$

where A is determined by the initial value of i_2. In this case, we have $A = 5/3$, since $i_2(0^+) = 0$. Hence, the current in the resistor is

$$i_R(t) = -i_2(t) = \frac{5}{3}(1 - e^{-3t})$$

5.4 Coupling Coefficient

We have shown in Section 5.2 that the value of M^2 is always less than the product of L_1 and L_2. For very tightly coupled coils, the value of M is close to $\sqrt{L_1 L_2}$. Conversely, the value of M approaches zero for very loosely coupled coils, and is identically zero for uncoupled coils. Thus, we may measure how tightly two coils are coupled in terms of the ratio $M/\sqrt{L_1 L_2}$. This ratio is known as the *coupling coefficient*, and is usually denoted by κ.

$$\kappa = \sqrt{\frac{M^2}{L_1 L_2}}$$

Remarks — L_1, L_2 and M share the same unit of inductance, i.e., *henry* (H), whereas κ is a dimensionless ratio.

Example 5.2: Coupling coefficient — For the coupled coils shown in figure 5.5, $L_1 = 3$H, $L_2 = 1$H and $M = 1$H. The coupling coefficient κ is given by

$$\kappa = \sqrt{\frac{M^2}{L_1 L_2}} = \sqrt{\frac{1}{3}} = 0.5774$$

5.5 The Transformer

Consider a pair of coupled inductors. If we apply a sinusoidal voltage to v_1 and open-circuit v_2, the coupling equations reduce to

$$v_1 = L_1 \frac{di_1}{dt} + 0$$

$$v_2 = M \frac{di_1}{dt} + 0$$

from which we have

$$v_2 = \frac{M}{L_1} v_1$$

Thus, at v_2, we obtain a replica of v_1. Also, by choosing M and L_1 we can select any voltage ratio as desired.

Similarly if we apply a sinusoidal current to i_1 and short-circuit i_2, we have

$$i_2 = -\frac{M}{L_2} i_1$$

which represents a current transformation. We may thus conclude that the pair of coupled inductors can be used to transform voltage or current, and at the same time maintain *galvanic isolation.*

5.5.1 The Ideal Transformer

In this subsection we consider a special application of the coupled coils, namely, changing the magnitude of a voltage or current. We start by considering a special case of the coupled inductors in which perfect coupling is maintained and the self-inductances are infinitely large. First, for a perfectly coupled pair of inductors, $M^2 = L_1 L_2$. The coupling equations become

$$v_1 = L_1 \frac{di_1}{dt} + M \frac{di_2}{dt}$$

$$v_2 = \frac{M}{L_1} \left(L_1 \frac{di_1}{dt} + M \frac{di_2}{dt} \right)$$

Hence, putting the first equation to the second equation gives

$$v_2 = \frac{M}{L_1} v_1$$

which does not require open-circuiting v_2 in order to achieve voltage transformation.

With perfect coupling, both coils have the same flux linkage per turn, since the flux that links one coil must also link the other coil. Suppose the flux linkage per turn is Φ, and inductors L_1 and L_2 have N_1 and N_2 turns respectively. We have, from Faraday's law,

$$v_1 = N_1 \frac{d}{dt} \Phi$$

$$v_2 = N_2 \frac{d}{dt} \Phi$$

which gives

$$v_2 = \frac{N_2}{N_1} v_1$$

Hence, we have the following relations:

$$\frac{v_2}{v_1} = \frac{N_2}{N_1} = \frac{M}{L_1}$$

Furthermore, if we let $L_1 \to \infty$, we have, from the first coupling equation,

$$\lim_{L_1 \to \infty} \frac{v_1}{L_1} = \frac{di_1}{dt} + \frac{M}{L_1} \frac{di_2}{dt}$$

$$\Rightarrow \quad 0 = \frac{di_1}{dt} + \frac{M}{L_1} \frac{di_2}{dt}$$

$$\Rightarrow \quad 0 = \frac{d}{dt} \left(i_1 + \frac{N_2}{N_1} i_2 \right)$$

$$\Rightarrow \quad K = i_1 + \frac{N_2}{N_1} i_2$$

where K is the integration constant. Choosing $K = 0$, we obtain a new element called the *ideal transformer,* which is defined by:

$$i_1 = -n i_2$$
$$v_1 = \frac{1}{n} v_2$$

with n replacing N_2/N_1 for the sake of brevity. Note that the above derivation does not involve any requirement of short-circuit or open-circuit terminals.

In summary, a pair of perfectly coupled inductors having infinite inductances is by definition an ideal tranformer. The ideal transformer may be viewed as a new circuit element, regardless of its physical origin, and is characterized by the following constitutive relation:

$$\begin{bmatrix} i_1 \\ v_1 \end{bmatrix} = \begin{bmatrix} -n & 0 \\ 0 & 1/n \end{bmatrix} \begin{bmatrix} i_2 \\ v_2 \end{bmatrix}$$

with n being a characteristic constant. The symbol for the ideal transformer is shown in figure 5.6. Usually, we refer to the left side as the *primary* side, and the right side as the *secondary* side. Nevertheless, such an assignment is arbitrary.

Remarks — We can consider an ideal transformer as a theoretical circuit element which is defined by the above matrix relation. It is worth noting that in modelling of abstract behaviour of many electrical devices and systems, the ideal transformer concept is often made use of. The following example illustrates the analysis of circuits that contain ideal transformers.

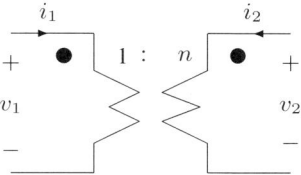

Figure 5.6: Symbol for the ideal transformer

Example 5.3: Circuit containing ideal transformers — Consider the circuit shown in figure 5.7. We will tackle this problem using the mesh method, with the transformer voltages as extra unknowns to be found. Moreover, in addition to the set of mesh equations, a set of ideal transformer equations come into play. Specifically, the usual set of mesh equations is

$$
\begin{aligned}
v_1 + v_3 + 5i_1 &= 10 \\
v_2 + 10i_2 + 8(i_2 - i_4) &= 0 \\
v_4 - 8(i_2 - i_4) &= 0 \\
i_1 - i_3 &= 0
\end{aligned}
$$

and the set of transformer equations is

$$
\begin{aligned}
2v_1 &= v_2 \\
i_1 &= -2i_2 \\
4v_3 &= v_4 \\
i_3 &= -4i_4
\end{aligned}
$$

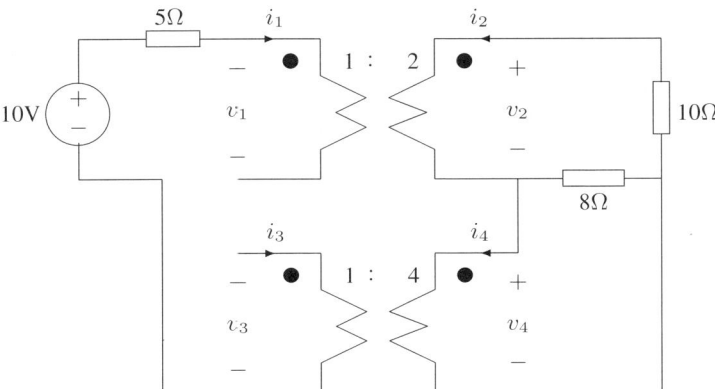

Figure 5.7: Circuit with ideal transformers

Putting the transformer equations into the above mesh equations gives

$$v_1 + v_3 + 5i_1 = 10$$

$$2v_1 - 5i_1 - 8\left(\frac{i_1}{2} - \frac{i_1}{4}\right) = 0$$

$$4v_3 + 8\left(\frac{i_1}{2} - \frac{i_1}{4}\right) = 0$$

Upon simplifying, we have the following set of equations:

$$v_1 + v_3 + 5i_1 = 10$$

$$2v_1 - 7i_1 = 0$$

$$2v_3 + i_1 = 0$$

The solution to this set of equations is $i_1 = 1.25$A, $v_1 = 4.375$V and $v_3 = -0.625$V. We may also calculate other voltages and currents. For example, $i_2 = -0.5i_1 = -0.625$A, $v_2 = 2v_1 = 8.75$V, $v_4 = 4v_3 = -2.5$V, $i_4 = -0.25i_1 = -0.3125$A, etc.

Remarks — We should stress that our aim is to study the method of analysis for circuits containing ideal transformers that are defined by the characteristic v-i relations. Although such a problem is important in the theoretical context, we should bear in mind that real transformers cannot transform DC voltages and currents which have zero change rate. Indeed, if we build the circuit using real transformers, we will get zero voltages and currents on both sides of the transformers. To make this particular problem meaningful (if we really wish to attach a physical meaning to the problem), we may consider the voltage source to be a time-varying function such as a sine wave. In particular, if the 10V source is actually a sinusoidal voltage source of amplitude 10V, then all voltages and currents are sinusoidal and their amplitudes are given by the values calculated above. Also, positive and negative values denote in-phase and anti-phase waveforms, respectively, with respect to the voltage source.

5.5.2 Practical Transformer Model

In practice, when transformers are constructed using coupled coils, the two assumptions made in deriving the ideal transformer are not valid. Let us now examine the effects of finite inductances and imperfect coupling on the transformer operation. Firstly, since L_1 is finite, we do not have $\frac{d}{dt}(i_1 + ni_2) = 0$, but rather have

$$\frac{d}{dt}(i_1 + ni_2) = \frac{v_1}{L_1}$$

This equation can be exactly modelled as an inductance of L_1 carrying a current of $i_1 + ni_2$, as shown in figure 5.8 (a). Since the current reflected from the secondary to the primary is ni_2, we obtain the model shown in figure 5.8 (b). Secondly, the equation for v_2 should be written as it originally is.

$$v_2 = M\frac{di_1}{dt} + L_2\frac{di_2}{dt}$$

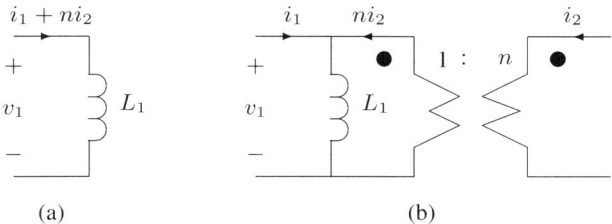

Figure 5.8: Practical transformer primary side (resistor for modelling loss omitted)

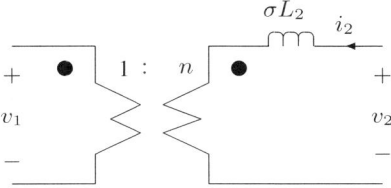

Figure 5.9: Practical transformer secondary side (resistor for modelling loss omitted)

In order to model the secondary side, we should try to eliminate di_1/dt. This can be done by expressing $\frac{di_1}{dt}$ as $\frac{v_1}{L_1} - n\frac{di_2}{dt}$ and putting it in the v_2 equation above.

$$
\begin{aligned}
v_2 &= M\left(\frac{v_1}{L_1} - n\frac{di_2}{dt}\right) + L_2\frac{di_2}{dt} \\
&= \frac{M}{L_1}v_1 - \frac{M^2}{L_1}\frac{di_2}{dt} + L_2\frac{di_2}{dt} \\
&= \frac{M}{L_1}v_1 + \left(L_2 - \frac{M^2}{L_1}\right)\frac{di_2}{dt} \\
&= nv_1 + L_2\left(1 - \kappa^2\right)\frac{di_2}{dt}
\end{aligned}
$$

Putting $\sigma = 1 - \kappa^2$ gives

$$
v_2 = nv_1 + \sigma L_2\frac{di_2}{dt}
$$

from which the secondary side can be modelled as shown in figure 5.9. Combining the models of the primary and secondary sides, we obtain a complete model for the practical transformer.[2] It should be noted that this practical model reduces to the ideal transformer when $L_1 \to \infty$ and $\sigma = 0$, i.e., $\kappa = 1$.

Remarks — Real practical transformers dissipate power in the windings due to I^2R heating, and in the magnetic cores due to hysteresis and eddy currents. These

---

[2]Engineers usually refer to L_1 as the magnetising inductance and to σL_2 as the leakage inductance.

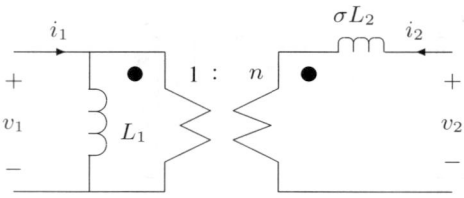

Figure 5.10: Practical transformer model (resistors for modelling losses are omitted)

losses can be modelled by inserting resistors in parallel with L_1 and also in series with σL_2.

Example 5.4: Deriving parameters for practical transformer model — Suppose a pair of coupled inductors has $L_1 = 3$H, $L_2 = 1$H and $M = 0.8$H. The physical turn ratio is given by

$$\frac{N_2}{N_1} = \frac{M}{L_1} = \frac{0.8}{3} = 0.2667$$

The coupling coefficient κ is

$$\kappa = \sqrt{\frac{0.64}{3}} = 0.4619$$

The equivalent inductance σL_2 placed on the secondary side, as in figure 5.10, is given by

$$\sigma L_2 = (1 - \kappa^2)L_2 = 0.7867\text{H}$$

5.6   Impedance Transformation

The complete transformer model described above is not the only way to represent a practical transformer. We may, for example, place the σL_2 inductance in the primary side, with its value multiplied by a factor of $1/n^2$. We may also place the L_1 in the secondary side with its value multiplied by a factor of n^2. In fact, it is possible to represent the practical transformer in a number of different ways.

Such different representations as the ones shown in figure 5.11 are possible because of an important property of the ideal transformer. Consider a simple situation where the secondary is terminated by an impedance Z. Here, the term *impedance* is used to distinguish circuits that contain capacitors and inductors from others that do not. Specifically we refer to an assembly of resistors, capacitors and inductors as impedance rather than resistance. Impedance, moreover, has the same unit as resistance. We will discuss impedance in detail in Chapter 7.

The equivalent impedance of Z_{in} observed from the primary side, as shown in figure 5.12, can be derived as follows. From the ideal transformer equations, i.e., $v_2 = nv_1$ and $i_2 = -i_1/n$, we have

$$\frac{v_2}{i_2} = \frac{nv_1}{\frac{-1}{n}i_1} = -n^2\frac{v_1}{i_1}$$

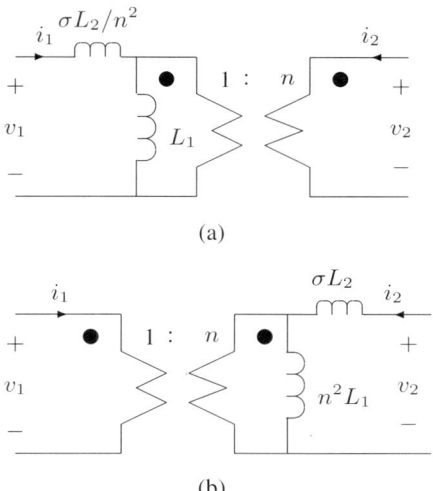

(a)

(b)

Figure 5.11: Possible practical transformer models

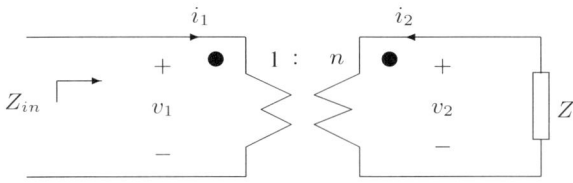

Figure 5.12: Impedance transformation by ideal transformer

Since $Z = -v_2/i_2$, the impedance seen from the primary side is given by

$$Z_{in} = \frac{v_1}{i_1} = \frac{1}{n^2}Z$$

Example 5.5: Impedance transformation using transformer — Impedance transformation of the transformer has many applications, e.g., in filter design and load matching. Suppose we have designed a filter for a 1Ω load, as shown in figure 5.13 (a). Also suppose the load is actually 50Ω. Instead of re-designing the filter for a 50Ω load, we may employ a transformer of turn ratio $1{:}\sqrt{50}$ to change the 50Ω load to 1Ω, as shown in figure 5.13 (b). Of course, in practice, the design must take into account the non-ideal property of the transformer.

5.7 Recapitulation — From a Physics Viewpoint

The foregoing discussion has relied entirely on the principle of superposition, by virtue of the linear property, the parameters being the terminal voltages and currents only.

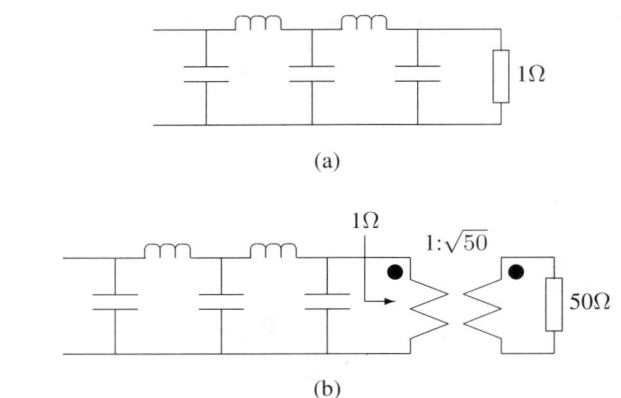

(a)

(b)

Figure 5.13: Application of impedance transformation

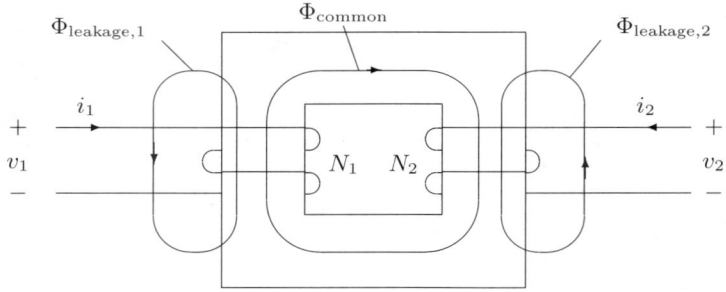

Figure 5.14: Coupled coils from a physics viewpoint

In this section we re-examine the coupled coils from a different viewpoint pertaining to the physics of magnetic coupling. The purpose is to improve our understanding of the basic equations developed in Section 5.1. Nevertheless, we should emphasize that the following treatment plays no role in the circuit theory of coupled coils, and the original coupling equations are already sufficient.

Now consider a pair of coupled coils wound on a magnetic core, as shown in figure 5.14. Under the excitation of i_1 and i_2, magnetic flux is induced in the core as well as through the air. Let Φ_{common} denote the flux confined in the core, and $\Phi_{\text{leakage},1}$ and $\Phi_{\text{leakage},2}$ denote the fluxes that leak through the air corresponding to excitation from i_1 and i_2 respectively. From Faraday's law, we can write

$$v_1 = N_1 \frac{d}{dt} \left(\Phi_{\text{common}} + \Phi_{\text{leakage},1} \right)$$

$$v_2 = N_2 \frac{d}{dt} \left(\Phi_{\text{common}} + \Phi_{\text{leakage},2} \right)$$

The inductance, by definition, is the ratio of the flux linking a coil (i.e., N times the total flux) and the current in that coil. Also, since Φ_{common} is due to the combined excitation from coils 1 and 2, we have

$$\Phi_{\text{common}} = \frac{L_{s1}}{N_1} i_1 + \frac{L_{s2}}{N_2} i_2$$

$$\Phi_{\text{leakage},1} = \frac{L_{\text{leakage},1}}{N_1} i_1$$

$$\Phi_{\text{leakage},2} = \frac{L_{\text{leakage},2}}{N_2} i_2$$

where L_{s1} and L_{s2} are simply the inductances that connect the flux in the core with the excitation i_1 and i_2 respectively, and, $\Phi_{\text{leakage},1}$ and $\Phi_{\text{leakage},2}$ are the inductances that connect the leakage fluxes with the excitations accordingly. Putting the above expressions in Faraday's equations, we have

$$
\begin{aligned}
v_1 &= N_1 \left(\frac{L_{s1}}{N_1} \frac{di_1}{dt} + \frac{L_{s2}}{N_2} \frac{di_2}{dt} \right) + L_{\text{leakage},1} \frac{di_1}{dt} \\
&= \left(L_{s1} + L_{\text{leakage},1} \right) \frac{di_1}{dt} + \frac{N_1 L_{s2}}{N_2} \frac{di_2}{dt} \\
v_2 &= N_2 \left(\frac{L_{s2}}{N_2} \frac{di_2}{dt} + \frac{L_{s1}}{N_1} \frac{di_1}{dt} \right) + L_{\text{leakage},2} \frac{di_2}{dt} \\
&= \frac{N_2 L_{s1}}{N_1} \frac{di_1}{dt} + \left(L_{s2} + L_{\text{leakage},2} \right) \frac{di_2}{dt}
\end{aligned}
$$

Comparing the above equations with the ones derived earlier, we can see that they are indeed the same. The above expressions, moreover, are able to provide physical insights into the coupling phenomenon. Essentially we see that the previously defined self-inductance (i.e., L_1 or L_2) can be decomposed into two inductances corresponding to a common and a leakage component, i.e.,

$$
\begin{aligned}
L_1 &= L_{s1} + L_{\text{leakage},1} \\
L_2 &= L_{s2} + L_{\text{leakage},2}
\end{aligned}
$$

Furthermore, the previously defined mutual inductance is

$$M = \frac{N_2 L_{s1}}{N_1} = \frac{N_1 L_{s2}}{N_2}$$

In the case of perfect coupling, we expect no leakage and hence both $L_{\text{leakage},1}$ and $L_{\text{leakage},2}$ are zero. This directly yields

$$\frac{N_2}{N_1} = \frac{M}{L_1} = \frac{L_2}{M}$$

and

$$M^2 = L_1 L_2$$

In the case of imperfect coupling, however, we have non-zero $L_{\text{leakage},1}$ and $L_{\text{leakage},2}$, and clearly we must have $M^2 < L_1 L_2$.

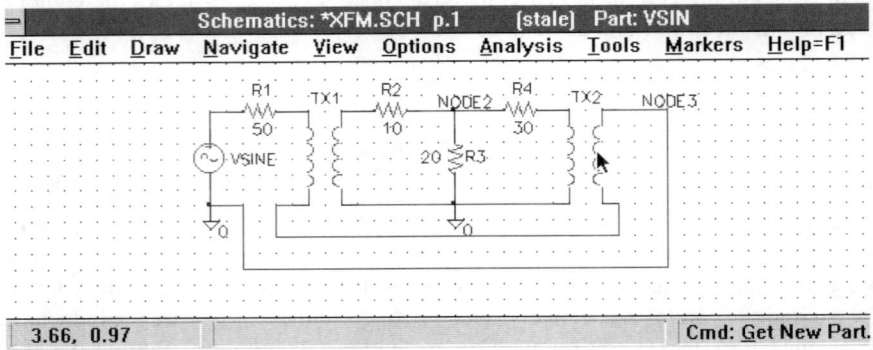

Figure 5.15: Circuit drawn in *Schematics* for PSPICE analysis

5.8 PSPICE Analysis of Circuits Containing Coupled Inductors

Circuits containing coupled inductors and transformers can be analyzed by PSPICE. To select a coupled inductor pair or transformer, the usual *"Get New Parts"* function is used. Specifically, the name of this element is XFRM_LINEAR which can be selected from the library "analog.slb". The parameters used to specify a transformer in PSPICE are L_1, L_2 and κ. Furthermore, since the transformer provides galvanic isolation, additonal ground nodes may be needed to avoid floating nodes, i.e., nodes having no DC path to ground, which are not permitted in PSPICE.

Example 5.6: PSPICE analysis of coupled inductors — The circuit to be analyzed is drawn in *Schematics,* as shown in figure 5.15. To specify the parameters for the transformers, we double-click on the transformer symbol to bring up a dialogue box, as shown in figure 5.16, in which we can enter the values of L_1, L_2 and κ. In this example, we enter $L_1 = 2H$, $L_2 = 1H$ and $\kappa = 0.8485$, which correspond to $M = 1.2H$. The sinusoidal voltage source used in this example is VSIN which is chosen from the library "source.slb". Double-clicking on the voltage source symbol will bring up the dialogue box for entering the desired amplitude, DC offset, and frequency. In this example, we set the amplitude to 10V and frequency 500Hz. Finally, to ensure all nodes have a DC path to ground, we need to define two ground nodes in this circuit.

A transient analysis is performed with a "no-print delay" of 44ms which effectively discards the initial transient phase and focuses on the steady-state waveforms. Figure 5.17 shows the steady-state waveforms of the voltage source and the voltage at NODE2. Note that the node "R1:1" refers to the left terminal of R1, which is simply the voltage source VSINE. (See also Example 4.6 for transient setup and the use of PROBE.)

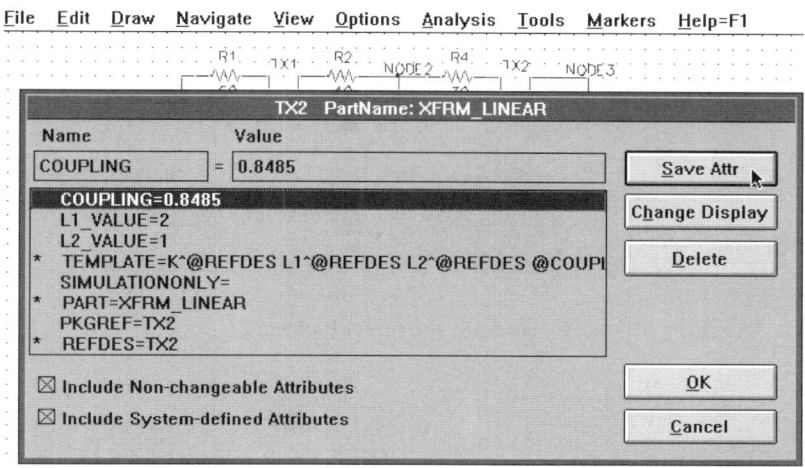

Figure 5.16: Dialogue box for specifying parameters for the transformer

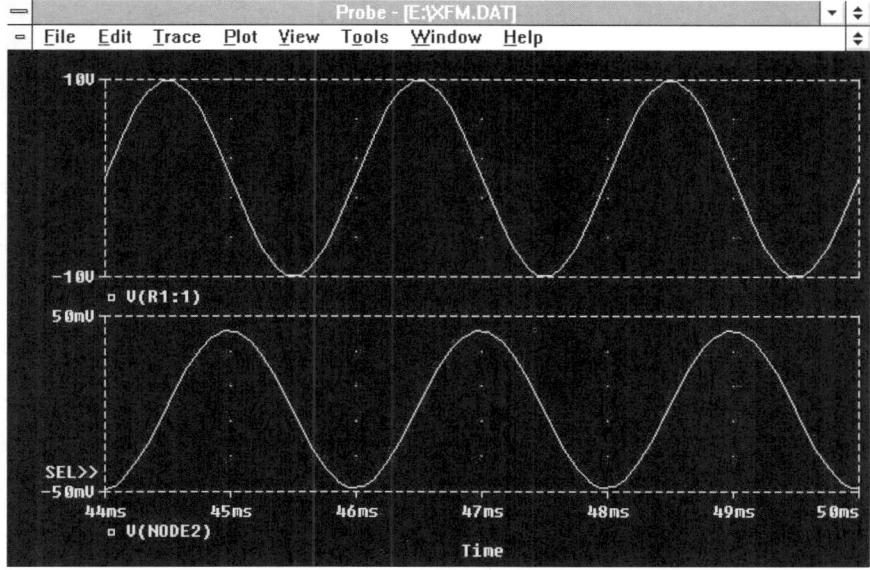

Figure 5.17: Transient waveforms from PSPICE. Upper graph: voltage source VSINE; Lower graph: voltage at NODE2

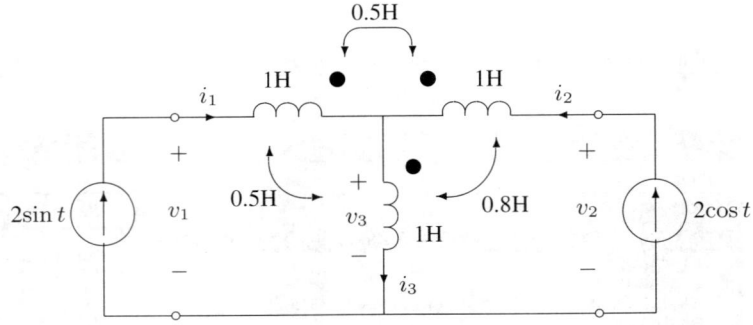

Figure 5.18: Circuit for Problem 2

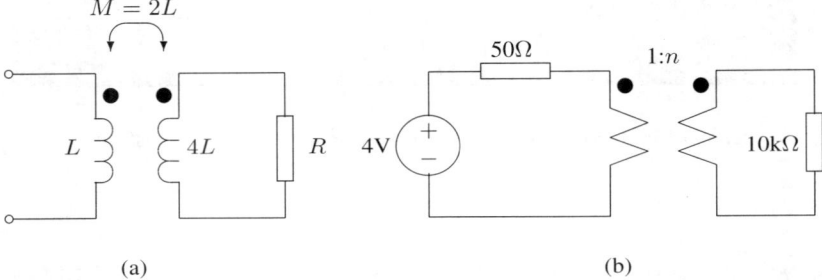

(a)   (b)

Figure 5.19: Circuits for Problems 3 and 4

5.9 Problems

1. The equations of coupled inductors have been derived for a pair of coupled inductors. It is in fact possible to extend the same concept to any arbitrary number of coupled coils. Assuming that linearity holds, write down a system of equations for n coupled inductors, $L_1, L_2, \ldots, L_n$. Denote the self-inductances by L_{11}, L_{22}, etc., and the mutual inductance between L_j and L_k by L_{jk}.

2. Consider the linearly coupled inductors in figure 5.18. Following the dot convention, write down the coupling equations for the circuit, and verify that

$$v_1 = 0.5\frac{di_1}{dt} - 0.3\frac{di_2}{dt} + 0.5\frac{di_3}{dt}$$

$$v_2 = 0.2\frac{di_2}{dt} + 0.2\frac{di_3}{dt}$$

$$v_3 = -0.5\frac{di_1}{dt} - 0.8\frac{di_2}{dt} + \frac{di_3}{dt}$$

Find v_1 and v_2.

3. Consider the circuit shown in figure 5.19 (a). Find the coefficient of coupling. Derive an equivalent representation of the circuit when observed from the left

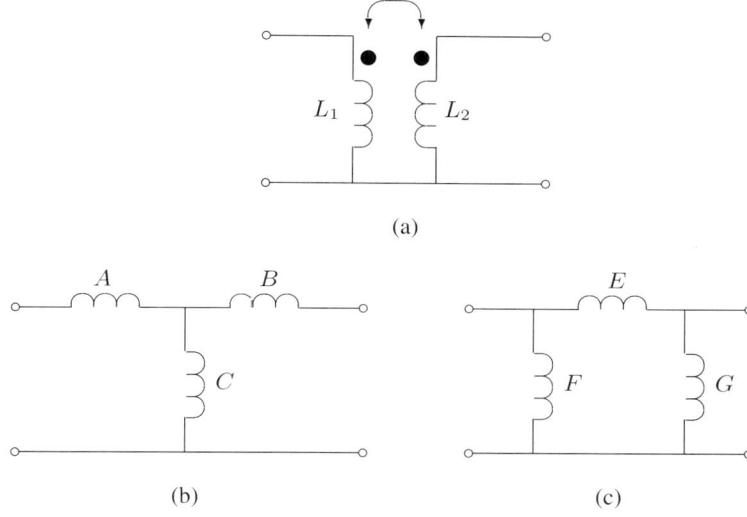

Figure 5.20: Circuits for Problem 5

terminals, i.e., input impedance. Comment on the limiting case as $L \to \infty$.

4. Referring to figure 5.19 (b), n is given a value such that maximum power transfer to the load is realized. Find n, and calculate the power dissipation in the whole circuit.

5. Determine A, B, C, E, F and G in figure 5.20 such that the three circuits are equivalent.

6. The circuit shown in figure 5.21 contains ideal transformers, resistors and a current source. Set up the necessary Kirchhoff's law equations and transformer equations, and solve the circuit completely. Bear in mind that this problem assumes the validity of the ideal transformer equations, which give no preference to AC or DC signals. In reality, an AC current source should be used since the transformer does not work with DC.

7. Determine $i_1(t)$ and $i_2(t)$ in the circuit of figure 5.22 in which the switches are thrown from left to right at $t = 0$.

8. The pair of coupled inductors shown in figure 5.19 (a) is used as a transformer. Develop a practical model for this transformer. Calculate the magnetizing inductance as observed from the primary (left side), and the leakage inductance as observed from the secondary (right side). Repeat the exercise with M reduced to L.

9. We have seen that when coupling is perfect and inductances are infinitely large, the coupled coils become an ideal transformer. Now, consider two uncoupled coils, i.e., $M = 0$. Starting from the transformer model, verify that the model reduces to two independent inductors as M tends to 0.

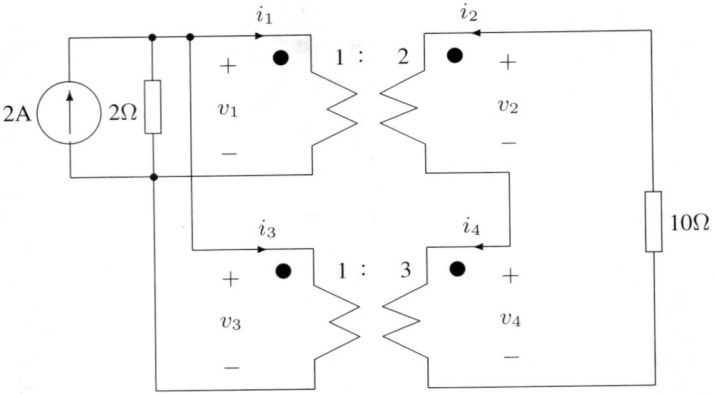

Figure 5.21: Circuit for Problem 6

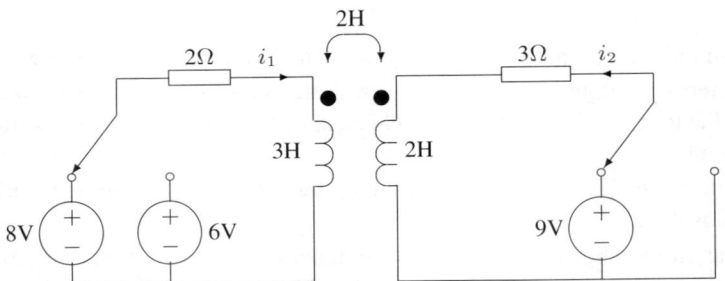

Figure 5.22: Circuit for Problem 7

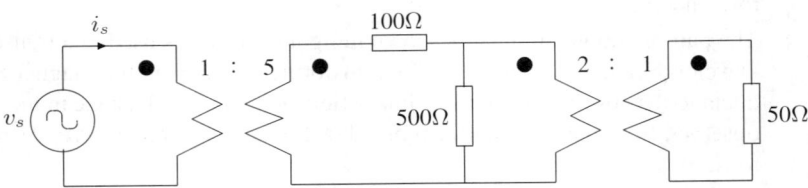

Figure 5.23: Circuit for Problem 10

10. Use PSPICE to analyze the circuit of figure 5.23 which contains ideal transform-
    ers. In particular find the input current amplitude, given that the input voltage
    is a sine wave of amplitude 10V and frequency 500Hz.

Chapter 6

Periodic Functions and Fourier Series Representations

An important class of signals that is commonly encountered in electrical and electronic systems is the class of *periodic signals*. In simple words, a periodic signal is a time-varying signal whose values at multiples of a fixed duration of time are identical. In this chapter we will examine the properties of periodic signals, and in particular of sinusoidal signals which are widely used in power transmission, communications and other applications.

6.1 Periodic Functions

A periodic function can be completely specified by a mathematical description that defines its value over an adequately long interval of time. The minimum interval of time for which a periodic signal must be known in order to define the entire signal is called the *period* of the signal, and is usually denoted by T. In mathematical terms, a function $x(t)$ is periodic if

$$x(t) = x(t + T) \quad \text{for all } t.$$

An example of a periodic function is shown in figure 6.1. This function repeats itself every T second. Sometimes we use *angular frequency* to describe how rapidly the function repeats itself. Denoted by ω, the angular frequency is related to the period by

$$\omega = \frac{2\pi}{T}$$

and its unit is *radians per second* (rad/s). In practice, the term *frequency* is more often used, and is defined as

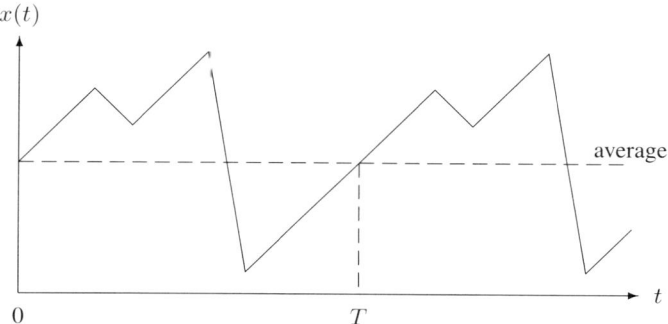

Figure 6.1: Periodic function

$$f = \frac{1}{T} = \frac{\omega}{2\pi}$$

The unit of frequency is the *hertz* (Hz) which is dimensionally equivalent to *cycles per second* (cycle/s).

Remarks — It is often misunderstood that the unit of f is *per second,* based on a wrong interpretation of the relation $f = 1/T$. In fact, the unit of ω is *per second* since radian is a real number with no unit, and the unit of f should be *cycles per second*. Also, since 1 cycle equals 2π, we have one *cycle per second* being equal to 2π *per second*, i.e., $\omega = 2\pi f$.

6.2 Average and Root-Mean-Square Values

Since the intensity of a time-varying signal changes from time to time, it is inappropriate and quite meaningless to quantify a signal by a value it assumes at an arbitrary instant of time. In some applications, however, we are interested in the average intensity of a signal rather than the full details of its waveform. We define the average value of a periodic signal by

$$x_{av} = \frac{1}{T} \int_0^T x(t)\, dt$$

Clearly, the average value provides no information about the amplitude of a periodic function. It merely tells the DC offset value. When amplitude information is required, it is more meaningful to talk about the root-mean-square value of the function, which is defined as the square root of the average of the square of the function over one period, i.e.,

$$x_{rms} = \sqrt{\frac{1}{T} \int_0^T x^2(t)\, dt}$$

Root-mean-square (rms) values are often used in electrical engineering to measure voltages and currents. Apart from providing amplitude information, use of rms values offer some computational advantages, especially when dealing with power. To see this, we compute the average power dissipation, p_{av}, in a resistor of resistance R. Assume that the voltage across this resistor is $v(t)$ and the current in it is $i(t)$. From the definition of average power, we have

$$p_{av} \;=\; \frac{1}{T}\int_0^T \frac{v(t)^2}{R}\,dt = \frac{v_{rms}^2}{R}$$

$$\text{or}\quad p_{av} \;=\; \frac{1}{T}\int_0^T i(t)^2 R\,dt = i_{rms}^2 R$$

Clearly, the above formulae resemble those of the DC case. A moment's thought will convince us that, as far as power is concerned, a circuit driven by AC sources can be analyzed as if it is driven by DC sources whose values equal those of the rms values of the corresponding AC sources.

Remarks — When $R = 1\,\Omega$, the power dissipation, known as *normalized power,* is exactly the mean-square value of the applied voltage or current. We will see later that the Fourier series representation can help compute the normalized power efficiently.

Example 6.1: Average and root-mean-square values — Consider two periodic functions $x_1(t) = \sin \omega t$ and $x_2(t) = 5 + \sin \omega t$. The period is $2\pi/\omega$ s. Using the above definition for the average value, we have

$$x_{1av} = \frac{\omega}{2\pi}\int_0^{2\pi/\omega} \sin \omega t\,dt = \frac{1}{2\pi}\int_0^{2\pi} \sin \theta\,d\theta = 0$$

and

$$x_{2av} = \frac{\omega}{2\pi}\int_0^{2\pi/\omega} (5 + \sin \omega t)\,dt = \frac{1}{2\pi}\int_0^{2\pi} (5 + \sin \theta)\,d\theta = 5$$

Obviously, for any periodic function having zero DC offset, the average value is zero, regardless of the amplitude of the function. The average value is thus unable to tell the amplitude of a periodic signal. Let us calculate the rms value of $x_1(t)$ and $x_2(t)$.

$$x_{1rms} = \sqrt{\frac{\omega}{2\pi}\int_0^{2\pi/\omega} \sin^2 \omega t\,dt}$$

$$= \sqrt{\frac{1}{2\pi}\int_0^{2\pi} \sin^2 \theta\,d\theta} = \frac{1}{\sqrt{2}} = 0.7071$$

$$x_{2rms} = \sqrt{\frac{\omega}{2\pi}\int_0^{2\pi/\omega} (5 + \sin \omega t)^2\,dt}$$

$$= \sqrt{\frac{1}{2\pi}\int_0^{2\pi} (25 + 10\sin \theta + \sin^2 \theta)\,d\theta} = \sqrt{25 + \frac{1}{2}} = 5.05$$

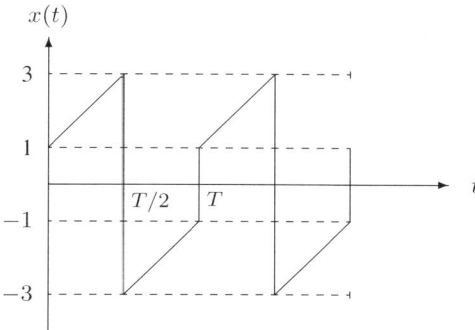

Figure 6.2: Piece-wise function

Furthermore, if $x_1(t)$ is the voltage (in volts) across a 50Ω resistor, and $x_2(t)$ is the current (in amperes) in a 100Ω resistor, we can calculate the power dissipations as 0.5/50W and 25.5×100W, respectively, for the 50Ω resistor and the 100Ω resistor.

Remarks — As shown above, the rms value of the sine function is equal to $1/\sqrt{2}$ times its peak value. In practice, voltages and currents in utility power lines are usually given in rms values. Their peak values are obtained by multiplying the rms values by $\sqrt{2}$.

Example 6.2: Root-mean-square value of a piece-wise function — The above definitions of average and rms values may be applied to any piece-wise function that is well defined over a period of repetition. Consider the periodic waveform shown in figure 6.2. This waveform is defined by

$$x(t) = \begin{cases} 1 + \dfrac{4t}{T} & \text{for } 0 \le t < \tfrac{T}{2} \\ -5 + \dfrac{4t}{T} & \text{for } \tfrac{T}{2} \le t < T \end{cases}$$

and $x(t + kT) = x(t)$ for all integers k. This waveform has a zero average value and an rms value given by

$$x_{rms} = \sqrt{\frac{1}{T}\left\{\int_0^{T/2}\left(1 + \frac{4t}{T}\right)^2 dt + \int_{T/2}^{T}\left(-5 + \frac{4t}{T}\right)^2 dt\right\}} = \sqrt{\frac{13}{2}}$$

Furthermore, if the above function is the voltage across a 1Ω resistor, the power dissipated in this resistor is 6.5W.

6.3 Sine Functions

Consider two sine functions $x_1(t)$ and $x_2(t)$, of amplitude X_1 and X_2, respectively. Suppose $x_2(t)$ is lagging $x_1(t)$ by a phase angle of ϕ. Choosing $x_1(t)$ as the reference

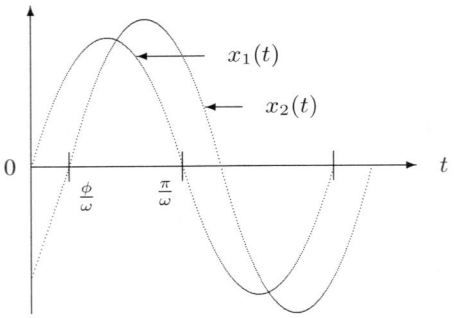

Figure 6.3: Sine functions

for phase comparison, we may write the two functions as

$$x_1(t) = X_1 \sin \omega t \quad \text{and} \quad x_2(t) = X_2 \sin(\omega t - \phi)$$

the sketches of which are shown in figure 6.3. Each of these waveforms is completely characterized by three parameters, namely frequency, amplitude, and phase angle difference from a chosen reference. Furthermore, if we limit ourselves to sine functions of one given frequency ω, then any such sine function can be specified by an amplitude and a phase angle. Hence, for brevity, we may denote a sine function by

$$x = X \angle \phi$$

where X is the peak or rms value of the sine function and ϕ is the phase angle relative to a chosen reference. For example, the above two sine functions may be represented by

$$x_1 = X_1 \angle 0 \quad \text{and} \quad x_2 = X_2 \angle -\phi$$

It should be noted that for any linear circuit driven by a sinusoidal voltage or current, all voltages and currents in the circuit will, in the steady state, be varying at the same frequency as the driving voltage or current. We will see in Chapter 7 that in analyzing linear circuits driven by a sinusoidal voltage or current, we need to deal only with the amplitudes and the phases of the voltages and currents in the circuit, while the frequency, being fixed, is unimportant.

6.4 Phasor Representation of Sine Functions

Consider the following five sine functions, all of frequency ω:

$$
\begin{aligned}
x_1(t) &= X_1 \sin \omega t & &\text{reference phase} \\
x_2(t) &= X_2 \sin(\omega t + \phi_2) & &\text{leading } x_1 \text{ by } \phi_2 \\
x_3(t) &= X_3 \sin(\omega t - \phi_3) & &\text{lagging } x_1 \text{ by } \phi_3 \\
x_4(t) &= X_4 \sin(\omega t + \phi_4) & &\text{leading } x_1 \text{ by } \phi_4 \\
x_5(t) &= X_5 \sin(\omega t - \phi_5) & &\text{lagging } x_1 \text{ by } \phi_5
\end{aligned}
$$

Although we may plot these waveforms against time as in figure 6.3, such a represen-

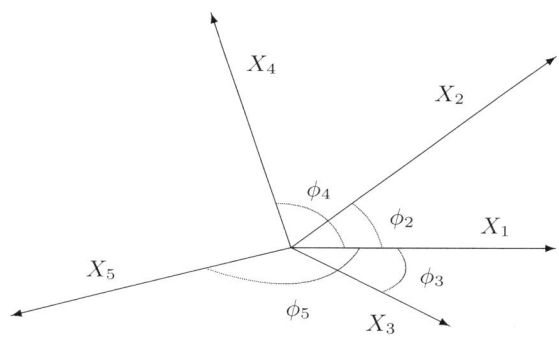

Figure 6.4: Phasor diagram of sine functions

tation is confusing and does not help us visualize the various waveforms conveniently. Since the parameters of interest are the amplitudes and phases, we may represent them on the complex Argand plane, treating x_1 as a real number of magnitude X_1, and x_2 as a complex number of magnitude X_2 and argument ϕ_2, etc., as shown in figure 6.4. Such a representation is known as *phasor diagram*.

The following alternative view may improve our understanding of the construction of a phasor diagram. Let us imagine that each sine function is represented by a rotating "vector" which rotates at an angular speed of ω rad/s around a centre in an anti-clockwise fashion, as shown in figure 6.5. The length of the vector is proportional to the amplitude of the sine function it represents, and the angular position of the vector indicates the phase angle of the sine function. Now all vectors are rotating, in accordance with the sine functions they represent. However, if the situation is viewed under a stroboscope which flashes on the rotating vectors every T seconds, where $T = 2\pi/\omega$, then all vectors will appear to be stationary. We use the term *phasors* to denote these apparently stationary "rotating vectors". We refer to this stroboscopic picture as a *phasor diagram*. Clearly, on the phasor diagram, the length of the phasor represents the magnitude of the sine function concerned, and the angular position specifies the phase angle.

Example 6.3: Phasor diagram — Consider the following sinusoidal functions: $x_1(t) = \sin \omega t$, $x_2(t) = 2\sin(\omega t - 60°)$ and $x_3(t) = \cos(\omega t + 45°)$. Suppose we want to draw a phasor diagram to represent these three functions. First, since $\cos \theta = \sin(\theta + 90°)$, we can write $x_3(t)$ as $\sin(\omega t + 135°)$. Thus, the three sinusoidal functions are

$$x_1 = 1\angle 0°, \quad x_2 = 2\angle -60° \quad \text{and} \quad x_3 = 1\angle 135°$$

Taking the phase of x_1 as the reference, the phasors for x_1, x_2 and x_3 are as shown in figure 6.6. We may also perform addition or subtraction of two functions on the phasor diagram. The rules follow the usual vector addition and substraction. For example, the sum of x_1 and x_2 is given by the diagonal of the parallelogram with x_1 and x_2 as

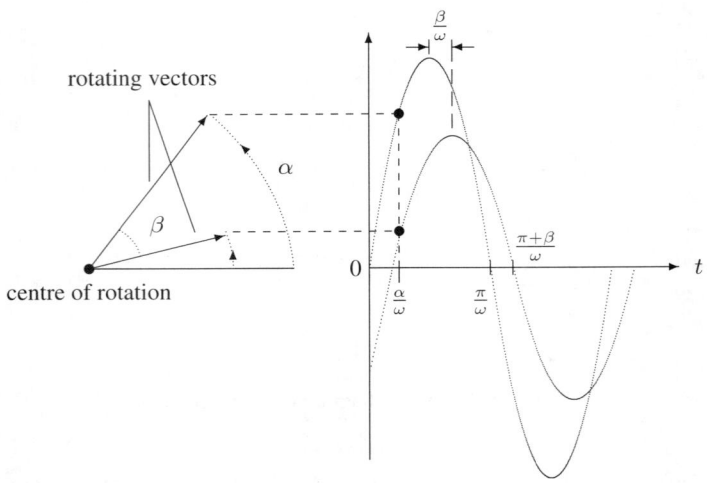

Figure 6.5: Rotating vector representation of sine functions

the adjacent sides, as shown in figure 6.6.

6.5 Fourier Series Representation of Periodic Functions

We can easily appreciate that when two sine functions $x_1(t)$ and $x_2(t)$ of period T_1 and T_2 add up, the resulting function $x_s(t)$ is periodic if the ratio of T_1 and T_2 is rational. In general, if

$$\frac{T_1}{T_2} = \frac{p}{q}$$

where p and q are integers having no common integer factor other than 1 (i.e., p/q is a simple fraction), then both x_1 and x_2 will repeat themselves after an interval T which is given by

$$T = qT_1 = pT_2$$

This value of T is clearly the period of x_s. Furthermore, from elementary arithmetic, any irrational number can be approximated by a rational number p/q, and the approximation improves as p and q become large. Thus, the limiting case of p and q tending towards infinity corresponds to an irrational ratio of T_1 and T_2. In this case, the period of $x_s(t)$ is infinite. We may now make the following conclusions regarding the sum of two sine functions:

1. The sum of two sine functions whose periods are in a rational ratio is periodic.

2. The sum of two sine functions whose periods are in an irrational ratio is non-periodic.

By induction, the sum of a number of sine functions whose periods are in rational ratios is periodic. In practice, however, the converse problem is important, and specifically,

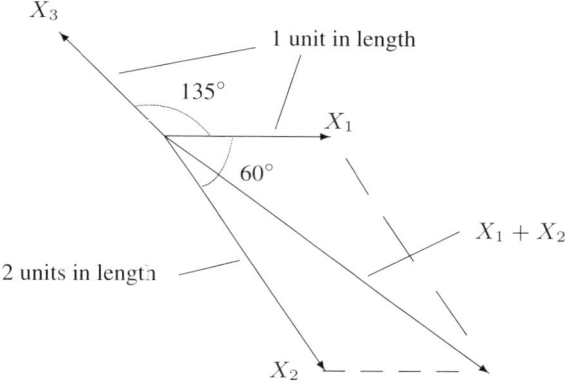

Figure 6.6: Phasor diagram for Example 6.3

we wish to know if a given periodic function can be constructed from such simple mathematical functions as the sine functions.

From the above discussion, if we choose $p = 1$ in particular, we see that the sum of any number of sine functions whose periods are T/q (where q is any positive integer) is periodic of period T. Let us now speculate that any periodic function $x(t)$ of period T can be constructed from a set of sine functions whose periods are T, $T/2$, $T/3$, $T/4$, etc. Thus, we have the following provisional representation for $x(t)$.

$$
\begin{aligned}
x(t) &= \frac{a_0}{2} + c_1 \sin(\omega t + \phi_1) + c_2 \sin(2\omega t + \phi_2) + \cdots \\
&= \frac{a_0}{2} + a_1 \cos \omega t + a_2 \cos 2\omega t + \cdots + b_1 \sin \omega t + b_2 \sin 2\omega t + \cdots \\
&= \frac{a_0}{2} + \sum_{n=1}^{\infty} \cos n\omega t + \sum_{n=1}^{\infty} \sin n\omega t
\end{aligned}
$$

where $\omega = 2\pi/T$. This infinite series is usually referred to as the *Fourier series,* and the coefficients $a_0, a_1, a_2, \ldots, b_1, b_2, \ldots$, are called the *Fourier coefficients* of $x(t)$. The problem now is to show that these coefficients exist and can be found for any given $x(t)$.

To determine whether $x(t)$ can be represented by the above Fourier series, we consider minimization of the following error function.

$$
\epsilon(t) = \frac{1}{T} \int_0^T \left(x(t) - \frac{a_0}{2} - \sum_{n=1}^{\infty} \cos n\omega t - \sum_{n=1}^{\infty} \sin n\omega t \right)^2 dt
$$

Also, the following relations hold for all integers m and n, with $m \neq n$.

$$
\int_0^T \cos n\omega t \sin m\omega t \, dt = 0
$$

$$\int_0^T \cos n\omega t \cos m\omega t \, dt = \int_0^T \sin n\omega t \sin m\omega t \, dt = 0$$

$$\int_0^T \cos^2 n\omega t \, dt = \int_0^T \sin^2 n\omega t \, dt = \frac{T}{2}$$

Upon expanding $\epsilon(t)$, with the help of the above relations, we get

$$
\begin{aligned}
\epsilon(t) \;=\;& \frac{1}{T}\int_0^T x(t)^2 \, dt - \frac{1}{T}\int_0^T a_0 x(t)\, dt - \frac{2}{T}\sum_{n=1}^{\infty}\left(\int_0^T a_n x(t)\cos n\omega t\, dt\right) \\
& -\frac{2}{T}\sum_{n=1}^{\infty}\left(\int_0^T b_n x(t)\sin n\omega t\, dt\right) + \left(\frac{a_0}{2}\right)^2 + \frac{1}{2}\sum_{n=1}^{\infty}a_n^2 + \frac{1}{2}\sum_{n=1}^{\infty}b_n^2
\end{aligned}
$$

which can be put as

$$
\begin{aligned}
\epsilon(t) \;=\;& \frac{1}{T}\int_0^T x(t)^2\, dt + \left(\frac{a_0}{2} - \frac{1}{T}\int_0^T x(t)\, dt\right)^2 \\
& + 2\sum_{n=1}^{\infty}\left(\frac{a_n}{2} - \frac{1}{T}\int_0^T x(t)\cos n\omega t\, dt\right)^2 \\
& + 2\sum_{n=1}^{\infty}\left(\frac{b_n}{2} - \frac{1}{T}\int_0^T x(t)\sin n\omega t\, dt\right)^2 \\
& - \left(\frac{a_0}{2}\right)^2 - \frac{1}{2}\sum_{n=1}^{\infty}\left(a_n^2 + b_n^2\right)
\end{aligned}
$$

The next step is to find a_0, a_n and b_n in order to minimize the above error function. Clearly, we must set the Fourier coefficients as

$$
\begin{aligned}
a_0 \;&=\; \frac{2}{T}\int_0^T x(t)\, dt \\[2mm]
a_n \;&=\; \frac{2}{T}\int_0^T x(t)\cos n\omega t\, dt \\[2mm]
b_n \;&=\; \frac{2}{T}\int_0^T x(t)\sin n\omega t\, dt
\end{aligned}
$$

With this choice, the error function becomes

$$\epsilon(t) = \frac{1}{T}\int_0^T x(t)^2\, dt - \left(\frac{a_0}{2}\right)^2 - \frac{1}{2}\sum_{n=1}^{\infty}\left(a_n^2 + b_n^2\right)$$

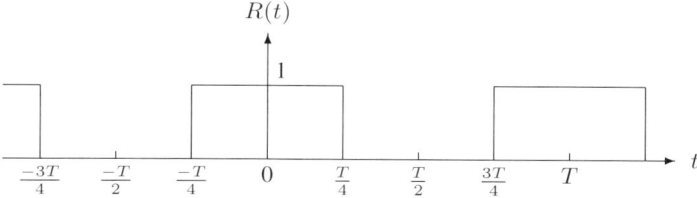

Figure 6.7: Rectangular waveform

Substituting $x(t)$ by its Fourier series, we can show that $\epsilon(t) \to 0$. In practice, if we use a finite number of terms to approximate $x(t)$, the error function becomes non-zero, but should be small as the number of terms increases.

Remarks — When a periodic function is represented by a Fourier series, the term $a_0/2$ corresponds to the DC offset or average value, the coefficients a_1 and b_1 correspond to the fundamental frequency, and the coefficients a_n and b_n (with $n > 1$) correspond to the nth harmonic. From elementary trigonometry, we have $a \sin\theta + b\sin\theta = \sqrt{a^2 + b^2}\sin(\theta + \phi)$. Thus, the amplitude of the fundamental component is $\sqrt{a_1^2 + b_1^2}$, and that of the nth harmonic component is $\sqrt{a_n^2 + b_n^2}$.

Example 6.4: Fourier series of a rectangular wave — The rectangular waveform shown in figure 6.7 is periodic with period T. The Fourier series for this waveform is

$$R(t) = \frac{a_0}{2} + \sum_{n=1}^{\infty} (a_n \cos n\omega t + b_n \sin n\omega t)$$

When evaluating the Fourier coefficients, we need to perform integration over a complete period. In this case, it is convenient to perform integration from $-T/2$ to $T/2$. Also, $R(t)$ is 1 between $-T/4$ and $T/4$, and 0 everywhere else in the interval of integration. Thus, the Fourier coefficients are given by

$$
\begin{aligned}
a_0 &= \frac{2}{T}\int_{-T/2}^{T/2} R(t)\, dt \\[2mm]
&= \frac{2}{T}\int_{-T/4}^{T/4} 1\, dt \\[2mm]
&= 1 \\[2mm]
a_n &= \frac{2}{T}\int_{-T/2}^{T/2} R(t)\cos n\omega t\, dt \\[2mm]
&= \frac{2}{T}\int_{-T/4}^{T/4} \cos n\omega t\, dt \\[2mm]
&= \frac{1}{\pi n}\left[\sin\left(\frac{\pi n}{2}\right) - \sin\left(\frac{-\pi n}{2}\right)\right]
\end{aligned}
$$

$$= \begin{cases} 0 & \text{for } n = 2, 4, 6, \ldots \\ \dfrac{2(-1)^{(n-1)/2}}{\pi n} & \text{for } n = 1, 3, 5, \ldots \end{cases}$$

$$b_n = \frac{2}{T} \int_{-T/2}^{T/2} R(t) \sin n\omega t \, dt$$

$$= \frac{2}{T} \int_{-T/4}^{T/4} \sin n\omega t \, dt$$

$$= 0$$

The first few terms of the Fourier series are

$$R(t) = \frac{1}{2} + \frac{2}{\pi} \cos \omega t - \frac{2}{3\pi} \cos 3\omega t + \frac{2}{5\pi} \cos 5\omega t - \frac{2}{7\pi} \cos 7\omega t + \cdots$$

6.6 Symmetry of Periodic Functions

The calculation of the Fourier coefficients for a given periodic function can be drastically simplified if the function is *symmetric* in one of the following ways.

1. Even symmetry, or mirror symmetry
2. Odd symmetry, or radial symmetry
3. Displacement symmetry

A periodic function $x_e(t)$ has *even symmetry* if $x_e(t) = x_e(-t)$. The waveform is symmetric about the y-axis. Figure 6.8 (a) shows a typical waveform having even symmetry. For this type of functions, all sine terms vanish in the Fourier series, i.e.,

$$b_n = 0 \quad \text{for all } n = 1, 2, 3, \ldots$$

Also, in computing a_n, we need only perform integration from 0 to $T/2$ because of the even symmetry of $x_e(t) \cos n\omega t$, i.e.,

$$a_n = \frac{4}{T} \int_0^{T/2} x_e(t) \cos n\omega t \, dt \quad \text{for all } n = 1, 2, 3, \ldots$$

A periodic function $x_o(t)$ has *odd symmetry* if $x_o(t) = -x_o(-t)$. The waveform exhibits a radial symmetry about the origin. Figure 6.8 (b) shows a typical waveform having odd symmetry. For this type of functions, the DC term and all cosine terms vanish in the Fourier series, i.e.,

$$a_n = 0 \quad \text{for all } n = 0, 1, 2, 3, \ldots$$

Also, in computing b_n, we need only perform integration from 0 to $T/2$ because of the even symmetry of $x_o(t) \sin n\omega t$, i.e.,

$$b_n = \frac{4}{T} \int_0^{T/2} x_o(t) \sin n\omega t \, dt \quad \text{for all } n = 1, 2, 3, \ldots$$

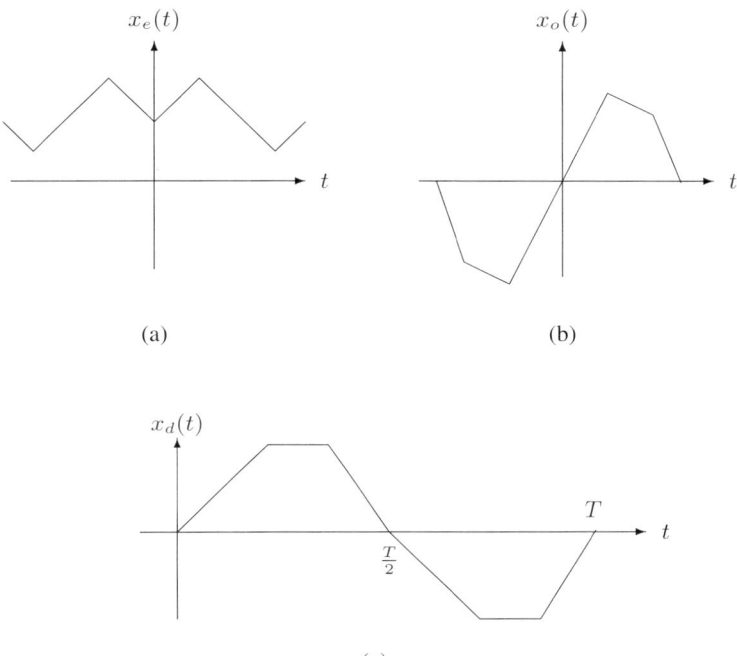

Figure 6.8: (a) Even symmetry; (b) odd symmetry; (c) displacement symmetry

A periodic function $x_d(t)$ has *displacement symmetry* if $x_d(t) = -x_d(t - T/2)$. The waveform of $x_d(t)$ in the first half of each period is exactly the same as that in the second half turned upside down. Figure 6.8 (c) shows a typical waveform having displacement symmetry. For this type of function, the DC offset is zero, i.e.,

$$a_0 = 0$$

Also, only odd harmonics exist in this type of function because if n is even, the integral of $x_d(t) \cos n\omega t$ or $x_d(t) \sin n\omega t$ over the first half period always cancels the integral over the next half period, i.e.,

$$a_n = b_n = 0 \quad \text{for all } n = 2, 4, 6, \ldots$$

Furthermore, integration from 0 to $T/2$ suffices to calculate the odd cosine and sine terms, i.e.,

$$a_n = \frac{4}{T} \int_0^{T/2} x_d(t) \cos n\omega t \, dt \quad \text{for all } n = 1, 3, 5, \ldots$$

$$b_n = \frac{4}{T} \int_0^{T/2} x_d(t) \sin n\omega t \, dt \quad \text{for all } n = 1, 3, 5, \ldots$$

Type of symmetry	Properties of Fourier coefficients
Even symmetry $x(t) = x(-t)$	$b_n = 0$ for all $n = 1, 2, 3, \ldots$ $a_n = \dfrac{4}{T} \displaystyle\int_0^{T/2} x(t) \cos n\omega t\, dt$ for all $n = 1, 2, 3, \ldots$
Odd symmetry $x(t) = -x(-t)$	$a_0 = 0$ $a_n = 0$ for all $n = 1, 2, 3, \ldots$ $b_n = \dfrac{4}{T} \displaystyle\int_0^{T/2} x(t) \sin n\omega t\, dt$ for all $n = 1, 2, 3, \ldots$
Displacement symmetry $x(t) = -x(t - T/2)$	$a_0 = 0$ $a_n = b_n = 0$ for all $n = 2, 4, 6, \ldots$ $a_n = \dfrac{4}{T} \displaystyle\int_0^{T/2} x(t) \cos n\omega t\, dt$ for all $n = 1, 3, 5, \ldots$ $b_n = \dfrac{4}{T} \displaystyle\int_0^{T/2} x(t) \sin n\omega t\, dt$ for all $n = 1, 3, 5, \ldots$

Table 6.1: Properties of Fourier coefficients for symmetric functions

Table 6.1 summarizes the properties of the Fourier coefficients for the afore-discussed symmetric waveforms.

Example 6.5: Fourier series of a symmetric function — Consider the triangular waveform shown in figure 6.9. Even symmetry as well as displacement symmetry are observed in this waveform. Thus, we expect all sine terms and even cosine terms to vanish, i.e.,

$$
\begin{aligned}
b_n &= 0 \quad \text{for all } n = 1, 2, 3, \ldots \\
a_n &= 0 \quad \text{for all } n = 2, 4, 6, \ldots
\end{aligned}
$$

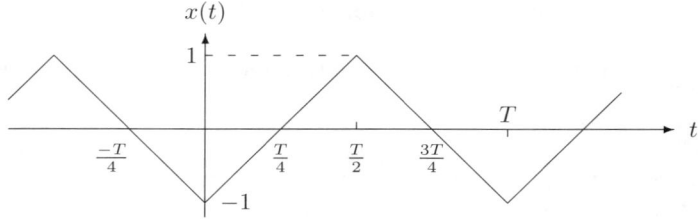

Figure 6.9: Triangular waveform with even and displacement symmetry

Also, the DC offset is zero, i.e., $a_0 = 0$. The waveform in the interval between 0 and $T/4$ is given by

$$x(t) = \frac{4}{T}\left(t - \frac{T}{4}\right)$$

The value of a_n for odd n is given by

$$
\begin{aligned}
a_n &= \frac{4}{T}\int_0^{T/2}\left[\frac{4}{T}\left(t - \frac{T}{4}\right)\cos n\omega t\right]dt \\
&= \frac{16}{T^2}\left[\frac{t - T/4}{n\omega}\sin n\omega t + \frac{1}{n^2\omega^2}\cos n\omega t\right]_0^{T/2} \\
&= \frac{-8}{n^2\pi^2} \qquad \text{for all } n = 1, 3, 5, \ldots
\end{aligned}
$$

The first few terms of the Fourier series are

$$x(t) = -\frac{8}{\pi^2}\cos\omega t - \frac{8}{9\pi^2}\cos 3\omega t - \frac{8}{25\pi^2}\cos 5\omega t - \cdots$$

6.7 Convergence of Fourier Series and the Gibbs Phenomenon

In this section we examine qualitatively the limitation of the Fourier series in representing a periodic function. First of all, the given function, $x(t)$, must satisfy

$$\int_{t_o}^{t_o+T}|x(t)|\,dt < \infty \qquad \text{for all } t_o$$

Otherwise, the error function $\epsilon(t)$ defined earlier cannot be minimized, and no Fourier coefficients can be found.

Secondly, if $x(t)$ has a rapidly changing segment within a period, then the Fourier series must contain very high harmonic terms. This is because the fast $dx(t)/dt$ can only be provided by sine functions of very high frequencies. The faster $dx(t)/dt$ is, the more higher harmonics are needed. In the limiting case, if $x(t)$ contains a step change in a period, i.e., $dx(t)/dt \to \infty$, no finite Fourier series can fully represent $x(t)$. Such a limitation is usually referred to as the *Gibbs phenomenon,* which describes the situation when a finite Fourier series is used to approximate a periodic function containing a simple discontinuity in a period. The so-called Gibbs phenomenon essentially refers to the oscillatory error, as depicted in figure 6.10, of the finite Fourier series approximation near the point of discontinuity. It has been shown by Gibbs that the overshoot is at maximum 9% of the step change. Moreover, if the number of terms tends towards infinity, the overshoot occurs for a vanishingly small range of time.

It should be noted that we can, in theory, represent a periodic function accurately by the infinite Fourier series, even though a discontinuity exists in the function. Moreover, the infinite Fourier series is still adequate if there are two, or even more, discontinuities in a period. But it will fail if there are an infinite number of discontinuities in a period.

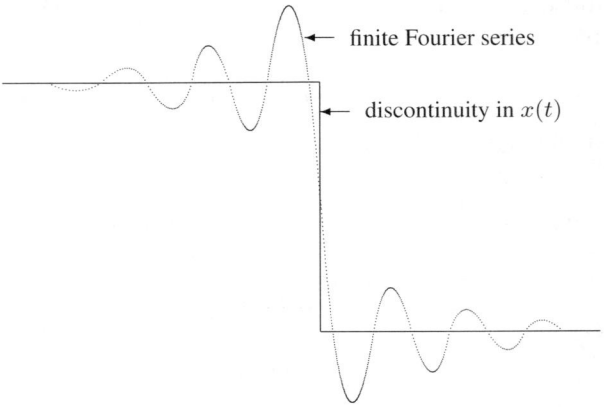

Figure 6.10: Gibbs phenomenon

A formal study of Fourier series representations for periodic functions was performed by Dirichlet, who proved rigorously that *a periodic function $x(t)$ can be accurately represented by the infinite Fourier series if it satisfies the following conditions:*

1. *It is absolutely integrable over a period, i.e.,* $\int_0^T |x(t)|\, dt < \infty$;
2. *It has a finite number of discontinuities in a period;*
3. *It has a finite number of maxima and minima in a period.*

Example 6.6: Illustration of the Gibbs phenomenon — Let us now plot the finite Fourier series for the rectangular waveform shown in figure 6.7. The Fourier series has been found previously as

$$R(t) = \frac{1}{2} + \frac{2}{\pi}\cos\omega t - \frac{2}{3\pi}\cos 3\omega t + \frac{2}{5\pi}\cos 5\omega t - \frac{2}{7\pi}\cos 7\omega t + \cdots$$

Suppose we approximate $R(t)$ with only the DC term and the first N cosine terms. Figure 6.11 shows the plots of the finite series with $N = 1$, 2, 3 and 4. Note the characteristic "ringing" as N increases.

6.8 Parseval's Theorem

As mentioned in Section 6.2, the mean-square (square of rms) value of voltage or current plays an important role in the description of power. When a periodic function $x(t)$ is represented by the infinite Fourier series, the mean-square value of $x(t)$ can be expressed in terms of the Fourier coefficients. Direct expansion of the mean-square expression gives

$$\frac{1}{T}\int_0^T x(t)^2\, dt = \frac{1}{T}\int_0^T \left[\frac{a_0}{2} + \sum_{n=1}^{\infty}(a_n\cos n\omega t + b_n\sin n\omega t)\right]^2 dt$$

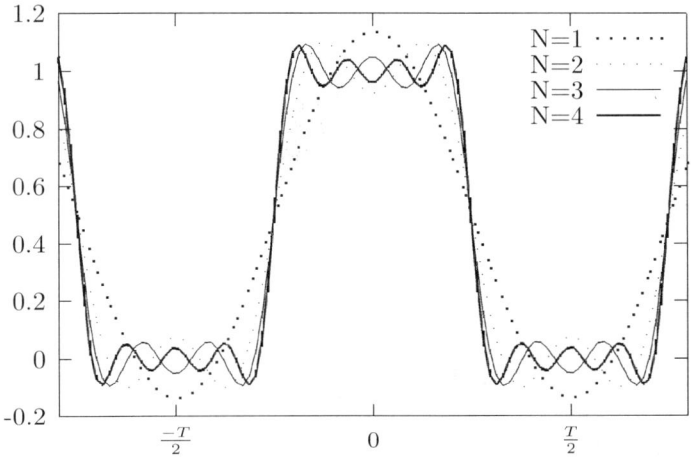

Figure 6.11: Finite Fourier series approximations of the rectangular waveform

Recall that $\int_0^T \sin n\omega t \cos m\omega t\, dt = 0$ for all n and m, $\int_0^T \sin n\omega t \sin m\omega t\, dt = \int_0^T \cos n\omega t \cos n\omega t\, dt = 0$ for $n \neq m$, and $\int_0^T \sin^2 n\omega t\, dt = \int_0^T \cos^2 n\omega t\, dt = T/2$ for all n. The above mean-square expression can be simplified to

$$\frac{1}{T}\int_0^T x(t)^2\, dt = \frac{a_0^2}{4} + \sum_{n=1}^{\infty} \frac{a_n^2}{2} + \sum_{n=1}^{\infty} \frac{b_n^2}{2}$$

This equation is known as the *Parseval's theorem* for real periodic functions. Essentially, Parseval's theorem provides an alternative means of evaluating the mean-square value, or the so-called *normalized power,* of a periodic signal. For example, if $x(t)$ is the voltage across a 1Ω resistor, then we can derive the power dissipation by adding up the terms $a_0^2/4, a_1^2/2, a_2^2/2, \ldots, b_1^2/2, b_2^2/2, \ldots$ In addition, we are able to tell the individual contributions from the DC component, the fundamental component, and other harmonic components.

Example 6.7: Application of Parseval's theorem — Suppose the voltage $v(t) = 10 + 5\cos\omega t + 2\sin\omega t + \cos 3\omega t$ is applied across a 50Ω resistor. The Fourier coefficients are $a_0 = 20$, $a_1 = 5$, $b_1 = 2$ and $a_3 = 1$. Using Parseval's theorem, we get the normalized power as

$$\text{Normalized power} = \left(\frac{20^2}{4} + \frac{5^2}{2} + \frac{2^2}{2} + \frac{1^2}{2}\right) = 115 \text{ W}$$

The power dissipated in the 50Ω resistor is $115/50 = 2.3$W.

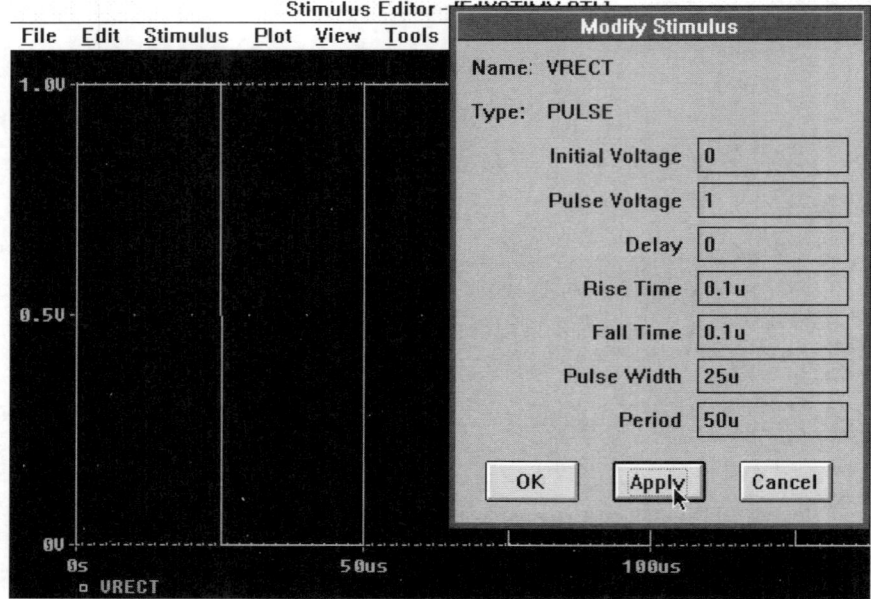

Figure 6.12: The "Modify Stimulus" dialogue box for specifying a pulse voltage (rectangular waveform)

6.9   Fourier Analysis with PSPICE

Two features of PSPICE are relevant to the subject of this chapter. The first one is the STIMULUS EDITOR, which can be used to create voltage sources of arbitrary waveform. The second one is *"Fourier Analysis"* which generates the Fourier coefficients of a specified voltage or current. It is also possible to use PROBE to perform a Fast Fourier Transform (FFT)[1] on any transient waveform generated from PSPICE analysis, and display the results directly within PROBE. In the following example we will focus on the use of the STIMULUS EDITOR and *"Fourier Analysis"* in PSPICE.

Example 6.8: Creating periodic functions and performing Fourier analysis in PSPICE — Voltage sources having such periodic waveforms as sawtooth and pulse are frequently encountered in electronics. In PSPICE analysis, we can create a voltage source of the required waveform by the usual *"Get New Part"* function. The name of the device to be selected is VSTIM, which can be found in the library "source.slb". Double-clicking on the VSTIM symbol will bring up the *"Edit Stimulus"* dialogue box for entering the desired name of the voltage source. Upon clicking "OK", the *"Stimulus Editor"* and the *"New Stimulus"* dialogue box pop up. From there, we can

---

[1]The Fast Fourier Transform is a fast discrete-time numerical algorithm for computing Fourier components of transient waveforms and is particularly efficient for periodic waveforms.

```
┌─────────────────────────────────────────────────────────────────────────┐
│ ▭                       Notepad - VSTIMX.OUT                          ▼  │
├─────────────────────────────────────────────────────────────────────────┤
│ File   Edit   Search   Help                                             │
│ DC COMPONENT =    5.000000E-01                                          │
│                                                                         │
│ HARMONIC    FREQUENCY     FOURIER     NORMALIZED    PHASE     NORMALIZED │
│    NO          (HZ)      COMPONENT     COMPONENT     (DEG)    PHASE (DEG) │
│                                                                         │
│     1      2.000E+04     6.367E-01    1.000E+00   -1.800E+00   0.000E+00 │
│     2      4.000E+04     3.422E-09    5.374E-09    1.238E+02   1.256E+02 │
│     3      6.000E+04     2.125E-01    3.338E-01   -5.400E+00  -3.600E+00 │
│     4      8.000E+04     2.434E-09    3.823E-09    1.538E+02   1.556E+02 │
│     5      1.000E+05     1.278E-01    2.008E-01   -9.000E+00  -7.200E+00 │
│     6      1.200E+05     1.334E-09    2.096E-09    1.700E+02   1.718E+02 │
│     7      1.400E+05     9.168E-02    1.440E-01   -1.260E+01  -1.080E+01 │
│                                                                         │
├─────────────────────────────────────────────────────────────────────────┤
│ ◄│ │                                                                  ► │
└─────────────────────────────────────────────────────────────────────────┘
```

Figure 6.13: The output file containing results from PSPICE's Fourier analysis of the rectangular waveform (up to the 7th harmonic)

select some commonly used waveforms, and modify their parameters to suit our own requirements. For example, we can select "PULSE" and click "OK" to bring up a *"Modify Stimulus"* dialogue box for entering parameters for a pulsating voltage. The modifiable parameters, as shown in figure 6.12, include "rise time", "fall time", "pulse width", "period", "pulsed voltage", and "initial voltage". In this example we wish to create a rectangular waveform similar to the one shown in figure 6.7. The parameters are:

$$\begin{aligned} \text{Rise time} &= \text{fall time} = 0.1\mu s \\ \text{Pulse width} &= 25\mu s \\ \text{Period} &= 50\mu s \\ \text{Initial voltage} &= 0V \\ \text{Pulse voltage} &= 1V \end{aligned}$$

Upon clicking "APPLY", the pulse waveform will be displayed for verification. We can modify as many times as we wish. Finally, we click "OK" to accept the parameters.

Our purpose here is to compute the Fourier coefficients for the rectangular waveform specified above by the STIMULUS EDITOR. In order for PSPICE to do the analysis, we have to complete the circuit. In this case we simply connect a 1kΩ resistor across the above pulse voltage source. Next, we have to select Fourier analysis. All we need to do here is to choose *"Setup"* from the **Analysis** menu, and click *"Transient"* to bring up a dialogue box in which we enter the name of the voltage and the number of harmonics to be analyzed, and click on the appropriate check box to enable Fourier analysis. Also, we must set the final time of the transient analysis to a value much larger than the repetition period of the rectangular function since PSPICE actually uses the transient waveform to evaluate Fourier components. Now we can invoke PSPICE analysis by choosing *"Simulate"* from the **Analysis** menu. PSPICE will save the results of the analysis to an output file which we can examine by choos-

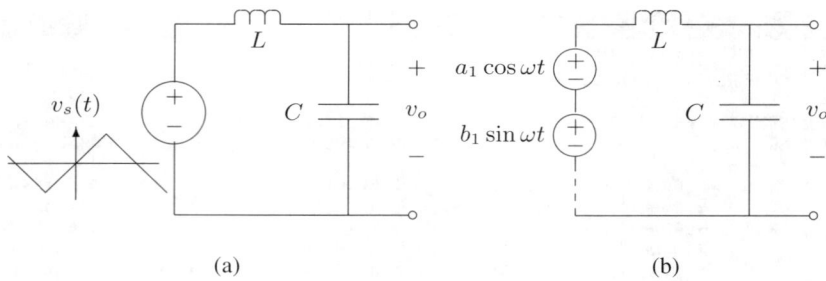

(a)   (b)

Figure 6.14: Circuit driven by a triangular voltage source

ing *"Examine Output"* from the **Analysis** menu. Figure 6.13 shows the part of the output file containing the result of the Fourier analysis. Note that PSPICE calculates $\sqrt{a_n^2 + b_n^2}$ and ϕ_n, instead of a_n and b_n. But we can always work out a_n and b_n using the following formulae from elementary trigonometry:

$$a_n = \sqrt{a_n^2 + b_n^2}\cos\phi_n \quad \text{and} \quad b_n = \sqrt{a_n^2 + b_n^2}\sin\phi_n$$

6.10 Using Fourier Series in Solving Circuit Problems

One final question before we close this chapter: What is the use of Fourier series in circuit analysis? There are indeed numerous applications of Fourier series. If we focus on circuit analysis, the use of a Fourier series representation for a complicated periodic function effectively translates a given circuit problem into a number of simpler problems that involve only sinusoidal functions. For example, in the circuit shown in figure 6.14, the input voltage is a triangular function. The problem is quite easy to solve if we decompose the triangular input voltage to a number of sinusoidal voltages, and use superposition to get the final answer. In the following two chapters, we will study the solution methods for circuit problems involving sinusoidal driving sources.

6.11 Problems

1. Calculate the average and rms values of the following functions. In each case, calculate the power dissipation in a 1Ω load whose terminal voltage is equal to the function $v(t)$.

   (i) $v(t) = \cos 3t$
   (ii) $v(t) = 10 + 2\sin 2t$
   (iii) $v(t) = 2 + \sin 2t + \cos 4t$
   (iv) $v(t) = \begin{cases} 0 & \text{for } nT \le t < (n + \frac{1}{2})T \\ 1 & \text{for } (n + \frac{1}{2})T \le t < (n + 1)T \end{cases}$

2. Find the average and rms values of the waveforms shown in figure 6.15. How are these values related to α?

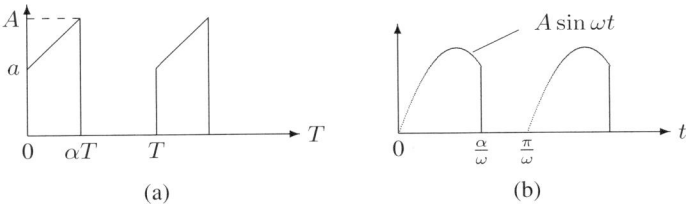

Figure 6.15: Waveforms for Problem 2

3. Suppose the input voltage and current to an electrical system are given by

$$v_i(t) \quad = \quad 300|\sin 100\pi t|$$

$$i_i(t) \quad = \quad \begin{cases} 10 & \text{for } 0 < t \le 0.002\alpha \text{ s} \\ 0 & \text{for } 0.002\alpha \text{ s} < t \le 0.002 \text{ s} \end{cases}$$

The voltage waveform repeats every $t = 1/100$s (i.e., at 100Hz), while the current waveform repeats every $t = 2$ms (i.e., at 500Hz). Calculate the average input power to the system. How does the input power depend on α? (Hint: Use the formula $p_{av} = \frac{1}{T} \int_0^T v(t)i(t)\, dt$, where T is the larger period.)

4. Draw a phasor diagram to represent the following sine functions, and hence find the sum of these functions.

$$v_1(t) = 10\sin \omega t, \quad v_2(t) = 15\cos(\omega t + 45°), \quad v_3(t) = -5\sin(\omega t - 60°)$$

5. Find the Fourier series for each of the waveforms shown in figure 6.16. State the type of symmetry in each case. For (c), assume that $T = 4$ for ease of calculation.

6. Verify that the rectangular waveform of figure 6.16 (a) can be obtained by differentiating the triangular waveform of figure 6.16 (b), with an appropriate adjustment factor.

7. Calculate the normalized power for the function $8 + 5\sin 100\pi t + 4\cos 400\pi t$. If this function is the voltage across a 50Ω load, calculate the power dissipation. Also, if the function is the current in a 50Ω load, calculate the power dissipation. (Hint: Use Parseval's theorem.)

8. Use Parseval's theorem to approximate the mean-square value of the waveforms shown in figure 6.16. Use terms up to the 5th harmonics. If each of these waveforms is the voltage (in volts) across a resistor of 10Ω, what is the power dissipation in each case?

9. Suppose a power supply is delivering current to a load, whose waveform is either one of those shown in figure 6.16. If the power supply provides a very well regulated voltage of 12V, what is the power dissipation in each case? What can you say about the relation between the current waveform and the power

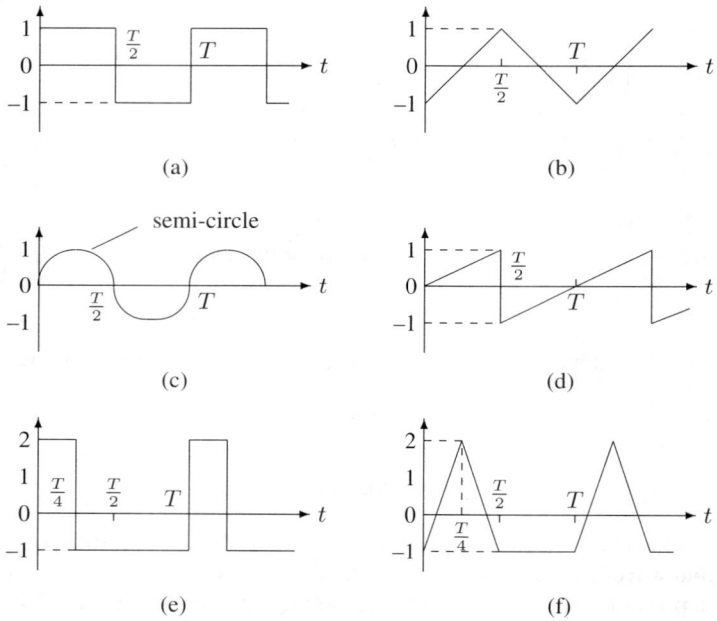

Figure 6.16: Waveforms for Problems 5, 6, 8 and 9

dissipation? If the power supply provides a sinusoidal voltage of frequency $1/T$, would you expect to draw the same conclusion as before regarding the power dissipation in each case?

10. Use PSPICE to study the transient waveform of the capacitor voltage in a simple RC circuit such as the one shown in figure 4.5 of Chapter 4. In particular, use VSTIM to create a pulse voltage source for this circuit, so that the exponential growth and decay can be repeatedly observed in the waveform of the capacitor voltage. Obtain the Fourier coefficients of the capacitor voltage waveform up to the 4th harmonic.

Chapter 7

Steady-State Analysis of AC Circuits

In linear resistive circuits, the voltage and current waveforms are exact images of the source that drives them. In the case of more than one source driving a circuit, superposition applies. Therefore, the methods used to analyze linear resistive circuits driven by DC sources can be direcly applied to linear resistive circuits driven by sinusoidal varying sources. Essentially we treat the sinusoidal varying sources as if they are DC sources. In this process we take either the peak value or the rms value as an equivalent DC value. The analysis goes exactly as in Chapters 1 to 3. The values of the voltages and currents obtained will then correspond consistently to either peak or rms values. All voltages and currents will be in phase with the driving source. In the case of more than one sinusoidal driving source of differing phases, we apply the principle of superposition. At the beginning, we treat each source separately. At the end, we add up the results from individual cases on the phasor diagram, with one source chosen arbitrarily as the reference phase. Alternatively we may perform the addition algebraically using the relation:

$$X_1 \sin(\omega t + \phi_1) + X_2 \sin(\omega t + \phi_2) = X_t \sin(\omega t + \phi_t)$$

$$\text{or} \qquad X_1 \angle \phi_1 + X_2 \angle \phi_2 = X_t \angle \phi_t$$

where

$$X_t = \sqrt{X_1^2 + X_2^2 + 2X_1 X_2 \cos(\phi_1 - \phi_2)}$$

$$\phi_t = \tan^{-1}\left(\frac{X_1 \sin \phi_1 + X_2 \sin \phi_2}{X_1 \cos \phi_1 + X_2 \cos \phi_2}\right)$$

When capacitors and/or inductors are present, however, the analysis in general requires setting up the describing differential equations, the solution of which reveals the transient motion as well as the steady state. In this chapter we focus on the steady-state solutions of circuits driven by sinusoidal varying sources.

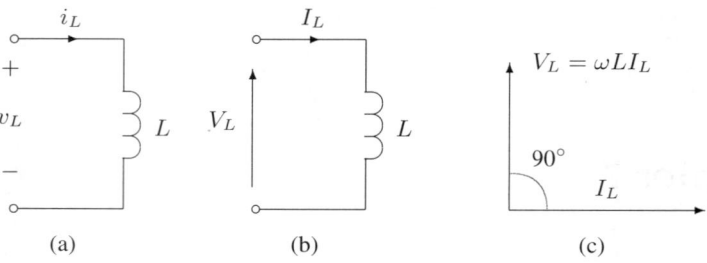

Figure 7.1: The inductor. (a) Sign convention; (b) using capital letters for steady-state values and an arrow for AC voltage; (c) phasor diagram showing current and voltage

7.1   Voltages and Currents in Capacitors and Inductors in the Steady-State

Let us now begin by examining the voltage and current relationship of a linear inductor. If the current through an inductor is

$$i_L(t) = \hat{i}_L \sin \omega t,$$

then the voltage across its terminal, using our usual sign convention as in figure 7.1 (a), is given by

$$v_L(t) = L\frac{di_L}{dt} = \omega L \hat{i}_L \cos \omega t$$

which may be written as

$$v_L(t) = \omega L \hat{i}_L \sin\left(\omega t + \frac{\pi}{2}\right)$$

Hence, the current is $\frac{\pi}{2}$ or 90° lagging the voltage, and the ratio of the magnitude of the voltage to that of the current is ωL. This may be represented by the phasor diagram shown in figure 7.1 (c).

In the case of a linear capacitor, if the voltage across its terminal is

$$v_C(t) = \hat{v}_C \sin \omega t,$$

then the current flowing through it is given by

$$i_C(t) = C\frac{dv_C}{dt} = \omega C \hat{v}_C \cos \omega t$$

or

$$i_C(t) = \omega C \hat{v}_C \sin\left(\omega t + \frac{\pi}{2}\right)$$

Hence, the current is $\frac{\pi}{2}$ or 90° leading the voltage, and the ratio of the magnitude of the voltage to that of the current is $1/\omega C$. This may be represented by the phasor diagram shown in figure 7.2 (c). Thus, the voltage-to-current ratio of the inductor and

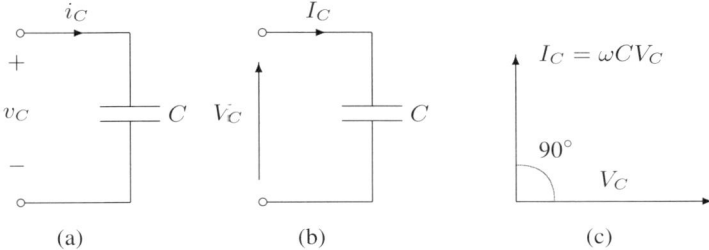

Figure 7.2: The capacitor. (a) Sign convention; (b) steady-state symbols; (c) phasor diagram

of the capacitor is frequency dependent. Specifically, for the inductor, the voltage is ωL times the current, and for the capacitor, the current is ωC times the voltage. Such v-i relations are similar to the case of resistance for which the voltage is R times the current. We may therefore regard the inductor or capacitor as a kind of frequency-dependent "resistance", with the additional feature of a fixed phase difference between its current and voltage, as described above. Normally, for the sake of clarity, we do not call it resistance; rather we refer to it as *reactance*.

Referring to the phasor diagram shown in figure 7.1 and treating the phasors as complex numbers on the Argand plane, the voltage–current relationship in the inductor may be represented by[1]

$$V_L = j\omega L I_L$$

and likewise for the capacitor,

$$V_C = \frac{I_C}{j\omega C}$$

Appendix E contains a summary of essential techniques of complex calculus that are relevant to the analysis of steady-state behaviour of AC circuits described in the subsequent sections.

7.2 Complex Representation of Impedance

We define, in general, the *impedance*, Z, of an element as the complex voltage-to-current ratio. Suppose V is the voltage across the element and I is the current through it. Both V and I are complex phasors. Then, in general, we have

$$Z = \frac{V}{I} = R + jX$$

---

[1]Recall from elementary complex calculus that if $\bar{a} = |a|\angle\theta_a$ and $\bar{b} = |b|\angle\theta_b$, then $\bar{a}.\bar{b} = |a|.|b|\angle(\theta_a + \theta_b)$. Thus, if $I_L = |I_L|\angle 0$, then $V_L = j\omega L.I_L = \omega L\angle 90°.|I_L|\angle 0 = \omega L|I_L|\angle 90°$.

Element	Z	R	X	Y	G	B
Resistance R	$R + j0$	R	0	$\frac{1}{R} + j0$	$\frac{1}{R}$	0
Inductance L	$0 + j\omega L$	0	ωL	$0 - \frac{j}{\omega L}$	0	$\frac{-1}{\omega L}$
Capacitance C	$0 - \frac{j}{\omega C}$	0	$\frac{-1}{\omega C}$	$0 + j\omega C$	0	ωC

Table 7.1: Impedance and admittance for resistor, inductor and capacitor

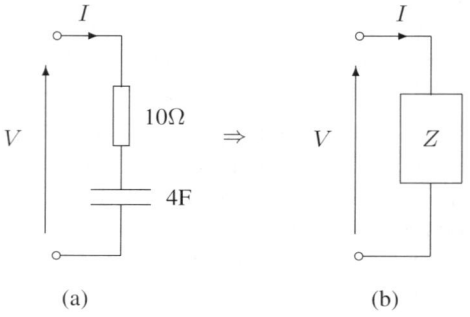

(a)   (b)

Figure 7.3: A capacitive load

where R is the real part and X is the imaginary part. Usually we refer to the real part as the *resistive part* (resistance), and to the imaginary part as the *reactive part* (reactance).

The dual of the impedance is the *admittance* which is simply the current-to-voltage ratio of a given element. The usual symbol for the admittance is Y.

$$Y = \frac{1}{Z} = \frac{1}{R + jX} = G + jB$$

where G is known as *conductance* and B as *susceptance*.

We now consider the impedance of three basic elements, namely the resistor, the inductor and the capacitor. For the resistor, voltage and current are in phase, and the impedance is a real number equal to the resistance R. For the capacitor and the inductor, we have shown in the preceding section that their impedances are respectively $j\omega L$ and $1/j\omega C$. Table 7.1 summarizes the impedance, admittance, resistance, conductance, reactance and susceptance values for these basic elements.

Example 7.1: Impedance of a capacitive load — Consider the series combination of a 10Ω resistor and a 4F capacitor shown in figure 7.3 (a). If the circuit that contains this combination of elements is driven by a 50Hz source, the impedance of this combination is

$$\begin{aligned} Z &= 10 + \frac{1}{j100\pi \times 4} \; \Omega \\ &= 10 - j0.8 \times 10^{-3} \; \Omega \end{aligned}$$

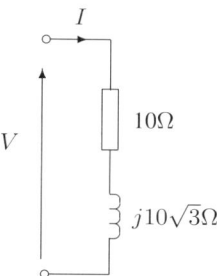

Figure 7.4: An inductive load

Sometimes it is more convenient to use the polar form, especially when performing division or multiplication. The polar form for $Z = R + jX$ is

$$Z = \sqrt{R^2 + X^2} \ \angle\tan^{-1}\left(\frac{X}{R}\right)$$

In this example, we have

$$Z = (10.00000003 \ \angle-0.004584°)\Omega$$

Clearly the polar form gives a convenient interpretation for the impedance. Since $Z = V/I$, the magnitude of Z is $|V|/|I|$ and the argument of Z is $\phi_v - \phi_i$ where ϕ_v and ϕ_i are the phase angles of the voltage and of the current, respectively, with respect to a common reference. In this example, the ratio of the magnitude of the voltage to that of the current is 10.00000003, and the voltage lags behind the current by a phase angle of 0.004584°.

Example 7.2: Impedance of an inductive load — Consider an impedance formed by a resistance and an inductance in series, as shown in figure 7.4. The impedance is equal to $10 + j10\sqrt{3}$ Ω, and in polar form it is $Z = 20 \ \angle60°$ Ω.
    If $V = \hat{v}\sin\omega t$, then the current I is

$$I = \frac{V}{Z} = \frac{V}{20 \ \angle60°} \ \text{A}$$

If we assume that the voltage has a zero phase angle, i.e., $V = \hat{v} \ \angle0$, then the current is

$$I = \frac{\hat{v}}{20} \ \angle-60° \ \text{A}$$

In the time domain, the above expression actually means

$$I = \frac{\hat{v}}{20} \sin(\omega t - 60°) \ \text{A}$$

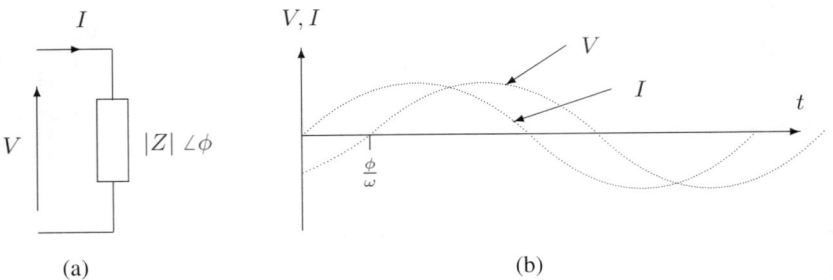

Figure 7.5: Phase angle ϕ

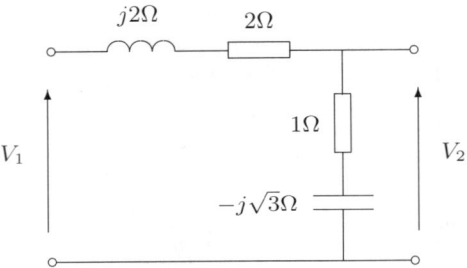

Figure 7.6: A simple voltage divider problem

7.3   The Impedance Angle

The impedance angle ϕ of an impedance Z is defined as the phase angle by which the voltage V leads the current I, using the usual sign convention as shown in figure 7.5.

Generally speaking, if Z is inductive, then $\phi > 0$ and V leads I, and if Z is capacitive, then $\phi < 0$ and V lags I.

7.4   Solving Circuit Problems

7.4.1   Using Complex Calculus

In solving circuits containing capacitors and inductors driven by a fixed frequency source, we may treat the capacitors and inductors as if they are "resistances" with imaginary values if we are only interested in the steady-state solution. Essentially the techniques for solving resistive circuits can be applied, but instead of solving real algebraic equations we have to solve complex equations. We use two simple examples to illustrate the procedure.

Example 7.3: Simple voltage divider problem — Suppose we wish to find the voltage V_2 in the circuit of figure 7.6. Given that $v_1 = \hat{v}_1 \sin \omega t$, or in polar form $V_1 = \hat{v}_1 \angle 0$. Obviously the problem reduces to solving a simple potential divider. The

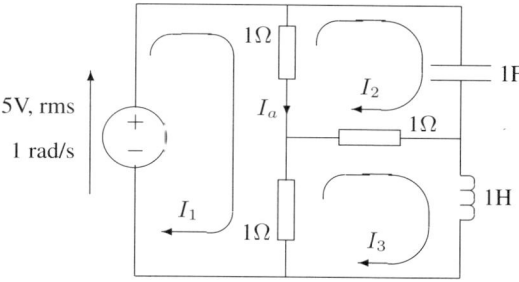

Figure 7.7: Application of the mesh method

transfer ratio is

$$
\begin{aligned}
\frac{V_2}{V_1} &= \frac{1 - j\sqrt{3}}{2 + j2 + 1 - j\sqrt{3}} \\
&= \frac{1 - j\sqrt{3}}{3 + j(2 - \sqrt{3})} \\
&= \frac{2\angle -60°}{3.012\angle 5.1°} \\
&= 0.664\ \angle -65.1°
\end{aligned}
$$

Thus, the time-domain expression for v_2 is $v_2 = 0.664\hat{v}_1 \sin(\omega t - 65.1°)$. It is worth noting that in manipulating complex number algebra we often interchange the numbers between rectangular form and polar form. The general rule is that the rectangular form is preferred when we do addition and subtraction, while the polar form is preferred when we do multiplication and division.

As mentioned before, the techniques for solving resistive circuits are applicable to steady-state analysis of dynamic circuits driven by sinusoidal sources. The following example illustrates the use of the mesh method.

Example 7.4: Steady-state solution via the mesh method — Consider the circuit shown in figure 7.7. Assume that $\omega = 1$ rad/s. We can write down the matrix equation involving the mesh currents as

$$
\begin{bmatrix}
2 & -1 & -1 \\
-1 & 2 - j & -1 \\
-1 & -1 & 2 + j
\end{bmatrix}
\begin{bmatrix}
I_1 \\
I_2 \\
I_3
\end{bmatrix}
=
\begin{bmatrix}
5 \\
0 \\
0
\end{bmatrix}
$$

Solving this equation gives

$$
I_1 =
\begin{vmatrix}
5 & -1 & -1 \\
0 & 2 - j & -1 \\
0 & -1 & 2 + j
\end{vmatrix}
/\Delta
$$

$$I_2 = \begin{vmatrix} 2 & 5 & -1 \\ -1 & 0 & -1 \\ -1 & 0 & 2+j \end{vmatrix} / \Delta$$

$$I_3 = \begin{vmatrix} 2 & -1 & 5 \\ -1 & 2-j & 0 \\ -1 & -1 & 0 \end{vmatrix} / \Delta$$

where

$$\Delta = \begin{vmatrix} 2 & -1 & -1 \\ -1 & 2-j & -1 \\ -1 & -1 & 2+j \end{vmatrix} = 2$$

Hence, we have

$$I_1 = 10\text{A}$$

$$I_2 = \frac{15 + 5j}{2} = \frac{5\sqrt{10}}{2} \angle 18.4° \text{ A}$$

$$I_3 = \frac{15 - 5j}{2} = \frac{5\sqrt{10}}{2} \angle -18.4° \text{ A}$$

We may find any branch current by suitably combining I_1, I_2 and I_3 vectorially. For example,

$$I_a = I_1 - I_2$$
$$= 2.5(1 - j) = 3.54 \angle -45° \text{ A}$$

7.4.2 Using Phasor Diagrams

Since only steady-state solutions are required, phasor diagrams may be used to derive the solution geometrically. The following rules should be observed when developing a phasor diagram.

1. Each voltage or current is represented by a phasor drawn on the complex Argand plane. The lengths of the phasors are drawn proportional to the magnitudes of the voltages or currents they represent, and their arguments must be consistent with the corresponding phase angle differences.

    (a) Phasors representing the voltage and the current of a resistor have the same argument, i.e., with the same orientation.

    (b) The phasor representing the current of a capacitor is drawn with an angle 90° leading the phasor of the voltage of the same capacitor.

    (c) The phasor representing the current of an inductor is drawn with an angle 90° lagging the phasor of the voltage of the same inductor.

2. Summation, substraction, multiplication and division are performed in accordance with the rules of elementary complex number manipulation.

    (a) Summation/subtraction of two phasors are performed vectorially according to the parallelogram law.

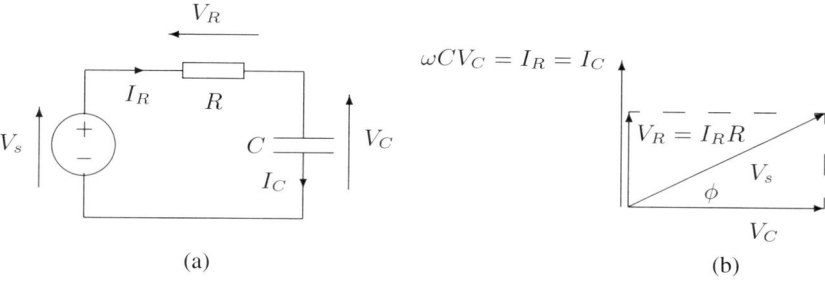

Figure 7.8: (a) Example circuit; (b) phasor diagram

(b) The product (quotient) of two phasors has magnitude equal to the algebraic product (quotient) of the individual magnitudes, and argument equal to the sum (difference) of the individual arguments.

Example 7.5: Steady-state solution via phasor diagram (capacitive circuit) — Consider the simple RC circuit shown in figure 7.8 (a). We may start arbitrarily by drawing V_C on the phasor diagram as a horizontal phasor pointing to the right. Then, I_C must be 90° leading V_C, and hence pointing upwards. Since the resistor and the capacitor are in series, I_R must be collinear with I_C, and so is V_R. Finally the vectorial sum of V_C and V_R is V_s. These are summarized in figure 7.8 (b).

Pythagoras's theorem gives $|V_R|^2 + |V_C|^2 = |V_s|^2$. Since $|V_R| = |I_R|.R = |I_C|.R = \omega C.|V_C|.R$, we obtain

$$|V_C| = \frac{|V_s|}{\sqrt{1 + \omega^2 C^2 R^2}}$$

Also, the angle between V_s and V_C is given by

$$\phi = \tan^{-1} \frac{|V_R|}{|V_C|} = \tan^{-1} \omega C R$$

Thus, if $v_s(t) = \hat{v}_s \sin \omega t$, then we have

$$
\begin{aligned}
v_C(t) &= \frac{\hat{v}_s}{\sqrt{1 + \omega^2 C^2 R^2}} \sin(\omega t - \phi) \\
v_R(t) &= \omega C R \hat{v}_C \sin(\omega t + \frac{\pi}{2} - \phi) \\
&= \frac{-\omega C R \hat{v}_s}{\sqrt{1 + \omega^2 C^2 R^2}} \cos(\omega t - \phi) \\
i_C(t) = i_R(t) &= \frac{-\omega C \hat{v}_s}{\sqrt{1 + \omega^2 C^2 R^2}} \cos(\omega t - \phi)
\end{aligned}
$$

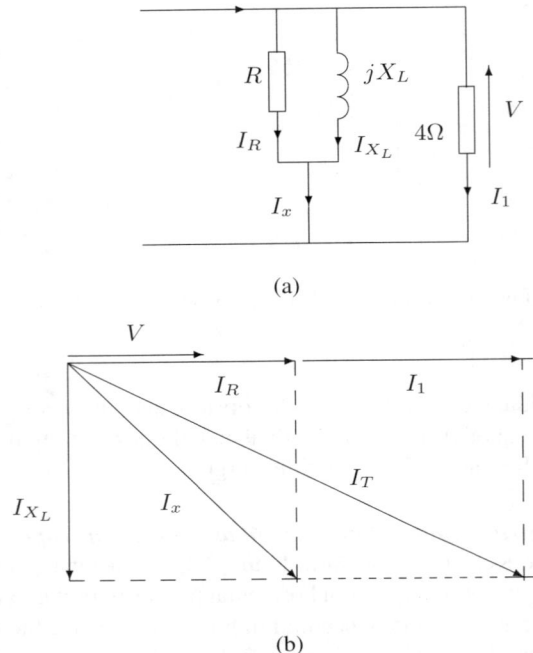

(a)

(b)

Figure 7.9: (a) Circuit for Example 7.6; (b) phasor diagram

Example 7.6: Steady-state solution via phasor diagram (inductive circuit) — We
consider another example. Suppose we wish to find the values of R and X_L in the
circuit of figure 7.9 (a) such that

$$I_x = 18\text{A}, \quad I_1 = 15\text{A}, \quad \text{and} \quad I_T = 30\text{A}$$

where all values are in rms. Since $I_1 = 15\text{A}$, we have $V = 4I_1 = 60\text{V}$, from which
the currents in R and L are found as

$$\text{|Current in } R| \quad = \quad \frac{V}{R} = \frac{60}{R}$$
$$\text{|Current in } L| \quad = \quad \frac{V}{X_L} = \frac{60}{X_L}$$

Inspection of the circuit gives the phasor diagram shown in figure 7.9 (b), from which
we have the following Pythagoras relations:

$$(|I_R| + |I_1|)^2 + |I_{X_L}|^2 \quad = \quad |I_T|^2$$
$$|I_{X_L}|^2 + |I_R|^2 \quad = \quad |I_x|^2$$

Substituting the values of the currents yields

$$\left(\frac{60}{R} - 15\right)^2 + \left(\frac{60}{X_L}\right)^2 = 30^2$$

$$\left(\frac{60}{X_L}\right)^2 + \left(\frac{60}{R}\right)^2 = 18^2$$

Letting $B = 1/X_L$ and $G = 1/R$, for convenience, we have

$$3600G^2 + 3600B^2 + 1800G = 675$$
$$3600G^2 + 3600B^2 = 324$$

which can be solved to give $G = 0.195$ and $B = 0.228$. Thus, the required values are

$$R = 5.13\Omega \quad \text{and} \quad X_L = 4.36\Omega$$

7.5   Power

7.5.1   Instantaneous Power

Power is an important quantity in practical circuits. It is a measure of energy dissipation in an apparatus. Since voltage and current can vary as functions of time, we immediately appreciate that an instantaneous value and an average value may be used to describe this dissipation. By definition, the instantaneous power is the product of the instantaneous voltage and current.

$$p(t) = v(t) \times i(t)$$

Thus we may use $p(t)$ to study the intensity of energy dissipation at any particular instant of time.[2]

7.5.2   Average Power

The average power dissipated in an element over an interval of time T is defined by the following integral.

$$P_{av} = \frac{1}{T} \int_0^T p(t) \, dt$$

In the case of voltage and current waveforms being sinusoidal, we can derive the average power dissipation as follows. For a resistance, denoting peak values with a

---

[2]This is useful, for example, when we wish to limit the maximum stress applied to a semiconductor device which can fail for a large vi product.

" ^ ", we have

$$P_{R_{av}} = \frac{1}{T} \int_0^T \hat{v}\hat{i} \sin^2 \omega t \, dt$$

$$= \frac{\hat{v}\hat{i}}{2\pi} \int_0^{2\pi} \sin^2 \theta \, d\theta$$

$$= \frac{1}{2}\hat{v}\hat{i}$$

$$= \frac{1}{2}\frac{\hat{v}^2}{R} = \frac{1}{2}\hat{i}^2 R$$

Alternatively we may express the average power in terms of the rms values and write $P_{R_{av}} = V_{rms}I_{rms} = V_{rms}^2/R = I_{rms}^2 R$. From this, we also see that the average power dissipated in a device is non-recoverable. It is *real* power dissipation.

For the inductor and the capacitor, the voltage and current are orthogonal. For this reason, the integral average of the vi product is identically zero. To see this, we let $x = \hat{x} \sin \omega t$ and $y = \hat{y} \cos \omega t$. Thus, x and y denote the voltage and current respectively for the case of a capacitor, whereas x and y denote the current and voltage respectively for the case of an inductor. For both cases, the average power is

$$P_{av} = \frac{1}{T} \int_0^T \hat{x}\hat{y} \sin \omega t \cos \omega t \, dt$$

$$= \frac{\hat{x}\hat{y}}{2\pi} \int_0^{2\pi} \sin \theta \, \cos \theta \, d\theta$$

$$= \frac{\hat{x}\hat{y}}{2\pi} \int_0^{2\pi} \frac{\sin 2\theta}{2} \, d\theta$$

$$= 0$$

Hence, we can conclude that the capacitor and the inductor do not dissipate any power in the average sense, or it has no real power dissipation.

7.5.3 Reactive Power

Although the capacitor and the inductor dissipate zero average power, they actually give and take power alternately, with the net intake over a complete period equal to zero. It would be useful to define another quantity which can reflect how large the "give" and the "take" are. A natural choice of a quantity that can reflect this is the $V_{rms}I_{rms}$ product. For the case of the capacitor and the inductor, such a quantity, denoted by Q, is called the *reactive power*.

$$Q = V_{rms}I_{rms} = \begin{cases} \omega L I_{rms}^2 & \text{for inductor} \\[2ex] \omega C V_{rms}^2 & \text{for capacitor} \end{cases}$$

In both cases, we have

$$Q = I_{rms}{}^2 X = \frac{V_{rms}{}^2}{X}$$

where X is the reactance.

7.6 Active Power, Reactive Power and Apparent Power

The terms *active power* and *real power* are often used interchangeably to mean the average power dissipated in an apparatus. For the case of a general impedance Z, we expect that the active power is non-zero while the total $V_{rms}I_{rms}$ product could be larger than the average power dissipation. The situation seems rather complicated. Let us now recall the instantaneous power expression, assuming that $i(t) = \hat{i}\sin\omega t$ and $v(t) = \hat{v}\sin(\omega t + \phi)$.

$$\begin{aligned} p(t) &= \hat{v}\sin(\omega t + \phi)\,\hat{i}\sin\omega t \\ &= \frac{\hat{v}\hat{i}}{2}\left[\cos\phi - \cos(2\omega t + \phi)\right] \end{aligned}$$

Writing $P = \dfrac{\hat{v}\hat{i}}{2}\cos\phi$ and $S = \dfrac{\hat{v}\hat{i}}{2}$, we have

$$p(t) = P - S\cos(2\omega t + \phi)$$

Obviously, P is the average power since the second term in the above expression has zero average. The quantity S is exactly the $V_{rms}I_{rms}$ product. This quantity does not give the real power dissipation, and we will call it *apparent power*. As shown in figure 7.10, S corresponds to the amplitude of the instantaneous power waveform, whereas P corresponds to the average value.

Let us now re-examine the situation from a different approach. With no loss of generality we may assume that any impedance Z is given by $Z = R + jX$. If the current that flows into Z is I, then the real power dissipation is

$$P = I_{rms}{}^2 R$$

whereas the reactive power is

$$Q = I_{rms}{}^2 X$$

It can be readily shown that the voltage across Z is $I_{rms}\sqrt{R^2 + X^2}$ and the impedance angle ϕ is $\tan^{-1}(X/R)$. We therefore have

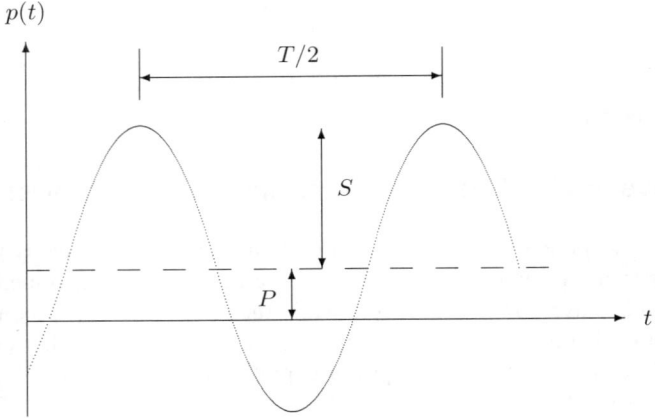

Figure 7.10: Instantaneous power waveform for a general impedance ϕ

$$\boxed{V_{rms}I_{rms} = \sqrt{P^2 + Q^2}}$$

the LHS of which is by definition the *apparent power*, S. Also, since $R = |Z|\cos\phi$ and $X = |Z|\sin\phi$, we obtain

$$\boxed{\begin{aligned} P &= S\cos\phi \\ Q &= S\sin\phi \end{aligned}}$$

Hence, for inductive impedance, $Q > 0$, and for capacitive impedance, $Q < 0$.

Remarks — The unit of real power is the *watt* (W). However, we do not use the *watt* for apparent power and reactive power since we would like to reserve the unit only for the dissipative VI product. For apparent power, we simply use the *volt-ampere* (VA) as the unit, whereas for reactive power we use the "var" as the unit. It should be borne in mind that all these units have the same dimension: the product of voltage and current.

7.7 Complex Power Notation

The afore-discussed concept of active, reactive and apparent powers may be conveniently described in complex number notation. If we define the complex apparent power as the multiplication of the voltage and the complex conjugate of the current,

$$\boxed{S = V.I^* = |V|.|I| \angle\phi}$$

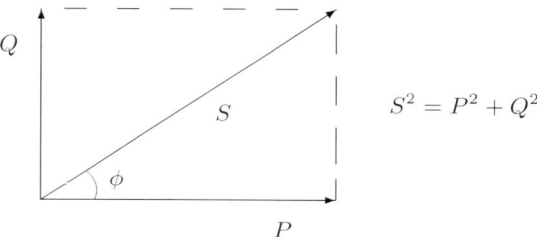

Figure 7.11: Power diagram

then the real power and reactive power can simply be extracted from the real part and imaginary part of S. Note here that all voltage and current values are rms values.

$$\begin{aligned} \text{Real Power } P &= \Re\{S\} = |V|.|I|\cos\phi \\ \text{Reactive Power } Q &= \Im\{S\} = |V|.|I|\sin\phi \end{aligned}$$

Figure 7.11 shows the complex apparent power on the Argand plane. Such a diagram is also known as the power diagram. As will be illustrated in the following examples, the concept of power often provides a fast solution to the steady-state problem.

Example 7.7: Complex power — If the input current to an impedance Z is $2\sin(\omega t + 30°)$ A and the voltage across Z is $5\sin(\omega t + 45°)$ V, then the complex apparent power is

$$S = 5\angle 45° \; 2\angle -30° = 10\angle 15° \text{ VA}$$

Also, the real power and reactive power are

$$P = \Re\{S\} = 9.6593\text{W} \quad \text{and} \quad Q = \Im\{S\} = 2.5882\text{var}$$

Example 7.8: Steady-state solution via the power concept — In solving circuit problems, sometimes, the use of the power concept can make the otherwise tedious problem very easy to handle. Referring to figure 7.12, the voltage source V_s having 10V rms is supplying current to a load Z via a transmission line which can be modelled as a resistance and an inductance in series. Suppose we want to calculate the current I_s.

The solution is tedious when resort to complex number algebra or phasor diagram. Let us consider the total real power and reactive power supplied by V_s. Bearing in mind that the total real power is the sum of all individual real powers, and the total reactive power is the sum of all individual reactive powers, we can write down the following almost by inspection, with all voltage and current magnitudes in rms.

$$\begin{aligned} P_T &= |I_s|^2(10 + 5\cos 30°) \\ &= 14.33\,|I_s|^2 \text{ W} \\ Q_T &= |I_s|^2(2 + 5\sin 30°) \\ &= 4.5\,|I_s|^2 \text{ var} \end{aligned}$$

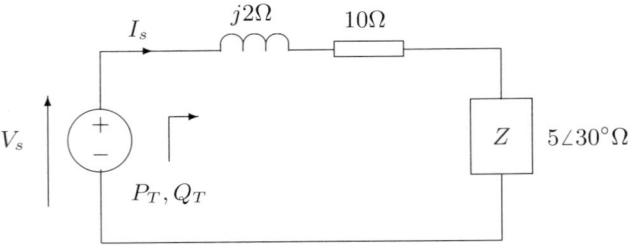

Figure 7.12: Solving circuits using power concepts

Hence, the apparent power S_T is

$$|S_T| = |I_s|^2 \sqrt{(14.33)^2 + (4.5)^2} = 15.02|I_s|^2 \text{ VA}$$

Since $|S_T| = |V_s|.|I_s|$, we obtain the current I_s as

$$|I_s| = \frac{10}{15.02} = 0.665 \text{ A}$$

and the phase angle lagging V_s is $\tan^{-1}(Q/P) = 17.43°$. Thus, we have

$$I_s = 0.665 \angle - 17.43° \text{ A}$$

7.8 Power Factor

From the foregoing discussion, the apparent power that is supplied to a load containing reactive elements is always larger than the real power actually dissipated in it. This is because $S^2 = P^2 + Q^2 > P^2$. Since the apparent power directly relates to the size and cost of the equipment,[3] it would be undesirable if much of S is due to Q. The ratio of real power to apparent power is therefore an important indicator of how "effective" (not efficient!) a load is performing its function in respect of power dissipation. This ratio is commonly known as the *power factor* (p.f.), which is defined as

$$\boxed{\text{p.f.} = \frac{P}{S}}$$

For instance, an electric induction motor is an inductive load, and hence can demand a high apparent power S while it actually produces much less real power P. In other words it has a power factor of less than 1.

For the case of sinusoidal voltages and currents, the power factor may be expressed as

---

[3]With a large $V_{rms}I_{rms}$, equipment of a sufficiently high rating is required, wires of sufficient thickness are needed, etc.

$$\boxed{\text{p.f.} = \frac{S \cos \phi}{S} = \cos \phi}$$

which is simply the cosine of the impedance angle. Thus, we see that inductive and capacitive loads have a power factor less than one while resistive loads have unity power factor.

$$0 \leq \text{p.f.} < 1 \quad \text{for reactive loads}$$
$$\text{p.f.} = 1 \quad \text{for resistive loads}$$

To distinguish between inductive and capacitive loads, we use the terms *lagging p.f.* and *leading p.f.* to describe respectively inductive and capacitive loads.

Example 7.9: Power factor and power factor correction — A load having a lagging p.f. of 0.8 is supplied by a sinusoidal voltage of rms value 220V and frequency 50Hz. Suppose it dissipates 80W of real power. In order to calculate the supply current, we must know the apparent power. Since p.f. = 0.8, we have $S = 80/0.8 = 100$ VA and $Q = \sqrt{100^2 - 80^2} = 60$ var. The supply current is therefore equal to $100/220$A or 0.4545A rms.

We now consider putting a capacitor C across the load to improve the power factor. This extra capacitor will demand a negative reactive power, thus cancelling out part of the positive reactive power absorbed by the inductive load. Suppose we wish to calculate the value of C in order to raise the p.f. to 0.95. Let P be the real power and Q be the total reactive power after capacitance is added. Obviously, Q now becomes $(60 - 220^2 \omega C)$ var. Thus, the value of p.f. becomes

$$\text{p.f.} = \frac{P}{\sqrt{P^2 + Q^2}} = \frac{80}{\sqrt{80^2 + (60 - 220^2 \omega C)^2}}$$

Putting p.f. = 0.95, we get $C = 2.2\mu$F. Furthermore, the total reactive power in the presence of the extra 2.2μF capacitance becomes 84.21 VA. Hence, the input current is reduced to

$$I_{in} = \frac{S}{220} = \frac{\sqrt{80^2 + 84.21^2}}{220} = 0.383\text{A rms}$$

7.9 Problems

1. Find the equivalent impedances of the circuits shown in figure 7.13, as observed from the end a-b, as functions of angular frequency ω. Indicate whether any singular point exists where the magnitude of the impedance is infinitely large or zero.

2. Find $v_o(t)$ at steady state in the circuit of figure 7.14, given that $v_s = 0.1 \sin \omega t$ V. Compare the cases when $\omega = 5000$ rad/s and $\omega = 300$ rad/s.

3. Use the nodal method to solve the circuit in figure 7.15 (a), and use the mesh method to solve the one in figure 7.15 (b). Finally convert all voltages and currents to the form $X \sin(t + \theta)$. You need only find the steady-state solutions.

4. Solve the circuit of figure 7.15 (a) using phasor diagram.

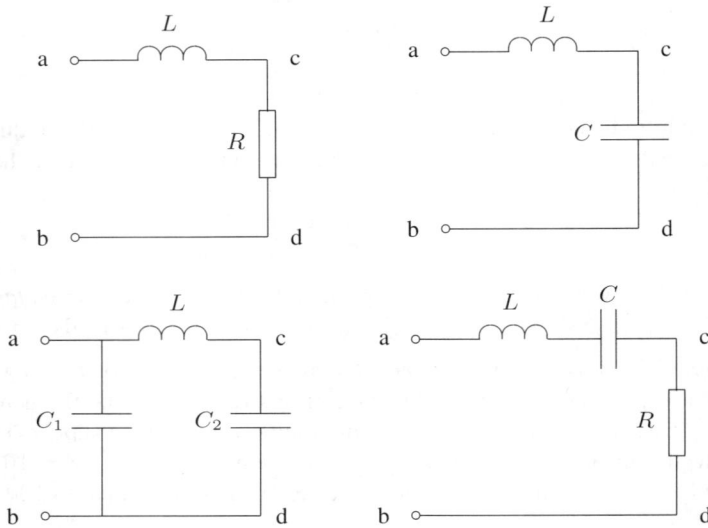

Figure 7.13: Circuits for Problem 1

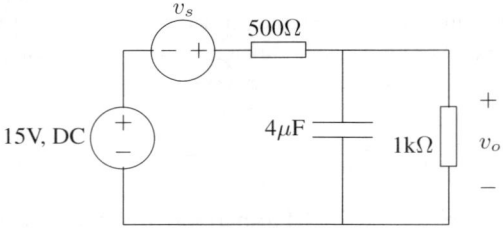

Figure 7.14: Circuit for Problem 2

5. The input current and voltage to a circuit are 3A and 110V rms. The measured power is 295W. Calculate the apparent power and the power factor. Draw the power diagram, assuming a lagging power factor.

6. The current flowing in an impedance, Z, is $10\sin(\omega t - 10°)$ A, and the voltage across it is $20\sin(\omega t + 15°)$ V. Calculate the complex power, real power and reactive power in the impedance. Find Z.

7. For the circuit shown in figure 7.16, the current through the resistance is 3A rms. What are the input current and voltage, assuming the driving angular frequency is 1 rad/s? Calculate the input active, reactive and apparent powers, and hence the input power factor.

8. Referring to figure 7.17, the load at the right end is an electric motor of power output 1kW, efficiency 85%, and lagging power factor 0.8. The supply voltage is

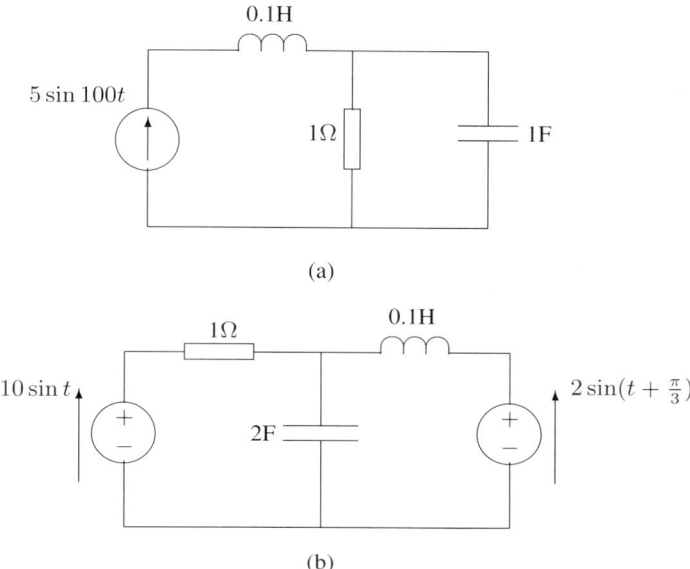

(a)

(b)

Figure 7.15: Circuits for Problems 3 and 4

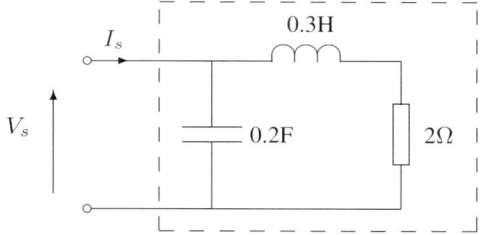

Figure 7.16: Circuit for Problems 7 and 9

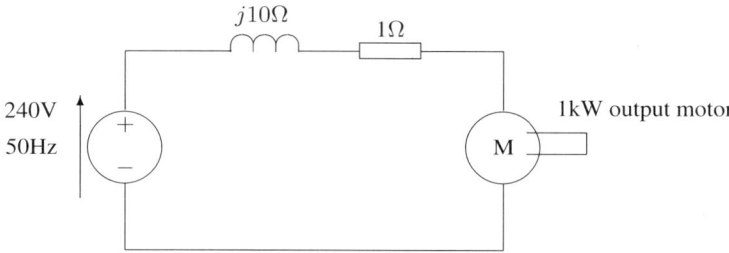

Figure 7.17: Circuit for Problem 8

240V rms, 50Hz. Calculate the input current, the input active power and power factor. (Hint: This problem can be solved easily with the power concepts.)

9. Consider the circuit of figure 7.16. Modify the value of the capacitor such that the power factor becomes 1. Verify that the input current is smaller than the one calculated in Problem 7.

10. Use PSPICE to verify answers to Problem 3. You may use ISIN and VSIN from the library "source.slb" for the sinusoidal sources. Perform a transient simulation. Make a sensible choice of final time so that enough time is allowed for the waveforms to reach the steady state.

Chapter 8

Free Oscillation and Complex Frequency Concepts

Steady-state analysis of linear dynamic circuits has been considered in Chapter 7, where the driving forces were sinusoidally varying functions of time. Two very handy methods of solution have been introduced, namely complex calculus and phasor diagrams. These methods work well for steady-state analysis but stop short of a general treatment that can give the transient behaviour, stability, and steady-state behaviour. In this chapter we will examine linear dynamic circuits from a more general perspective. We will begin with a review of the classical approach of differential equation solving, and see how qualitative behaviour of a circuit can be observed from its time-domain solution. We will also introduce a "new" domain — *the frequency domain* — in which the form of the solution allows much simpler derivation and the system behaviour is more readily observed. A number of important concepts will be introduced in this chapter, including free oscillation, stability, transfer function, poles and zeros, frequency response of circuits, and Bode plots. We will conclude this chapter with an illustrative example of deriving frequency response curves using PSPICE.

8.1 Differential Equation and Time-Domain Interpretation

Any dynamic circuit can be described by a system of differential equations, the form and complexity of which depend on the circuit configuration and the component's constitutive relations.[1] Linear dynamic circuits, in particular, are described by ordinary differential equations (ODE) with constant coefficients.

---

[1] The constitutive relation of an element refers to the relationship between the current through it and the voltage across it. For instance, for a linear resistance, the constitutive relation is simply the Ohm's law equation.

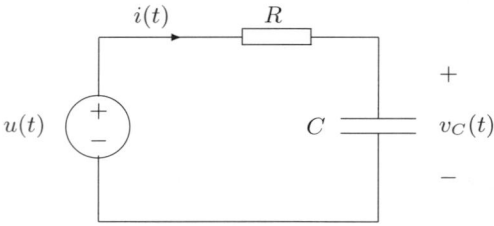

Figure 8.1: Simple RC circuit

8.1.1 Differential Equation for the Simple RC Circuit

Consider the simple RC circuit shown in figure 8.1. The capacitor's constitutive relation gives

$$i(t) = C \frac{dv_C(t)}{dt}$$

The circuit configuration implies that

$$u(t) = v_C(t) + i(t)R$$

Combining the above two equations yields the following first-order ODE:

$$\frac{dv_C(t)}{dt} + \frac{v_C(t)}{CR} = \frac{u(t)}{CR}$$

which can be solved to obtain $v_C(t)$ and $i(t)$, provided that $u(t)$ is given. Solutions thus obtained, being expressed as functions of time, are called *time-domain* solutions.

8.1.2 Free Oscillation and Natural (Transient) Response

If we solve the ODE for the above RC circuit, we will find that the solution consists of two terms: one is independent of $u(t)$ and dies out as time progresses; the other depends on $u(t)$ and has a more prolonged effect. Indeed, the former term governs the transient behaviour, whereas the latter is an "image" of $u(t)$ and constitutes the steady-state behaviour, as we have seen in Chapter 7.

It should be stressed that the circuit itself governs its own transient behaviour, irrespective of what $u(t)$ is. Thus, it would suffice, if the transient behaviour is what we want to know, to study the circuit with all driving sources reduced to zero. This involves short-circuiting all voltage sources and open-circuiting all current sources. The resulting circuit is known as the *free-oscillating* version of the given circuit, or simply the *free-oscillating circuit.*

Let us now put $u(t) = 0$ into the ODE of the RC circuit. This gives

$$\frac{dv_C(t)}{dt} + \frac{v_C(t)}{CR} = 0$$

which is the ODE of free oscillation. Two solutions are possible. A trivial solution is $v_C(t) \equiv 0$ for all t. Another solution (more general), which involves a non-zero initial condition, is given by

$$v_C(t) = v_C(0)e^{-t/CR}$$

The above solution is also called the *natural response* of the circuit, since it describes the *natural flow* of the circuit when it is let go with some charge initially deposited on the capacitor.

8.1.3 Stability

Another important property that can be derived from the free-oscillating circuit is *stability*. In simple terms, *stability refers to the ability to converge to a steady state under all possible initial conditions.* In the presence of bounded driving sources, the circuit, if stable, tends towards a steady state which depends on these driving sources. In the absence of a driving source, however, a stable circuit simply tends to a zero state where all voltages and currents are zero. Thus, we can determine whether a circuit is stable by inspecting its free-oscillating version. The essential test consists in determining whether the free-oscillating circuit tends eventually to a zero state.

Referring to the same RC circuit, we see that, provided CR remains positive, the value of v_C tends towards zero as $t \to \infty$. Thus, we say that the RC circuit is stable.

8.2 Natural Response as a Test for Stability

Clearly, from the foregoing discussion, we need only consider the circuit in its free-oscillating state when we examine its stability. Since driving sources are eliminated, the ODE will have zero RHS, i.e.,

$$\frac{d^n}{dt^n}x(t) + a_1\frac{d^{n-1}}{dt^{n-1}}x(t) + \cdots + a_{n-1}\frac{d}{dt}x(t) + a_nx(t) = 0 \qquad (*)$$

where $x(t)$ is a chosen variable.[2] The general solution is of the form

$$x(t) = A_1e^{\lambda_1 t} + A_2e^{\lambda_2 t} + \cdots + A_ne^{\lambda_n t}$$

where λ_i can be real or complex, and the As are found from the initial condition. Complex λs always come in conjugate pairs.

Obviously, $x(t)$ tends towards zero as $t \to \infty$ if and only if $\Re(\lambda_i) < 0$ for all i. Thus, stability, as defined previously, requires that *all λs be either negative real numbers or complex numbers with negative real parts.*

---

[2]The variable can be an inductance current or a capacitance voltage. We shall postpone the exact procedure for choosing the variables to be used in formulating differential equations to a later chapter.

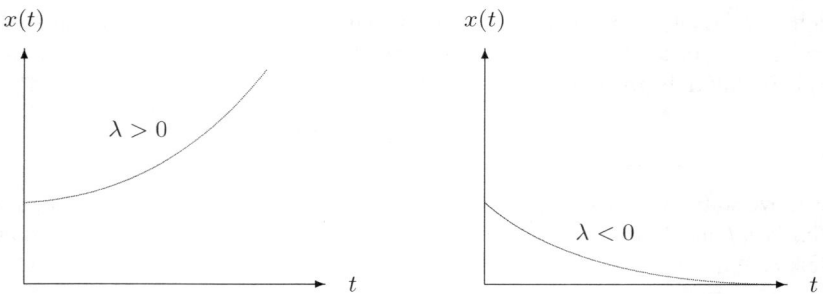

Figure 8.2: First-order natural responses

We take the general first-order case as an example to illustrate the test for stability. The solution of the free-oscillating circuit is

$$x(t) = Ae^{\lambda t}$$

The two natural responses corresponding to $\lambda > 0$ and $\lambda < 0$ are sketched in figure 8.2. Obviously, in order to guarantee convergence to zero, λ must be negative. The case where $\lambda = 0$ corresponds to a so-called marginally stable circuit.

To illustrate the case where some λs are complex, we consider the general solution of a second-order circuit.

$$x(t) = A_1 e^{\lambda_1 t} + A_2 e^{\lambda_2 t}$$

Three cases are possible.

Case 1 — *λ_1 and λ_2 are real.* In this case, both λ_1 and λ_2 must be negative for the solution to converge to zero.

Case 2 — *λ_1 and λ_2 are complex conjugates.* In this case, we can write λ_1 and λ_2 as

$$\lambda_1 = \alpha + j\omega \quad \text{and} \quad \lambda_2 = \alpha - j\omega$$

The solution can then be written as

$$x(t) = e^{\alpha t}(A\cos\omega t + B\sin\omega t)$$

which clearly demonstrates that convergence to zero requires α to be negative. Figures 8.3 (a) and (b) depict the cases for $\alpha > 0$ and $\alpha < 0$ respectively.

Case 3 — *λ_1 and λ_2 are pure imaginary.* This case corresponds to marginal stability or sustained oscillation. As shown in figure 8.3 (c), the oscillation amplitude depends on the initial value.

Thus, for all cases, we need

$$\boxed{\Re(\lambda_i) < 0 \quad \text{for all } i}$$

to guarantee convergence to zero, as $t \to \infty$.

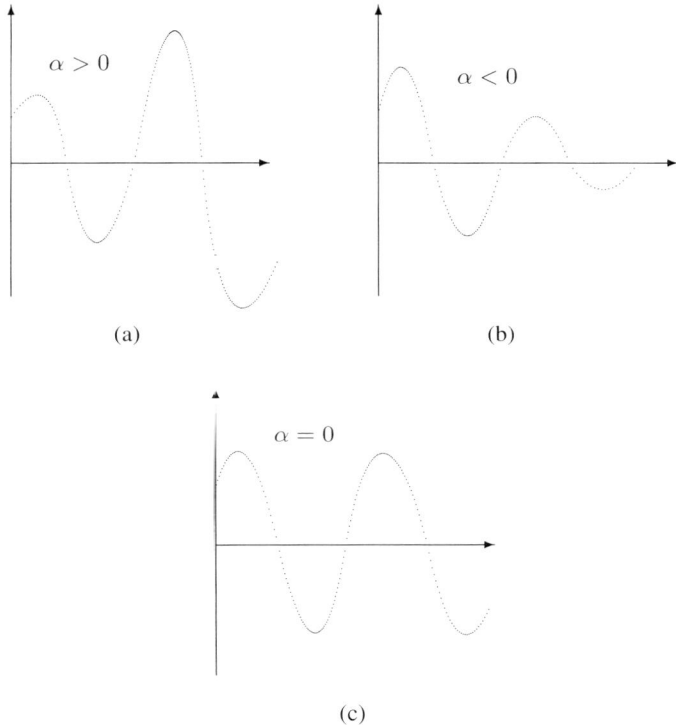

Figure 8.3: Second-order natural responses

8.3 The Characteristic Equation

The problem now reduces to finding the λs. Suppose the describing ODE is given by equation (*). Since Ae^{st} is a solution to this equation, we can substitute Ae^{st} for $x(t)$ in equation (*) to get

$$s^n Ae^{st} + s^{n-1}a_1 Ae^{st} + s^{n-2}a_2 Ae^{st} + \cdots + a_n Ae^{st} = 0$$
$$\Rightarrow \quad s^n + a_1 s^{n-1} + a_2 s^{n-2} + \cdots + a_n = 0$$

Clearly the roots of the last equation will give all the λs. In other words, once we have obtained this equation, we are able to tell whether the circuit is stable provided we can solve for its roots. This polynomial equation in s, which contains important information about stability, is given the name *characteristic equation*.[3] We may

---

[3]Literature abounds with techniques for detecting roots with a positive real part of a given characteristic equation. These techniques are beyond the scope of this book, and are usually treated in a systems theory text. It is, however, important to appreciate the important role the characteristic equation plays in determining the stability of a given circuit.

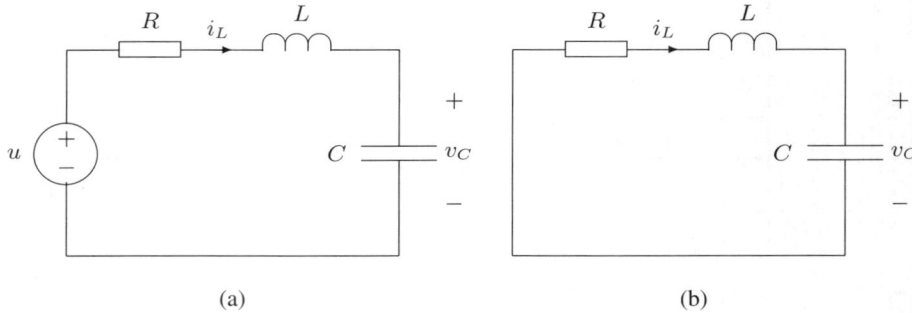

Figure 8.4: (a) Series LRC circuit; (b) free-oscillating circuit

summarize that *a circuit is stable if and only if all roots of its characteristic equation have a negative real part.*

Example 8.1: Simple LRC circuit — Figure 8.4 (a) shows a series LRC circuit, and figure 8.4 (b) shows its free-oscillating circuit. The constitutive relations for the capacitor and the inductor, together with Kirchhoff's law applied to the free-oscillating circuit, give

$$C\frac{dv_C}{dt} = i_L$$
$$L\frac{di_L}{dt} = -v_C - i_L R$$

Differentiating the first equation and eliminating i_L from these equations yield the describing ODE as

$$\frac{d^2}{dt^2}v_c + \frac{R}{L}\frac{d}{dt}v_C + \frac{1}{LC}v_C = 0$$

The characteristic equation is

$$s^2 + \frac{R}{L}s + \frac{1}{LC} = 0$$

from which we can tell whether the circuit is stable or not. Essentially the roots are

$$\lambda = -\frac{R}{2L} \pm j\sqrt{\frac{1}{LC} - \frac{R^2}{4L^2}} \quad \text{if} \quad \frac{1}{LC} > \frac{R^2}{4L^2}$$
$$\text{or} \quad \lambda = -\frac{R}{2L} \pm \sqrt{\frac{R^2}{4L^2} - \frac{1}{LC}} \quad \text{if} \quad \frac{1}{LC} < \frac{R^2}{4L^2}$$

Thus, the circuit is always stable if $R, L, C > 0$.

It is interesting to note that we can obtain another second-order ODE in terms of i_L if we choose to differentiate the second differential equation instead of the first, and eliminate v_C. The ODE in i_L is

$$\frac{d^2}{dt^2}i_L + \frac{R}{L}\frac{d}{dt}i_L + \frac{1}{LC}i_L = 0$$

This ODE gives exactly the same characteristic equation as the other ODE does. This is in fact a very important observation. *The characteristic equation for a given circuit is unique.* In other words, regardless of the method of derivation, the characteristic equation is always the same for a given circuit.

Remarks — λ is sometimes called the natural frequency of the circuit. A more common term for λ is, however, the *eigenvalue* which is widely used in systems theory and mathematics. In the case of first-order circuits, the reciprocal of λ is exactly the time constant τ.

8.4 The Complex Frequency Concept

8.4.1 Complex Natural Frequencies

We have seen that any nth order linear circuit is associated with a set of n natural frequencies (also called eigenvalues) which are generally complex numbers having the same dimension as the angular frequency (rad/s). From these numbers we can tell about the natural response of the circuit, based on the general solution expression.

$$x(t) = \sum_{k=1}^{n} X_k e^{\lambda_k t}$$

1. If a negative real eigenvalue exists, the natural response has a mode which is *exponentially decaying.*
2. If a positive real eigenvalue exists, the natural response has a mode which is *exponentially growing.*
3. If a pair of pure imaginary eigenvalues exist, the natural response has a mode which is *oscillatory.*
4. If a pair of complex eigenvalues with negative real parts exist, the natural response has a mode which is *oscillatory with exponentially decaying amplitude.*
5. If a pair of complex eigenvalues with positive real parts exist, the natural response has a mode which is *oscillatory with exponentially growing amplitude.*

In the following we will extend the complex frequency concept to sources and impedances.

8.4.2 Extension to Driving Source

Now let us use our imagination to create a voltage source which varies at a complex frequency s. By using the same complex frequency concept as applied to the natural response, we may express this voltage as

$$v(t) = V_o e^{st}$$

For example, if $s = -10$ /s, then this voltage source has a magnitude which is exponentially decaying with time constant 0.1s. If $s = -10 \pm j10$ rad/s, the source is a $10/2\pi$ Hz sinusoidal voltage whose amplitude decays with time constant 0.1s. Also, a pure sinusoidal voltage would have a frequency of $s = \pm j\omega$ rad/s since

$$\cos \omega t = \frac{e^{j\omega t} + e^{-j\omega t}}{2}$$

Remarks — Using the complex frequency notation, *sinusoidal signals have pure imaginary frequencies.* Thus, if we drive a circuit with a sinusoidal signal of ω rad/s, we are driving it at $\pm j\omega$ rad/s in the complex frequency domain.

8.4.3   Extension to Impedance

Extension of the concept of complex frequency to impedance can be made conveniently by considering the constitutive relations of the inductor and capacitor.

$$\text{Inductor:} \qquad v_L = L\frac{di_L}{dt}$$

$$\text{Capacitor:} \qquad i_C = C\frac{dv_C}{dt}$$

Substituting the general expressions $v_L = V_L e^{st}$, $i_L = I_L e^{st}$, $v_C = V_C e^{st}$ and $i_C = I_C e^{st}$ gives

$$\begin{aligned} V_L &= sLI_L \\ I_C &= sCV_C \end{aligned}$$

Thus, the impedance of an inductor is

$$\boxed{Z_L = sL}$$

and that of a capacitor is

$$\boxed{Z_C = \frac{1}{sC}}$$

They are functions of the complex frequency s. In general, s is complex, and so are Z_L and Z_C. For the special case of a circuit driven by a sinusoidal source, the impedance of an inductor is $j\omega L$ and that of a capacitor is $1/j\omega C$. This is consistent with our discussion of reactance in Chapter 7.

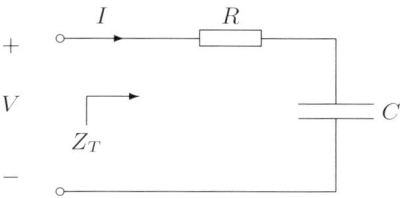

Figure 8.5: Simple series RC combination

Example 8.2: Simple series RC circuit — Consider a simple series combination of a resistance and a capacitor, as shown in figure 8.5. The general equivalent impedance Z_T is given by

$$Z_T = R + \frac{1}{sC} = \frac{1 + sCR}{sC}$$

We can thus write

$$\frac{I}{V} = \frac{sC}{1 + sCR}$$

The free-oscillating circuit corresponds to $V = 0$, i.e.,

$$(1 + sCR)I = 0$$

This gives the characteristic equation directly as

$$1 + sCR = 0$$

In this example, we have shown how the characteristic equation can be obtained very quickly using the complex frequency notation in conjunction with the usual circuit analysis techniques.

Remarks — As is customary in circuit theory, upper-case letters will be used for variables in the frequency domain, and lower-case letters for the time domain.

8.5 Transfer Function

In general, a transfer function is a ratio of two quantities (either voltage or current) measured at two different locations of a circuit. Usually one quantity is referred to as *input,* and the other as *output.* Depending on the types of the involving quantities, four forms of transfer function are possible.

1. *Voltage gain* — Both input and output are voltage.

2. *Current gain* — Both input and output are current.

3. *Trans-admittance* — Input is voltage, output is current.

4. *Trans-impedance* — Input is current, output is voltage.

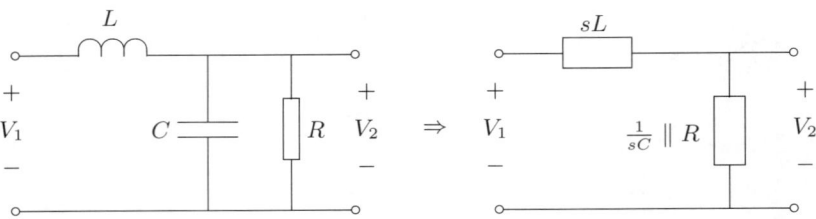

Figure 8.6: Transfer function example

Remarks — When input and output are taken from the same location in a circuit, we use the terms *driving-point admittance* and *driving-point impedance* in lieu of trans-admittance and trans-impedance. For instance, in the simple RC circuit shown in figure 8.5, if we take V as the input and I as the output, the transfer function V/I is a driving-point impedance.

Example 8.3: Voltage transfer function — Referring to the circuit shown in figure 8.6, V_1 and V_2 are the input and output respectively. Using the complex impedance concept, we can write down the transfer function V_2/V_1 as

$$\frac{V_2}{V_1} = \frac{\dfrac{1}{sC} \| R}{sL + \left(\dfrac{1}{sC} \| R \right)} = \frac{1}{1 + s\dfrac{L}{R} + s^2 CL}$$

It is also possible to obtain the transfer function directly from the differential equation. For the above circuit, it can be shown, using the nodal method for instance, that the differential equation involving v_2 is

$$LC \frac{d^2}{dt^2} v_2(t) + \frac{L}{R} \frac{d}{dt} v_2(t) + v_2(t) = v_1(t)$$

To obtain the complex-frequency domain equation, we simply replace d/dt by s. This results in an algebraic equation

$$\left(LC s^2 + \frac{L}{R} s + 1 \right) V_2 = V_1$$

which gives the same transfer function as before.

8.5.1   Obtaining the Characteristic Equation

Without probing too deeply, one thing we can do at this stage is to read out the characteristic equation from the denominator of the transfer function, and hence obtain the natural response. In general, the transfer function is of the form

$$\frac{X_o}{X_i} = \frac{N(s)}{D(s)}$$

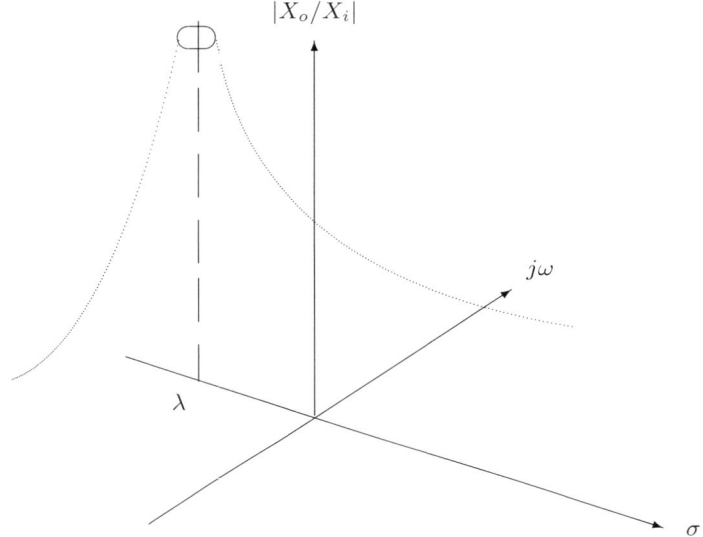

Figure 8.7: Transfer function magnitude plotted on the s-plane

where X_o and X_i are output and input respectively, and $N(s)$ and $D(s)$ are polynomials in s. The characteristic equation is simply given by

$$D(s) = 0$$

For the above circuit, $N(s) = 1$ and $D(s) = s^2LC + sL/R + 1$. The characteristic equation is thus

$$s^2LC + s\frac{L}{R} + 1 = 0$$

solving which gives the eigenvalues or natural frequencies for the circuit.

8.5.2 Plotting Transfer Function Magnitude

The transfer function can be regarded as the ratio of the output, X_o, to the input, X_i, at complex frequency s. It seems a bit difficult to comprehend its meaning when the driving frequency is a complex number. Let us now focus on the magnitude of the transfer function, $|X_o/X_i|$. As s varies, $|X_o/X_i|$ varies. Thus, if we plot $|X_o/X_i|$ against s, we obtain the transfer function magnitude as a "surface" over the complex s-plane. At those values of the eigenvalues, the transfer function magnitude goes to infinity since $D(s) = 0$ for $s = \lambda_i$. For this reason, the set of eigenvalues or natural frequencies are often referred to as a set of *poles* of the circuit. Figure 8.7 illustrates the transfer function magnitude on the complex s-plane.

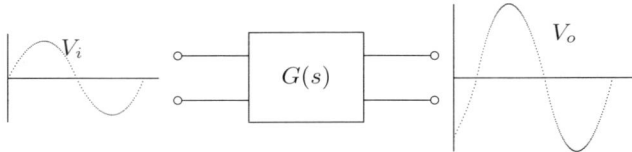

Figure 8.8: Block diagram representation of circuit with transfer function $G(s)$

8.6   Frequency Response – A Realistic Scenario

In practice, we drive the circuit with a sinusoidal source for which the frequency is pure imaginary, i.e., $\pm j\omega$, and the magnitude of the transfer function gives the magnitude ratio of $|X_o|$ to $|X_i|$ while the argument of the transfer function gives the phase shift of X_o from X_i.

Consider the block diagram shown in figure 8.8, which represents a linear circuit. The transfer function V_o/V_i is $G(s)$. That means, if we drive the circuit with V_i, it gives $V_o = G(s)V_i$. Suppose V_i is a sinusoidal voltage at angular frequency ω rad/s. We would like to see how V_o/V_i varies as we vary ω from zero to infinity. If we perform measurements in the laboratory, we can plot $|V_o|/|V_i|$ against ω. In addition, we observe a phase shift between V_i and V_o, from which we can produce another graph with the phase angle between V_i and V_o plotted against ω. Contained in these two plots (magnitude and phase angle) is what engineers commonly refer to as the *frequency response* of the circuit.

In fact, the frequency response can be observed from the complex frequency domain. Since sinusoidal signals have pure imaginary frequencies $j\omega$, we can concentrate on the imaginary axis, and look at the variation of $|V_o|/|V_i|$ for $s = j0$ to $s = j\infty$. Thus, instead of inspecting the whole "surface" over the s-plane, we focus on the curve where the surface cuts the cross-section $\sigma = 0$, as shown in figure 8.9. Similar reasoning applies to the argument of V_o/V_i in the s-plane.

8.7   Time Domain Versus Frequency Domain

It must be borne in mind that the frequency response obtained above corresponds to steady-state behaviour of the given circuit. This means that the magnitude ratio so derived is the ratio of the output to input at steady state, and the phase shift also corresponds to the steady-state waveforms. In other words we have a complete picture of the steady-state response of the circuit driven by sinusoidal sources for the whole range of frequencies. Such information is also extractable from the time-domain description, though it is a rather clumsy business, since we can generate the steady-state response for as many frequency values as we desire. But can we get the transient response information from the frequency-domain description? Indeed it is possible. The link is via the characteristic equation, which is simply the denominator of the transfer function equated to zero. Thus, we observe an amazing correspondence between the time-domain and frequency-domain descriptions.

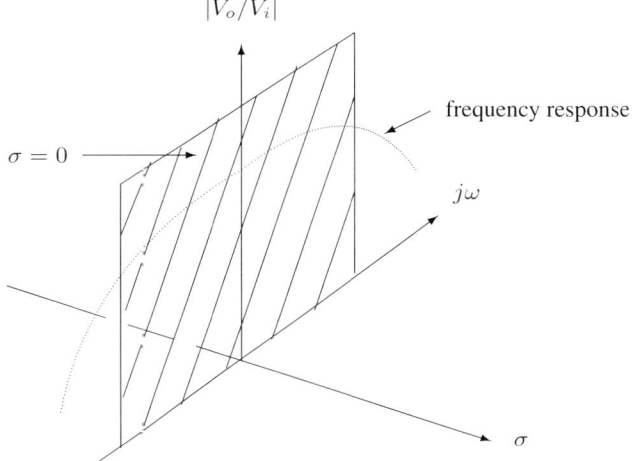

Figure 8.9: Viewing frequency response on the s-plane

Let us take a closer look. When we observe the form $\frac{1}{s+a}$ in the frequency-domain transfer function, we know immediately that the circuit has a natural response typified by e^{-at}. In other words, the frequency-domain description is mapped uniquely to the type of time-domain natural response. We may collect these mappings in the form of a translation table, e.g.,

$$Ke^{-at} \longleftrightarrow \frac{1}{s+a}$$

$$K\sin\omega t \longleftrightarrow \frac{1}{s^2+\omega^2}$$

$$Ke^{-at}\sin\omega t \longleftrightarrow \frac{1}{s^2+2as+a^2+\omega^2} \quad \text{etc.}$$

The formal mathematical treatment for translating between the two domains is known as Laplace Transform, which we will briefly introduce in Chapter 11 as a tool for solving differential equations. In fact, the use of Laplace Transform not only provides a linkage between the form of transfer function and the type of natural response, it is also able to provide complete solution when the exact forms of the driving sources and the initial condition are given. This branch of mathematics is widely known in the mathematical literature as *initial value problems*.[4]

---

[4]In Appendix C, we discuss the Laplace transform method for solving differential equations in some depth. A formal treatment of the subject should be found in an engineering mathematics text.

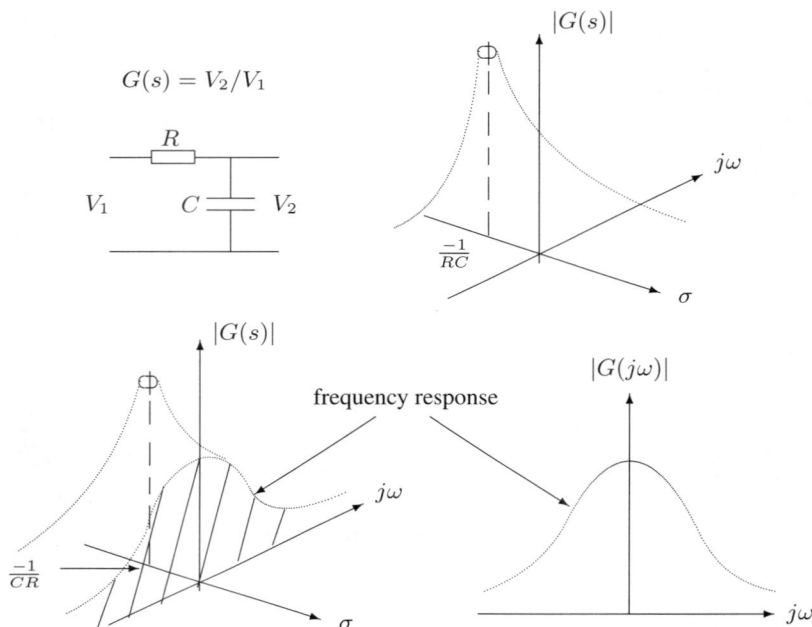

Figure 8.10: RC circuit and its frequency response

8.8 Frequency Response of Low-Order Circuits

8.8.1 Frequency Response of First-Order Circuits

A first-order circuit has one eigenvalue, which is the reciprocal of the time constant. Any transfer function from a first-order circuit has a first-degree denominator. For example, the simple RC circuit shown in figure 8.10 has a transfer function given by

$$\frac{V_2}{V_1} = \frac{\frac{1}{sC}}{R + \frac{1}{sC}} = \frac{1}{1 + sCR}$$

In order to derive the frequency response for this circuit, we first notice that $V_2/V_1 \rightarrow \infty$ as $s \rightarrow -1/CR$, and $V_2/V_1 \rightarrow 0$ as $|s| \rightarrow \infty$. Plotting the transfer function magnitude against s, we expect to see a surface on the s-plane which tends to infinity at $s = -1/CR$ and goes to zero everywhere else that is far from the origin, as shown on the top right of figure 8.10. For sinusoidal signals, $s = j\omega$. Thus, if we cut the surface along the $j\omega$ axis as shown in the bottom left of the figure (see also figure 8.11), the curve intersecting the surface and the cutting plane gives the frequency response. Several points are worth noting here.

    1. The point $s = -1/CR$ is called the *pole* of the circuit. Intuitively speaking, if the pole is close to the origin, the frequency response of $|V_2/V_1|$ will have a narrow top and drop off at a relatively lower frequency. If, however, the pole

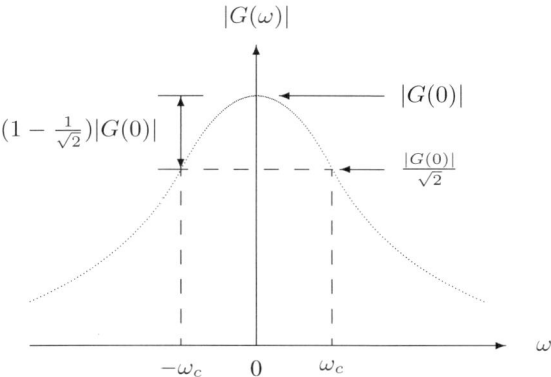

Figure 8.11: First-order low-pass frequency response

is farther from the origin, the frequency response will have a wider flat top and drop off at a higher frequency.

2. The frequency at which the magnitude response drops to $1/\sqrt{2}$ of the highest value is called the *3dB corner frequency* or simply *corner frequency*, because $(1 - \frac{1}{\sqrt{2}})$ is 3dB.

3. It can be shown that the corner frequency is given by $\omega_c = \frac{1}{CR}$ rad/s. This agrees with our intuition as mentioned above.

Example 8.4: First-order response — Consider a first-order system whose transfer function is given by

$$\frac{Y(s)}{X(s)} = \frac{10}{1 + \frac{s}{10000}}$$

The eigenvalue is -10000. Thus, we expect the natural response to take the form $y(t) = Ae^{-10000t}$. In the frequency domain, if we measure the ratio of the magnitude of $y(t)$ to that of $x(t)$ for the entire frequency range (i.e., $x(t)$ is a sine function), we will see that the gain $|X(j\omega)/Y(j\omega)|$ decreases as frequency increases. At very low frequencies, the gain is 10 or 20dB, and at very high frequencies it is 0. In particular, the 3dB corner frequency as defined above is

$$\omega_c = 10000 \text{ rad/s} \quad \text{or} \quad f_c = 1.59 \text{ kHz}$$

8.8.2 Frequency Response of Second-Order Circuits

The transfer function

$$G(s) = \frac{1}{\dfrac{s^2}{\omega_n^2} + \dfrac{2\zeta}{\omega_n}s + 1}$$

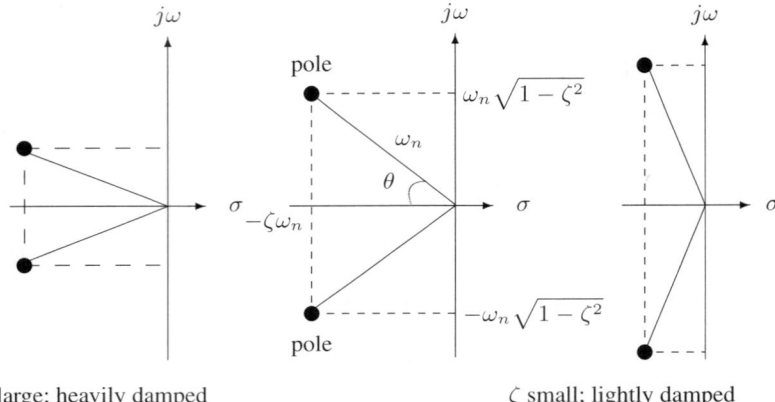

ζ large; heavily damped ζ small; lightly damped

Figure 8.12: Poles of second-order circuits

describes a general second-order low-pass function. Assuming complex conjugate eigenvalues, the positions of the poles on the s-plane are $s = -\zeta\omega_n \pm j\omega_n\sqrt{1-\zeta^2}$. The parameter ζ is commonly called the *damping factor*. As we can see in figure 8.12, the smaller the value of ζ, the closer the poles to the $j\omega$ axis, hence a larger peak in the frequency response. Using the same procedure as for the first-order case, the frequency response for the second-order low-pass function can be obtained, as shown in figure 8.13. The following points concerning the frequency response curves are noteworthy.

1. For circuits with $\zeta < 0.7071$, i.e., light damping, the frequency response shows a peak near ω_n. (See figure 8.14.)
2. For circuits with $\zeta > 0.7071$, i.e., heavy damping, the frequency response shows no peak.
3. The bandwidth is approximately ω_n where the peak occurs.

The above description corresponds exactly to the low-pass filter function realized by the circuit shown earlier in figure 8.6. Note that the numerator in this particular case is 1, and hence the magnitude of the transfer function tends to zero as s tends to ∞, while being finite for $s = 0$. The low-pass characteristic is thus apparent. In the following example, we illustrate a second-order band-pass function in which the numerator is also a polynomial in s. This example will serve as an introduction to the concept of pole and zero, which will be discussed in the next section.

Example 8.5: Second-order band-pass filter — Consider the circuit shown in figure 8.15 (a) which realizes a band-pass filter function. We may apply the nodal analysis technique to find the transfer function V_2/V_{in}. The KCL equations for this circuit are summarized in the following matrix equation.

$$\left[\begin{array}{cc} G_1 + G_2 + \frac{1}{sL} & -G_2 \\ -G_2 & G_2 + G_3 + sC \end{array} \right] \left[\begin{array}{c} V_1 \\ V_2 \end{array} \right] = \left[\begin{array}{c} G_1 V_{in} \\ 0 \end{array} \right]$$

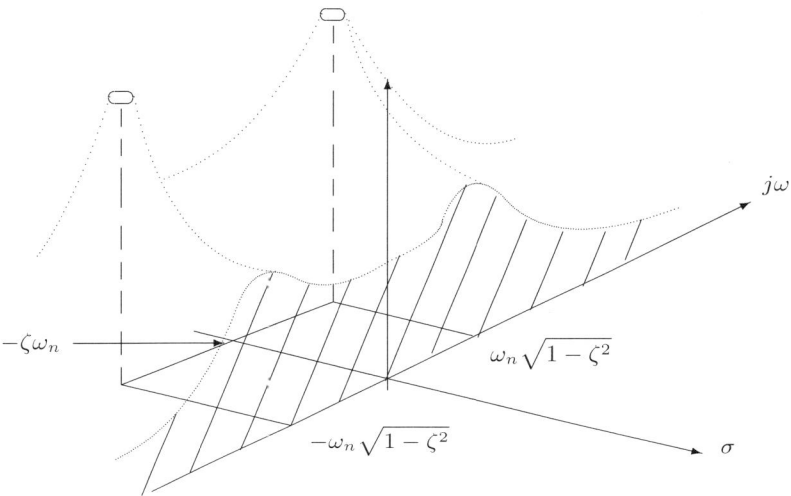

Figure 8.13: Second-order frequency response with complex poles

The required transfer function is

$$
\frac{V_2}{V_{in}} = \frac{sL\dfrac{G_1}{G_3}}{s^2LC\left(\dfrac{G_1+G_2}{G_2G_3}\right) + sL\left(1 + \dfrac{G_1}{G_3} + \dfrac{G_1}{G_2}\right) + sC\dfrac{1}{G_2G_3} + 1}
$$

Clearly, when $s = 0$ as well as when $s \to j\infty$, the magnitude of the transfer function is zero. Also, from what we know about the characteristic of the second-order denominator, there should be a maximum somewhere around $s = j\omega_n\sqrt{1 - \zeta^2}$, where

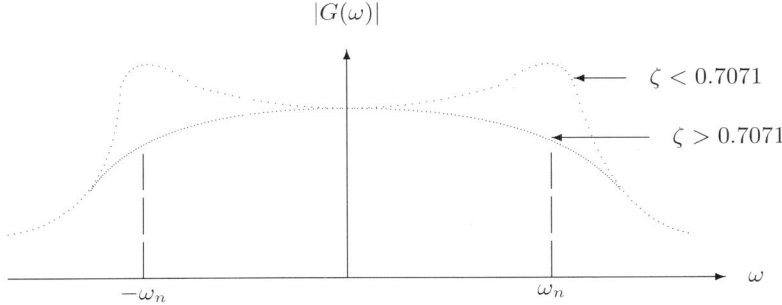

Figure 8.14: Second-order frequency responses for different damping factors

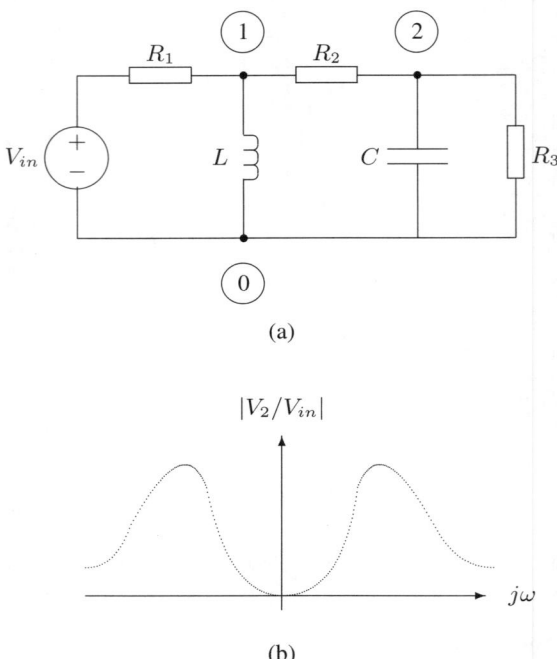

(a)

(b)

Figure 8.15: (a) Second-order band-pass filter; (b) frequency response

ω_n and ζ can be identified by comparing terms. A sketch of the resulting magnitude response is shown in figure 8.15 (b).

8.8.3 Poles and Zeros

The general form of the transfer function is

$$G(s) = \frac{N(s)}{D(s)}$$

where $N(s)$ and $D(s)$ are polynomials in s.

For realistic circuits, the degree of $D(s)$ is always higher than or equal to that of $N(s)$, so that $\lim_{s \to \infty} |G(s)|$ is bounded. The roots of $D(s) = 0$ are known as the *poles* of the circuit, and those of $N(s) = 0$ are known as the *zeros* of the circuit. Thus, at the poles, the magnitude of $G(s)$ tends to ∞, and at the zeros, it goes to 0.

Example 8.6: Rubber sheet analogy — Consider the following transfer function

$$G_1(s) = \frac{s(s^2 + 100)}{(s^2 + 4s + 13)(s + 10)}$$

It is easy to verify that

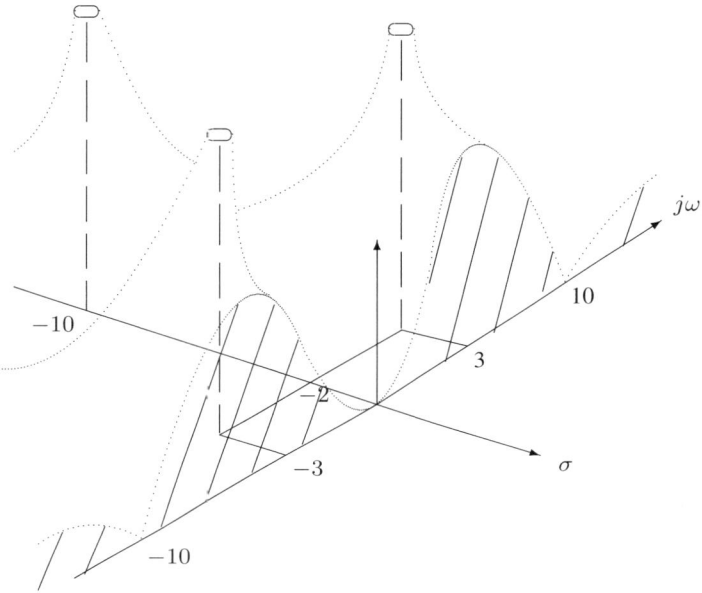

Figure 8.16: Rubber sheet analogy for sketching $G_1(s)$

1. $\lim_{s \to \infty} |G_1(s)| = 1$.
2. The poles are $-1 \pm j3$ and -10.
3. The zeros are 0 and $\pm j10$.

The above information can be used to obtain a sketch of the frequency response. At the poles the $|G(s)|$ surface goes to infinity, and at the zeros it touches the zero horizontal plane.

Since $\lim_{s \to \infty} |G_1(s)| = 1$, we expect the magnitude of $G_1(s)$ to approach 1 for points on the s-plane far away from the origin. We may imagine the top of a very large table to be the s-plane. A rubber sheet is lying horizontally at a distance 1 unit from the table. At each pole, a long vertical stick is inserted underneath the rubber sheet and pushes the sheet very high up. At each zero, the sheet is fixed to the table top. Figure 8.16 illustrates this rubber sheet analogy.

8.8.4 Phase Shift

So far, we have been concentrating on the magnitude of the frequency response. In fact, any given frequency response $G(j\omega)$ is characterized by two quantities. The magnitude is one of them. The other quantity is the phase shift, which is also a function of ω. Our discussion of frequency response cannot be complete if the phase shift is ignored.

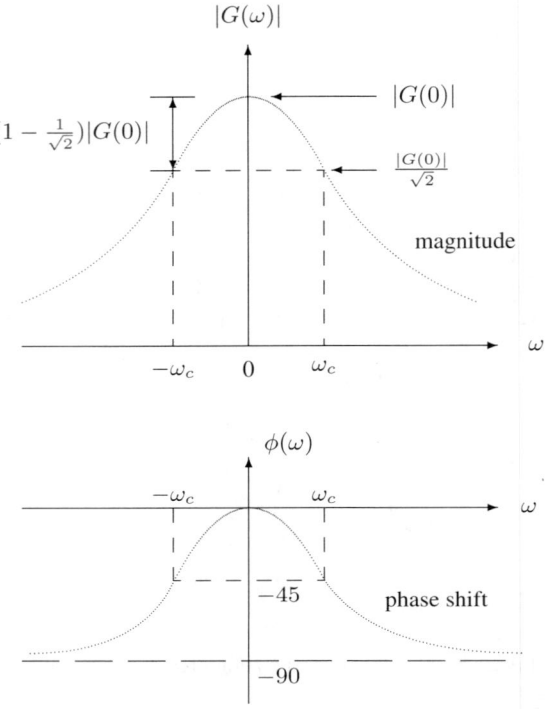

Figure 8.17: Complete frequency response of an RC circuit

Consider the simple RC circuit discussed in Section 8.8.1. The output V_2 and the input V_1 have a phase difference which is given by the argument of the transfer function $G(s)$. More precisely, the output V_2 leads the input V_1 by the argument of $G(s)$, i.e.,

$$
\begin{aligned}
\text{Phase shift } \phi(\omega) \;&=\; \arg\{G(s)\} \\
&=\; \arg\left\{\frac{1}{1+j\omega CR}\right\} \\
&=\; -\tan^{-1}\omega CR
\end{aligned}
$$

Thus, we see that the phase shift for this first-order circuit is always negative. In other words, the output always lags the input. At low frequencies, the phase shift between the input and output is small. At the corner frequency, the output lags the input by exactly $45°$. At high frequencies, the phase shift tends towards $-90°$. Indeed, the effect of poles is to contribute a lagging phase shift, and for each pole, a maximum of $90°$ phase shift is expected which occurs at high frequencies. The complete frequency response is shown in figure 8.17.

For the case of the second-order circuit discussed previously, the phase shift can

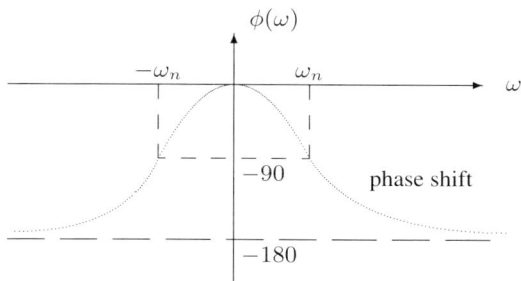

Figure 8.18: Phase shift of second-order circuit with complex poles

likewise be derived from the argument of the transfer function.

$$\text{Phase shift } \phi(\omega) = \arg\left\{\frac{1}{\frac{(j\omega)^2}{\omega_n^2} + \frac{2\zeta}{\omega_n}j\omega + 1}\right\}$$

$$= -\tan^{-1}\left\{\frac{2\zeta\omega_n}{\omega_n^2 - \omega^2}\right\}$$

Hence, for low frequencies, $\phi(\omega)$ is small. At $\omega = \omega_n$, $\phi = -90°$. At high frequencies, $\phi(\omega)$ tends towards $-180°$. This is a typical scenario for second-order circuits with complex poles, and is illustrated in figure 8.18.

For the case of second-order circuits with two negative real poles, such as $G(s) = 1/(s+1)(s+1000)$, the phase shift can be obtained by combining two first-order phase shifts, one corresponding to $1/(s+1)$ and the other corresponding to $1/(s+1000)$. In this particular case, the result is that the phase shift at low frequecies remains small. At the first pole, i.e., $\omega = 1$ rad/s, the phase shift is about $-45°$, since the effect of the other pole is small. At the second pole, i.e., $\omega = 1000$ rad/s, the phase shift is about $-135°$, since the first pole contributes nearly $-90°$. Eventually, at high frequencies, the phase shift tends to $-180°$. Figure 8.19 illustrates this case.

8.9 Direct Methods for Determining Frequency Response

8.9.1 Direct Substitution

The foregoing section has introduced a rubber sheet analogy to sketch the magnitude frequency response. Such an approach provides quick visualization of the response shape, but does not quantitatively plot the response as a function of angular frequency. In fact we can directly put $s = j\omega$ into the transfer function and plot the magnitude as well as the phase shift directly as functions of ω. For example, putting $s = j\omega$ into the first-order low-pass filter discussed earlier gives

$$G(j\omega) = \frac{1}{1 + j\omega CR}$$

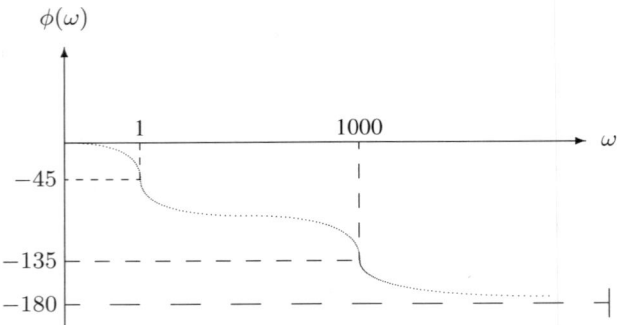

Figure 8.19: Phase shift of second-order circuit with negative real poles

The magnitude and phase shift are

$$|G(\omega)| = \sqrt{\frac{1}{1 + \omega^2 C^2 R^2}}$$

$$\theta(\omega) = -\tan^{-1} \omega C R$$

from which we can plot the responses precisely as functions of ω. The result will be similar to figure 8.17.

8.9.2 Log Scale and the Bode Technique

The frequency response curves plotted previously are non-linear since $|G(\omega)|$ varies non-linearly with ω (actually a reciprocal of a square inside a square root). This is not very convenient to derive and to use by engineers. Can we make a better plot which is easier to sketch and can give more analytically useful clues? Usually, a straight line function is more desirable. Thus, we should try to convert the transfer function to a form that can be represented by a straight line function like $y = mx + c$. Consider the above first-order transfer function. The square of the magnitude is

$$|G(\omega)|^2 = \frac{1}{1 + \omega^2 C^2 R^2}$$

If we take the logarithm of both sides and multiply them by 10, we have

$$20 \log |G(\omega)| = -10 \log(1 + \omega^2 C^2 R^2)$$

Now we define

$$y = 20 \log |G(\omega)|$$

$$x = \log \omega$$

The original transfer function can be written as

$$y = -10 \log(1 + \omega^2 C^2 R^2)$$

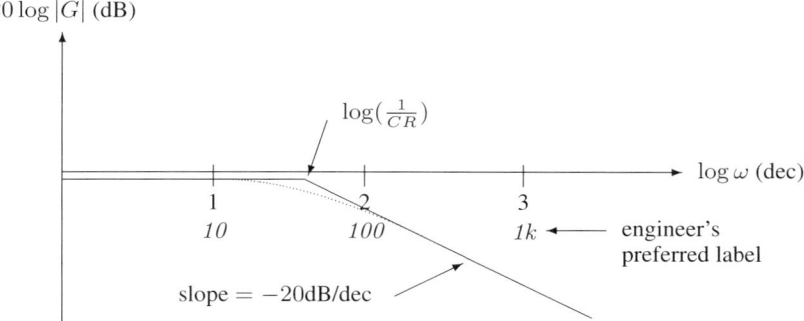

Figure 8.20: Asymptotic approximation of first-order magnitude response

Let us make some approximation of y for small ω and large ω.

$$y \approx \begin{cases} -10 \log \omega^2 C^2 R^2 & \text{for } \omega \gg \frac{1}{CR} \\ -10 \log 1 & \text{for } \omega \ll \frac{1}{CR} \end{cases}$$

$$\Rightarrow \quad y \approx \begin{cases} -20 \log \omega - 20 \log CR & \text{for } \omega \gg \frac{1}{CR} \\ 0 & \text{for } \omega \ll \frac{1}{CR} \end{cases}$$

$$\Rightarrow \quad y \approx \begin{cases} -20x + K & \text{for } \omega \gg \frac{1}{CR} \\ 0 & \text{for } \omega \ll \frac{1}{CR} \end{cases}$$

where $K = 20 \log \left(\frac{1}{CR}\right)$. Hence, we obtain a piece-wise straight-line (asymptotic) approximation for the frequency response magnitude, as shown in figure 8.20. Note that in this diagram, we are plotting $20 \log |G|$ against $\log \omega$ (y against x). The unit of $20 \log |G|$ is the *decibel (dB)*, and the unit of $\log \omega$ is the *decade (dec)*. In particular we observe that

1. The response, originally at 0dB, starts to drop off at $x = \log(\frac{1}{CR})$.

2. The slope of the straight line for the high-frequency approximation (right side) is -20 dB/dec.

3. The approximation is poor in the vicinity of $\omega = \frac{1}{CR}$. The actual response is shown as a dotted line in the diagram, and at $\omega = \frac{1}{CR}$ the response is exactly -3dB.

It should be noted that the *x*-axis is $\log \omega$ on which a "1" corresponds to "10rad/s", a "2" corresponds to "100rad/s", and so on. Although it is mathematically correct to label "1", "2", "3", ..., on the axis, it is rather inconvenient for practical engineers who are used to dealing with actual frequencies in rad/s or Hz. Therefore, we sometimes deliberately label "10", "100", "1k", ..., in lieu of "1", "2", "3", ..., as shown in the diagram.

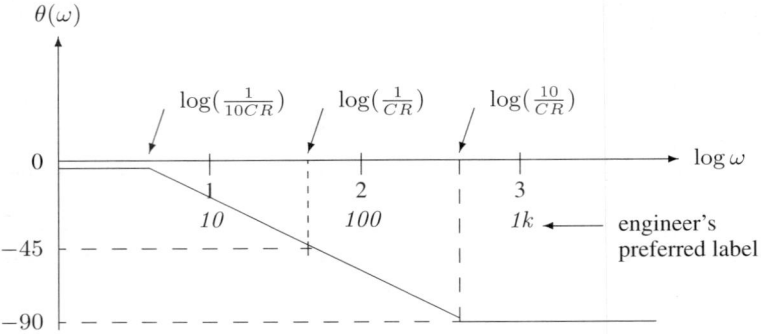

Figure 8.21: Asymptotic approximation of first-order phase response

In order to produce the complete frequency response, we must consider the phase shift as well as the magnitude. We again examine the asymptotic values of the phase shift at very low and very high frequencies.

$$\theta(\omega) = -\tan^{-1}\omega CR \begin{cases} \approx 0 & \text{for } \omega \ll \frac{1}{CR} \\ = 45° & \text{for } \omega = \frac{1}{CR} \\ \approx 90° & \text{for } \omega \gg \frac{1}{CR} \end{cases}$$

We now make two additional assumptions. First, the phase shift is nearly zero for frequencies one decade below the corner frequency $1/CR$. Second, the phase shift is nearly 90° for frequencies one decade above $1/CR$. This results in an asymptotic plot of the phase shift, as shown in figure 8.21. This turns out to be a reasonable approximation.

The afore-discussed technique of asymptotic approximation for plotting frequency response is very popular in electronic engineering. The resulting asymptotic plots for the magnitude and phase shift are widely known as the *Bode plots*. It should be noted that since the basic principle behind this technique is the logarithmic operation, the Bode plots of any transfer function with several poles and/or zeros can be derived by adding up Bode plots of individual poles and zeros. Figure 8.22 summarizes the Bode plots of various types of poles and zeros. Note that we have labelled the *x*-axis with the actual angular frequency values.

Example 8.7: Illustration of Bode plots construction — We illustrate in this example how we can obtain the Bode plots for a given transfer function by adding up individual responses, with the help of figure 8.23. Suppose we wish to obtain the asymptotic plots of the magnitude and phase shift for the following transfer function.

$$H(s) = \frac{1000\left(1 + \dfrac{s}{200\pi}\right)}{\dfrac{s}{2\pi}\left(1 + \dfrac{s}{60000\pi}\right)}$$

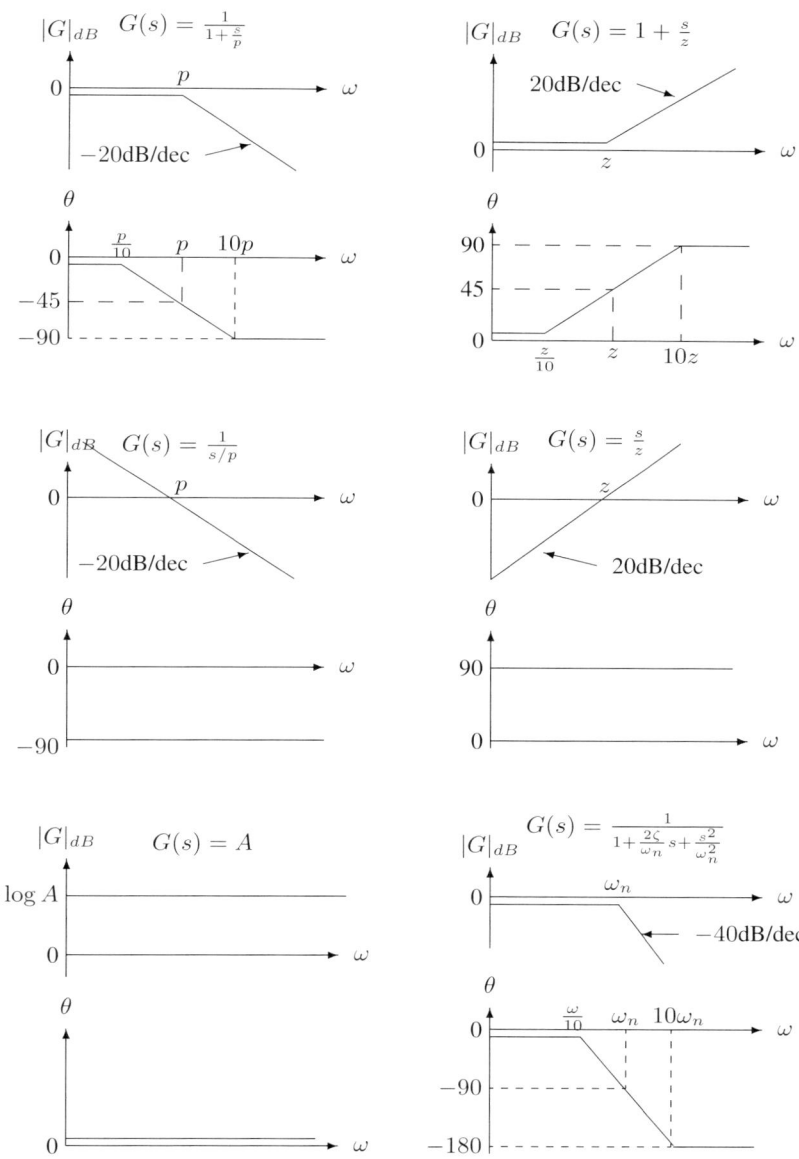

Figure 8.22: Standard Bode plots

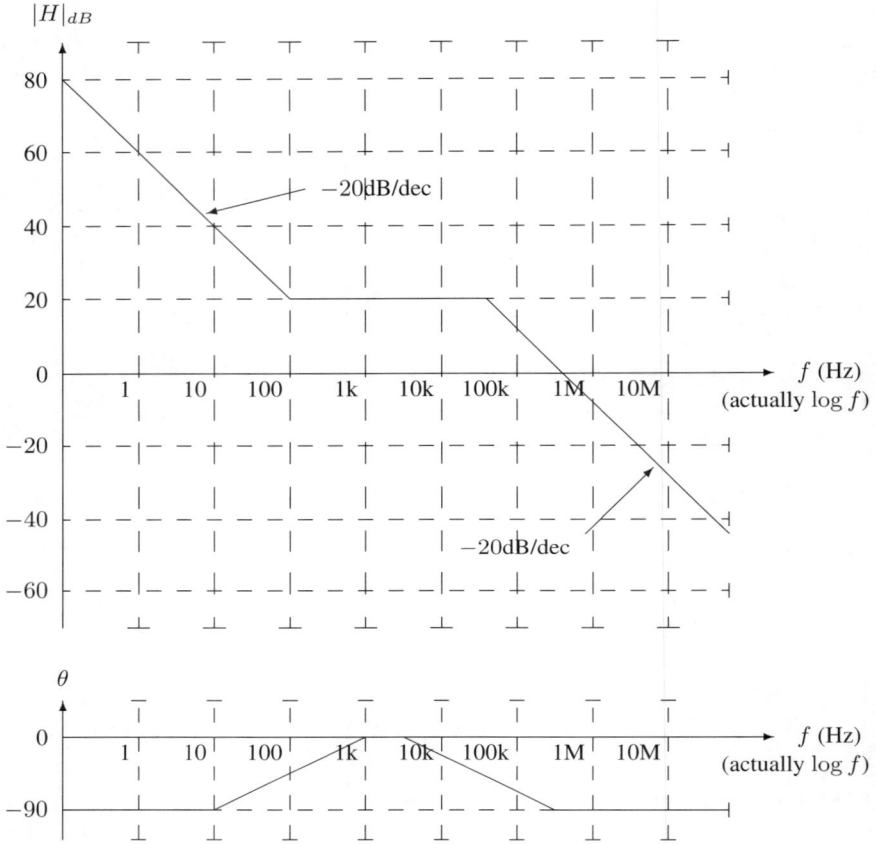

Figure 8.23: Bode plots for $H(s)$

Note that this function has a zero at 200π rad/s or 100 Hz, a pole at 60000π rad/s or 30kHz, and an integrating pole. However, the gain 1000 does not reflect the low-frequency or DC gain due to the presence of the integrating pole. In fact, at zero frequency, the gain is infinitely large. Nevertheless, we can still use the Bode technique to derive the overall response of $H(s)$. Essentially, we add up all the individual responses corresponding to $(1 + s/200\pi)$, $\frac{1}{s/2\pi}$, and $\frac{1}{1+s/6000\pi}$. Finally, due to the constant gain of 1000, we need to shift up the resulting response by 60dB. A likewise procedure can be applied to derive the phase shift response. The result is shown in figure 8.23. Note that we have plotted the response against frequencies (in Hz) instead of angular frequencies (in rad/s).

8.9.3 Bode Plots Based on Inverted Poles and Zeros

In the previous example, we see that the gain 1000 is neither the low-frequency gain, the mid-band gain, nor the high-frequency gain. Thus, the constant factor 1000 in $H(s)$ seems to be rather meaningless, at least not pointing to a direct physical attribute. The question is what form of the transfer function would preserve a greater amount of information that can be retrieved readily by inspection.

The constant factor in the transfer function usually refers to the gain in a certain range of frequency. For instance, the transfer function $\frac{100}{1+s}$ has a low-frequency gain of 100, i.e., 40dB. However, in some cases, such as the function $H(s)$ in the previous example, the constant factor has no direct physical relevance. This is rather undesirable, since we cannot make good use of the form of the transfer function to predict the characteristic. Guilty of the destruction of information, in this case, is the presence of an integrating pole, i.e., $1/s$. In fact, it is not difficult to see the same undesirable hiding of the gain information when the transfer function contains a differentiating zero, i.e., s. For example, the transfer function

$$F(s) = \frac{100s}{1 + \dfrac{s}{2000}}$$

actually has a high-frequency gain of 200000 or 106dB, but such information is not readily told from the constant factor 100.

In order to allow maximum preservation of gain information, we need to eliminate $1/s$ and s from the transfer function that contains them. The following possibilities are readily observed from simple algebraic manipulations:

1. We assume that the transfer function containing a differentiating zero s also contains a pole at p. This case is typified by the following transfer function:

$$F_1(s) = \frac{As}{1 + \dfrac{s}{p}}$$

We may multiply both the numerator and the denominator by p/s to give

$$F_1(s) = \frac{Ap}{1 + \dfrac{p}{s}}$$

Immediately, we see that the high-frequency gain is Ap, as $s \to \infty$. Note that the denominator does not represent a normal pole. Instead, we shall refer to it as an *inverted pole*.

2. Similarly, we assume that the transfer function containing an integrating pole $1/s$ also contains a zero at z. This case is typified by the following transfer

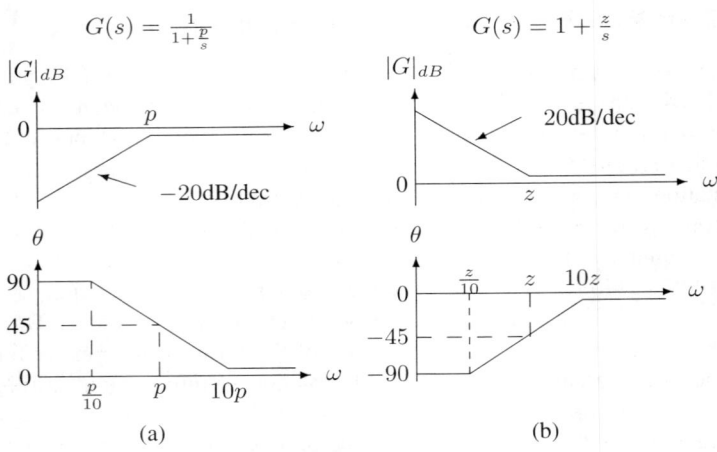

Figure 8.24: Bode plots for (a) inverted pole; (b) inverted zero

function:

$$F_2(s) = \frac{A\left(1 + \dfrac{s}{z}\right)}{s}$$

We may multiply both the numerator and the denominator by z/s to give

$$F_2(s) = \frac{A\left(1 + \dfrac{z}{s}\right)}{z}$$

Immediately, we see that the high-frequency gain is A/z, as $s \to \infty$. Again, note that the numerator does not represent a normal zero; we shall refer to it as an *inverted zero*.

Now, applying the asymptotic approximation technique as outlined in the previous sub-section, we obtain the Bode plots for the inverted pole and zero functions as shown in figure 8.24.

Example 8.8: Application of inverted poles and zeros — In the previous example, $H(s)$ has a constant factor of 1000 which has no apparent physical meaning. Let us now rearrange it as

$$H(s) = \frac{10\left(\dfrac{200\pi}{s} + 1\right)}{1 + \dfrac{s}{60000\pi}}$$

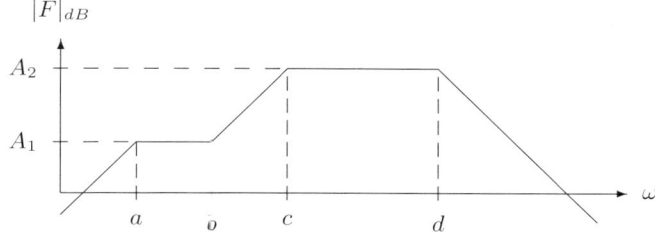

Figure 8.25: Magnitude plot of example transfer function

Here, we have a normal pole and an inverted zero. The Bode plots can be easily obtained by adding up the individual plots corresponding to the normal pole and the inverted zero, and shifting the resulting plot up by 20dB. The result is identical to figure 8.23 found previously. But, this time, the constant factor 10 (i.e., 20dB) is clearly identified as the mid-band gain. Thus, expressing the transfer function in terms of inverted poles and zeros restores the identity of the constant factor, and hence makes the transfer function more meaningful.

An additional merit of expressing transfer functions in terms of inverted poles and zeros is that we can select the most meaningful constant factor for the transfer function according to our design objective. The following example illustrates this interesting application.

Example 8.9: Choosing the right form of transfer function — Referring to the magnitude plot in figure 8.25, we can write down the transfer function as

$$F(s) = \frac{A_1(1 + s/b)}{(1 + a/s)(1 + s/c)(1 + s/d)}$$

corresponding to an inverted pole at a, a normal zero at b, and two normal poles at c and d. With this choice of poles and zeros, the gain A_1 for the frequency range a to b is highlighted in the transfer function. Moreover, if we wish to highlight A_2 instead, we can choose c to be an inverted pole and d to be a normal pole. In this case, b will be an inverted zero and a an inverted pole. The transfer function can be written alternatively as

$$F(s) = \frac{A_2(1 + b/s)}{(1 + a/s)(1 + c/s)(1 + s/d)}$$

It is interesting to compare the above two forms of $F(s)$ with the one containing only normal poles and zeros.

$$F(s) = \frac{A_2 s(1 + s/b)}{(1 + s/a)(1 + s/c)(1 + s/d)}$$

Having no obvious physical meaning for the constant factor A_2, this form of transfer function has limited usefulness for design. For instance, if we wish to alter the gain for the frequency range c to d, we should arrange $F(s)$ in the above second form.

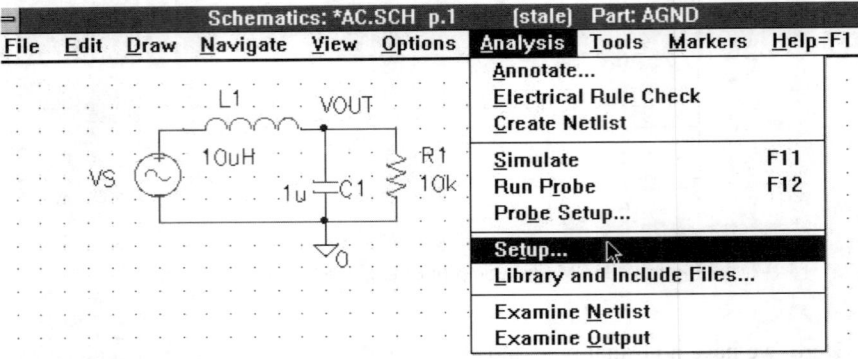

Figure 8.26: Circuit drawn in *Schematics* for PSPICE frequency-domain analysis

Suppose we do so and end up with A_2 equal to $g_m R_L R_1 / (R_1 + R_2)$. Then, we know exactly which component values to be tuned in order to give the required gain.

8.10 Frequency-Domain Analysis with PSPICE

In this section we introduce the "AC sweep" feature of PSPICE for studying the frequency response of dynamic circuits. Essentially we make use of the sweep feature of PSPICE to "sweep" the frequency through a range of values, and of PROBE to display the frequency response curves corresponding to both magnitude and phase. The following example illustrates the procedure.

Example 8.10: Using AC sweep in PSPICE for frequency-domain analysis — As usual, we start by telling PSPICE what circuit to analyze by drawing it in *Schematics*. Figure 8.26 shows a simple second-order low-pass filter which is to be analyzed by PSPICE. In this example we wish to obtain the magnitude and phase response curves for the transfer function VOUT/VS.

1. We use the part VSIN for the input voltage source, which can be obtained from the library "source.slb". Double-clicking it will bring up the dialogue box for entering parameters such as AC value, DC value, offset, amplitude and frequency. Here we enter 1V for the AC value and 0V for the DC value. Since we are interested in the frequency response only, the values of offset, amplitude and frequency are not important. (Warning: PSPICE will not run if any of these values is not specified.)

2. In the **Analysis** menu, we choose *"Setup"* and click *"AC sweep"* to bring up a dialogue box, as shown in figure 8.27, in which we enter the range of frequencies to be swept and set the "AC sweep type" to "Decade".

3. After PSPICE has finished the analysis, we invoke PROBE to examine the transfer function VOUT/VS. First of all, we must set the *x*-axis to logarithmic scale using the **Plot** menu in the PROBE window. If the gain is to be plotted

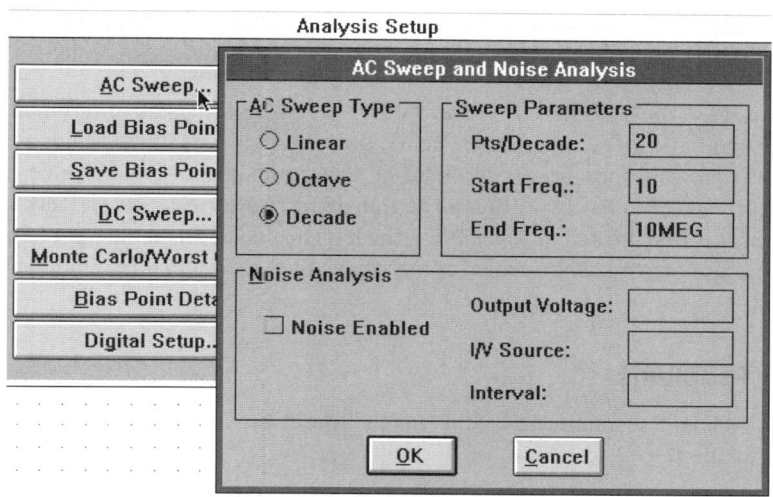

Figure 8.27: Dialogue box for specifying parameters for AC analysis

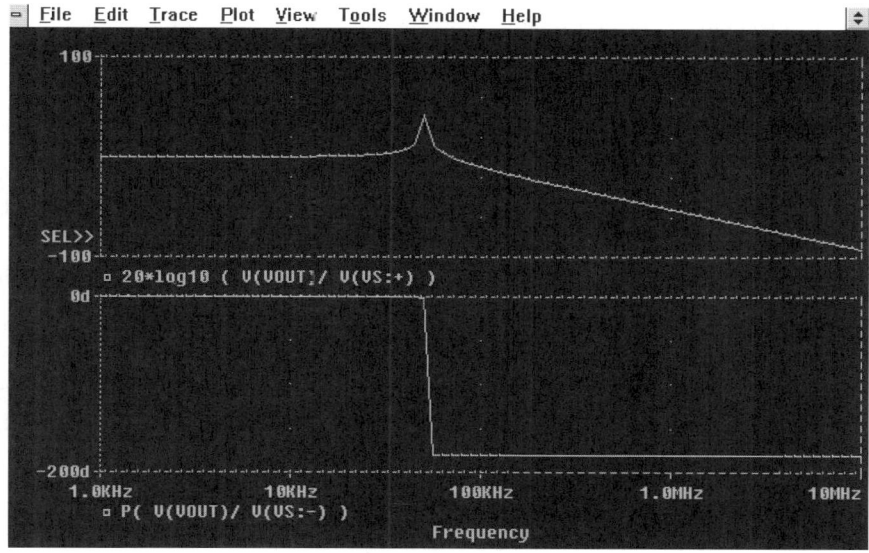

Figure 8.28: Frequency response curves displayed by PROBE

in dB, we should enter "20*log10 (V(VOUT)/(V(VS:+)))" as the variable to be displayed. Also, to display the phase shift, we enter "P (V(VOUT)/(V(VS:+)))" as the displaying variable in PROBE. (See also Examples 2.8 and 4.6 for using PROBE.) Results are shown in figure 8.28.

Remarks — In PROBE, the node "VS:+" refers to the positive terminal of the voltage source VS. For other components, the terminals are suffixed with a number 1 or 2 to indicate the specific terminal of that component. For instance, "R1:1" and "R1:2" refer to the two different terminals of resistor R1. By default, when a component is first created in *Schematics,* the left terminal is 1 and the right terminal is 2. In the case of vertically oriented components, the top terminal is 1 and the bottom one is 2.

8.11 Problems

1. Consider a second-order circuit which is described by the following differential equation:

$$\frac{d^2}{dt^2}v_o(t) + 2005\frac{d}{dt}v_o(t) + 10000v_o(t) = 10000\left(\frac{d^2}{dt^2}v_i(t) + 100v_i(t)\right)$$

Find the characteristic equation and the eigenvalues. Hence, determine the natural response of the free-oscillating circuit.

2. The corresponding complex-frequency domain transfer function for the circuit of Problem 1 is given by

$$\frac{V_o}{V_i} = \frac{10000(s^2 + 100)}{s^2 + 2005s + 10000}$$

Determine the limiting values of $|V_o/V_i|$ as s tends to 0 and infinity. Find also the poles and zeros of the transfer function. Hence, or otherwise, sketch the frequency response using the rubber sheet analogy.

3. What values of ζ will give an oscillatory transient response for the following system?

$$\frac{d^2}{dt^2}x(t) + 20\zeta\frac{d}{dt}x(t) + 100x(t) = y(t)$$

Sketch the frequency responses of X/Y for $\zeta < 0.7071$ and $\zeta > 0.7071$.

4. Consider a fourth-order circuit with differential equation:

$$\frac{d^4}{dt^4}v_o(t) + 1004\frac{d^3}{dt^3}v_o(t) + 4009\frac{d^2}{dt^2}v_o(t) + 9000\frac{d}{dt}v_o(t)$$

$$= 9000\left(\frac{d}{dt}v_i(t) + 100v_i(t)\right)$$

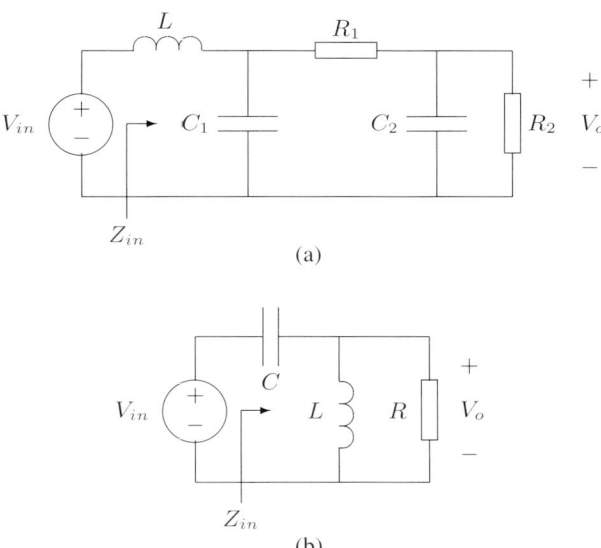

Figure 8.29: Circuits for Problem 6

Verify the following transfer function for V_o/V_i:

$$\frac{V_o}{V_i} = \frac{100\left(1 + \dfrac{s}{100}\right)}{s + \dfrac{4009s^2}{9000} + \dfrac{1004s^3}{9000} + \dfrac{s^4}{9000}}$$

Find the Bode plots for this transfer function.

5. Find all eigenvalues and the natural response of the free-oscillating version of the circuit described by the equation of Problem 4, i.e., with RHS set to 0.

6. Find the transfer function V_o/V_{in} and the input impedance Z_{in} for the circuits shown in figure 8.29. State the poles and zeros in each transfer function, assuming that all capacitors are 1F, all inductors are 1H and all resistors are 1Ω.

7. Find a suitable circuit that can realize the following input impedance function. (Hint: A 1H inductor has an impedance of s, and a 1F capacitor has an impedance of $1/s$.)

$$Z_{in} = \frac{1}{s} + \frac{1}{\dfrac{s}{s^2+1} + \dfrac{1}{1 + \dfrac{1}{s}}}$$

8. Rearrange the following transfer function such that the constant factor refers to the gain in the range 100kHz to 50MHz. Find the value of this gain and sketch the magnitude Bode plot.

$$F(s) = \frac{10s(1 + \frac{s}{600\pi})}{(1 + \frac{s}{20\pi})(1 + \frac{s}{200\pi \times 10^3})(1 + \frac{s}{100\pi \times 10^6})}$$

9. Consider the transfer function of Problem 8. Suppose the gain in the range 10Hz to 300Hz is a design parameter. Rearrange the transfer function so that this gain value can be readily observed.

10. Use PSPICE to plot the frequency response curves for V_o/V_{in} for the circuits of figure 8.29. Assume that all resistors are 1kΩ, all capacitors are 10μF, and all inductors are 500μH. Use VSIN for the voltage source and set AC sweep appropriately.

Chapter 9

Circuit Topology

In Chapters 1, 2 and 3, we introduced some useful theorems and methods of analysis, the basis of which are essentially Kirchhoff's two laws and the linearity assumption. In particular, for circuits having some relatively simple connection styles (e.g., series, parallel, ladder, "delta" and "star"), we formulate the Kirchhoff's law equations by inspection, and for circuits of arbitrary configuration, we resort to the mesh and nodal methods. However, neither the mesh method nor the nodal method takes advantage of the circuit topology of the particular circuit under study, although more efficient solutions can be obtained if the circuit topology is exploited in the formulation of Kirchhoff's law equations.[1] The branch of mathematics that is relevant to the description of interconnection of electric components in a circuit is known as *graph theory*. In this chapter, the basics of graph theory are introduced. We will revisit Kirchhoff's laws and explain how the graph helps formulate independent Kirchhoff's law equations for a given circuit, which is the key to formulating systematic solutions for general circuit analysis problems. This chapter and the following two chapters will discuss systematic and efficient approaches to solving circuit analysis problems based on graph theoretic concepts.

9.1 The Graph and the Di-Graph

The two basic elements of a graph are the *branch* and the *node*. The graph describes how the nodes are interconnected or linked up by the branches. Any given circuit is associated with a unique graph. As shown in figure 9.1, the graph on the right corresponds uniquely to the circuit shown on the left.

It should be noted that, in translating from a given circuit to its graph, each component in the circuit gives one branch. Thus, for example, two components connected in series will give two branches with a node in between. Furthermore, if we assign

---

[1] The term "topology" refers to the way in which devices are interconnected to form an electric circuit.

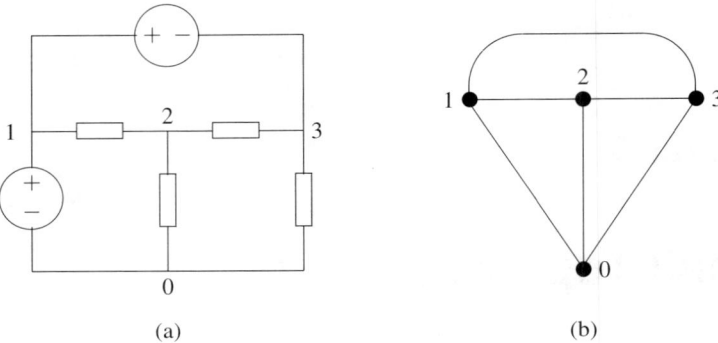

(a)   (b)

Figure 9.1: (a) A circuit; and (b) its graph

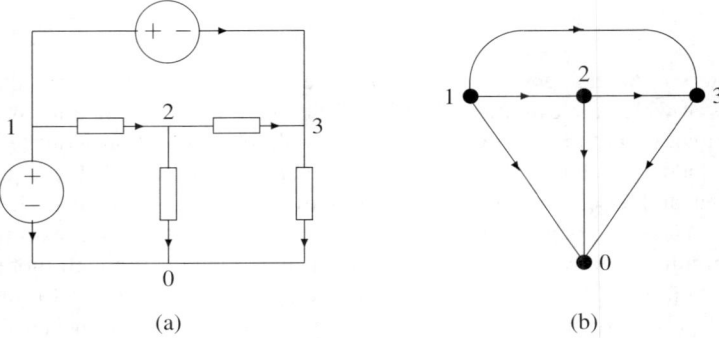

(a)   (b)

Figure 9.2: (a) A circuit; and its (b) di-graph

direction to each branch in the graph according to the polarity of the corresponding component in the circuit, the resulting graph is called a *di-graph*. Figure 9.2 shows the di-graph of a circuit.

Note that the direction of the arrow on a given branch is consistent with the assigned current direction through the corresponding component of the circuit, and in the case of components labelled with voltage polarity, the arrow direction is consistent with the convention that positive current enters the positive terminal and emerges from the negative terminal.

9.2   Loops and Cutsets

The two fundamental concepts in graph theory that are crucial to the formulation of independent Kirchhoff's law equations are the *loop* and the *cutset*. The concept of a loop has been introduced previously as a closed path in a circuit, whereas that of a cutset is new. In the following we will formally define the loop and the cutset in a given circuit graph in terms of *a set of branches* of the graph.

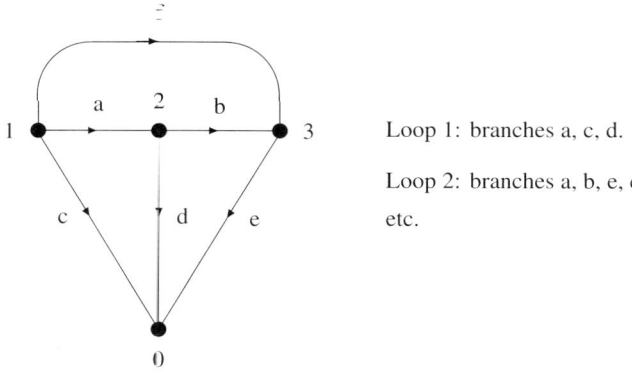

Loop 1: branches a, c, d.

Loop 2: branches a, b, e, c.

etc.

Figure 9.3: Loops

9.2.1 Definition of Loop

A *loop* is a set of branches of a graph forming a closed path. In the graph shown in figure 9.3, branches a, c and d form a loop, and branches a, b, e and c also form a loop.

9.2.2 Definition of Cutset

A *cutset* is a set of branches of a graph which, upon removal, will cause the graph to be separated into two disconnected sub-graphs, and upon resumption of any one branch of the cutset, the graph will be reconnected. For example, as shown on the upper left of figure 9.4, branches f, b, d and c form a cutset. Note that a cutset represents a minimal set that separates the graph. Thus, branches a, f, b, d, c do not form a cutset since they are more than necessary for the purpose of separating the graph. Figure 9.4 shows several cutsets of the same graph.

Remarks — A special case is shown on the lower right of figure 9.4. If we consider a node as a sub-graph, then it is always true that *branches emerging from a node form a cutset.*

9.3 Kirchhoff's Laws Revisited

In Chapter 1, Kirchhoff's voltage law (KVL) and Kirchhoff's current law (KCL) are stated as follows:

1. Kirchhoff's voltage law (KVL) states that the sum of voltages across branches along a loop is equal to zero, assuming the polarities of all voltages are chosen in the same sense.

$$\sum_{loop} v_k = 0$$

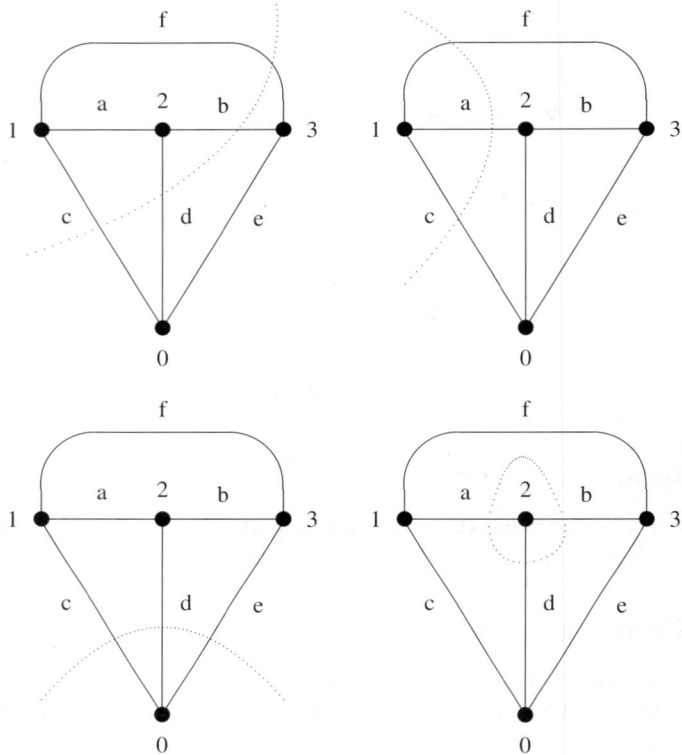

Figure 9.4: Cutsets

where k is the index of the summation.

2. Kirchhoff's current law (KCL) states that the sum of currents in branches emerging from a node is equal to zero.

$$\sum_{node} i_k = 0$$

where k is the index of the summation.

While the above version of KVL is general, the version of KCL stated above is only a special case of the general statement of KCL, which should be read as *the sum of currents in the branches of a cutset is equal to zero*, i.e.,

$$\boxed{\sum_{cutset} i_k = 0}$$

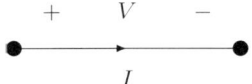

Figure 9.5: Sign convention for voltage and current in a branch

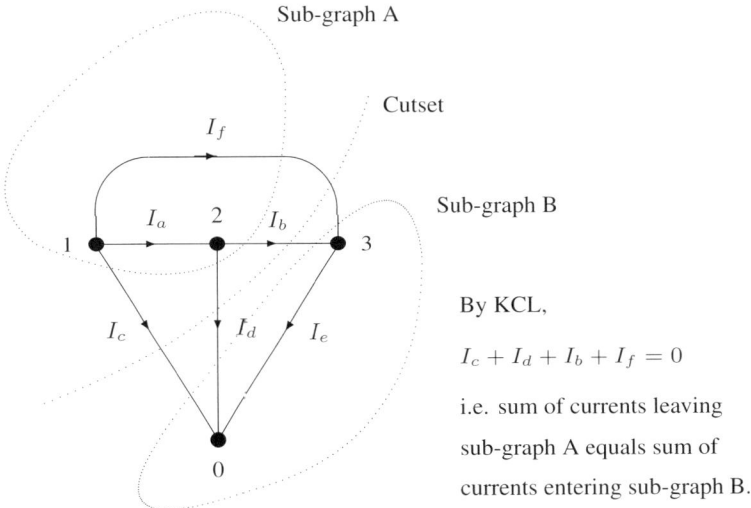

By KCL,

$$I_c + I_d + I_b + I_f = 0$$

i.e. sum of currents leaving sub-graph A equals sum of currents entering sub-graph B.

Figure 9.6: Kirchhoff's current law and cutsets

with the current directions chosen such that all currents flow from one sub-graph to another across the cutset. Since the cutset separates the graph into two sub-graphs, we may say that the sum of currents going from one sub-graph to the other is zero.

Remarks — Before going further and to avoid possible confusions, we should clarify our assignment of voltage polarity and current direction in a branch. The voltage across a branch is such that the current goes into the positive end and emerges from the negative end, as shown in figure 9.5.

Example 9.1: Kirchhoff's law based on loops and cutsets — Let us now write down some KVL and KCL equations based on loops and cutsets, and the sign convention defined above.[2] Application of KVL to the di-graph of figure 9.6 gives $V_a + V_d - V_c = 0$, $V_a + V_b + V_e - V_c = 0$, etc., and of KCL gives $I_a - I_b - I_d = 0$, $I_f + I_a - I_d - I_e = 0$, etc.

---

[2] In practice, if the polarity/direction is already assigned in a particular problem, then we have to adjust the sign of the value to suit the variable concerned.

9.4   Tree and Co-Tree

A *tree* is a set of branches of a graph, i.e., a sub-set of the given graph, which contains no loop, and upon addition of any other branch in the graph to this sub-set will make a loop. In simple words, a tree is a maximal sub-set of branches of a graph containing no loops.

After a tree is chosen, the remaining branches of the graph form another sub-set called *co-tree*. The union of the tree and the co-tree is exactly the given graph, and the intersection is a null set.

> ***Example 9.2: Construction of trees*** — The construction of a tree for any given graph involves, in practice, a very straightforward procedure. Referring to the same graph as in figure 9.6, we first include in the tree, for instance, branch a. Then, with no loop formed, we can take branch f as well. Furthermore, we can take branch c, and still no loop is formed. However, if we go on and take branch d, we would close a loop. Thus, branch d should not be included. If we try taking branch b or e, a loop would be closed as well. The tree can therefore contain only branches a, c and f. Figure 9.7 shows this tree and two other possible trees.

As seen from the above example, for any given graph, there exists more than one tree. Although the choice of tree is not unique, the total number of tree branches is always the same for a given graph. We will summarize the expressions, without proofs, relating the number of nodes n, the number of branches b, the number of tree branches t, and the number of co-tree branches l.

$$
\begin{aligned}
t &= n - 1 \\
b &= t + l \\
l &= b - n + 1
\end{aligned}
$$

The first equation above is valid for planar graphs only, and is the most important since it provides a means for detecting any missing or excess tree branches.

> ***Example 9.3: Number of tree branches in planar graphs*** — If a graph has 5 branches and 4 nodes, then the number of tree branches t is equal to $4 - 1 = 3$, and the number of co-tree branches l is equal to $5 - 3 = 2$.

9.5   Basic Cutsets and Independent KCL Equations

For a given circuit, we can find many cutsets, and since each cutset gives a KCL equation, we will have many KCL equations. However, not all KCL equations are independent, meaning that some KCL equations could simply be linear combinations of others. Hence, it is important to identify a set of independent KCL equations so that we need not do any unnecessary work. For example, in the graph given previously, there are only three independent KCL equations. (We will explain why later on.) Hence, it is meaningless to set up more than three KCL equations for this graph. The question now is how to identify a set of independent KCL equations.

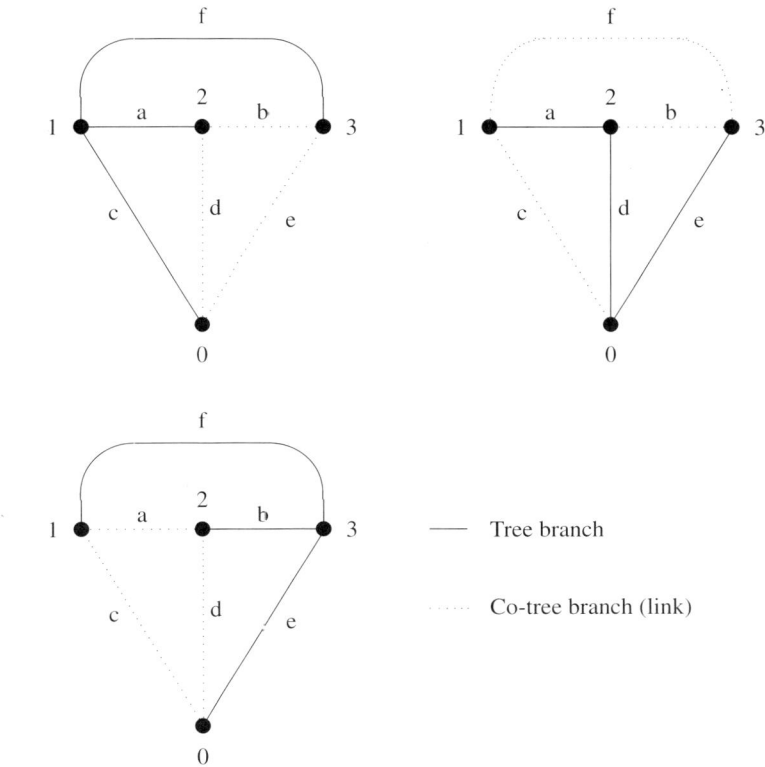

Figure 9.7: Tree and co-tree

We define a *basic cutset* as *a cutset that contains only one tree branch.*[3] Thus, there are exactly t basic cutsets for a given graph. If we write KCL equations for all basic cutsets, we then have exactly t KCL equations, and it can be shown that these t equations are independent.

Example 9.4: Independent KCL equations from basic cutset — Referring to figure 9.8, three independent KCL equations are

$$
\begin{aligned}
I_1 + I_3 - I_6 &= 0 \\
I_2 - I_3 - I_5 &= 0 \\
I_4 + I_5 + I_6 &= 0
\end{aligned}
$$

Note that each one of these equations corresponds to one basic cutset. Since there are 3 tree branches, we expect 3 independent KCL equations. It should be noted that

---

[3]Basic cutsets are also known as fundamental cutsets in the literature.

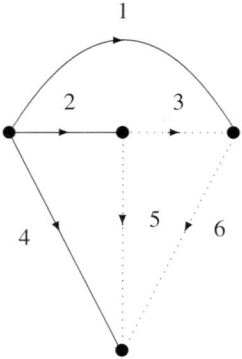

Figure 9.8: Basic cutsets: $\{1,3,6\}$, $\{2,3,5\}$ and $\{4,5,6\}$

1. we may end up with a different set of independent equations if we choose a different tree;
2. any set of independent equations gives essentially the same information, and hence the choice of tree is arbitrary;
3. any KCL equation derived from a non-basic cutset is simply a linear combination of the above set of equations.

9.6   Basic Loops and Independent KVL Equations

Similar to basic cutsets, basic loops give independent KVL equations. While basic cutsets are based on tree branches, basic loops are based on co-tree branches.

We define a *basic loop* as *a loop that contains only one co-tree branch (link).*[4] There are $b - t$ basic loops for a given graph, and hence $b - t$ independent KVL equations.

Example 9.5: Independent KVL equations from basic loop — For the same graph of Example 9.4, using the same choice of tree, the set of basic loops is $\{1,2,3\}$, $\{2,5,4\}$, and $\{1,4,6\}$, and the corresponding set of independent KVL equations is

$$\begin{aligned}
-V_1 + V_2 + V_3 &= 0 \\
V_2 + V_5 - V_4 &= 0 \\
V_1 - V_4 - V_6 &= 0
\end{aligned}$$

The sign convention is as described in figure 9.5. For instance, V_1 is the voltage across branch 1 with the left end being positive and the right end negative. We note again that we may end up with a different set of independent equations if we choose a different tree, and that any KVL equation derived from a non-basic loop is simply a linear combination of the above set of equations.

---

[4] Basic loops are also known as fundamental loops.

9.7 Matrix Representation of Circuit Graphs

In the development of general algorithms for systematic analysis of large-scale circuits, the graph plays an important role. In order to allow efficient implementation in the form of computer-aided analysis programmes, circuit topologies are described in matrix form and the Kirchhoff's law equations are formulated on the basis of the describing topological matrices. We will discuss here three important matrices.

1. Node Incidence Matrix (A-matrix).

2. Basic Cutset Matrix (Q-matrix).

3. Basic Loop Matrix (B-matrix).

9.7.1 The A-Matrix

The A-matrix describes the topology of a graph. Each column of the A-matrix corresponds to a branch, and each row corresponds to a node. We illustrate the construction of this matrix with an example

Example 9.6: Construction of A-matrix — Consider the graph shown in figure 9.9 (a). For row 0, for example, either a "1", "−1" or "0" is assigned to each column to indicate whether the branch corresponding to that column leaves from, enters, or is not connected to node 0. Applying this procedure to all nodes gives the following A-matrix.

$$A = \begin{array}{c} \\ 0 \\ 1 \\ 2 \\ 3 \end{array} \begin{array}{cccccc} 1 & 2 & 3 & 4 & 5 & 6 \\ \left[\begin{array}{cccccc} 0 & 0 & 0 & -1 & -1 & -1 \\ 1 & 1 & 0 & 1 & 0 & 0 \\ 0 & -1 & 1 & 0 & 1 & 0 \\ -1 & 0 & -1 & 0 & 0 & 1 \end{array}\right] \end{array}$$

9.7.2 The Q-Matrix

Once a tree is selected, the choice of basic cutsets is fixed. We can represent these basic cutsets by the *basic cutset matrix,* i.e., the Q-matrix. Again, we illustrate the construction with an example.

Example 9.7: Construction of Q-matrix — Referring to the graph shown in figure 9.9 (b), the following steps are taken to construct the Q-matrix.

1. Each column corresponds to a branch, and each row corresponds to a basic cutset. The columns are arranged such that tree branches are on the left and co-tree branches on the right, i.e., 1, 2, 4, 3, 5, 6, as shown below.

2. For each basic cutset (row), we put either a "1", "−1" or "0" in each column to indicate whether the branch corresponding to that column belongs to the cutset. (Note that there is exactly one tree branch in a basic cutset.)

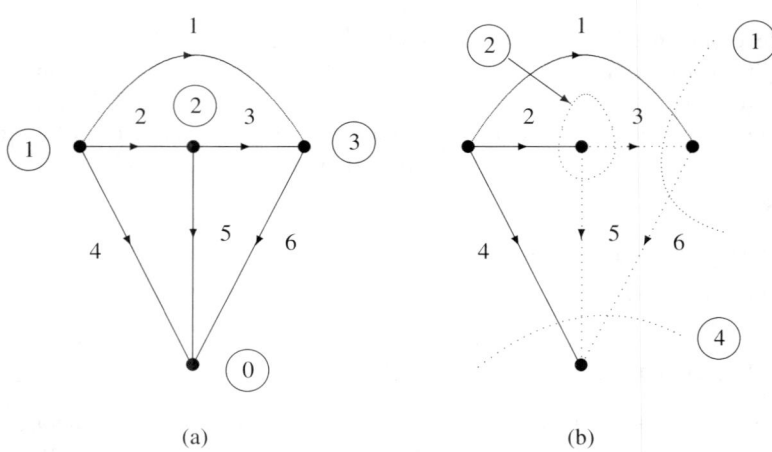

Figure 9.9: (a) A-matrix construction; (b) Q-matrix and B-matrix constructions

3. For branches of the cutset, either "1" or "−1" is assigned. In order to distinguish the direction of the branches, "1" is always assigned to the tree branch, and likewise to other co-tree branches in the cutset that have the same direction as the tree branch, and otherwise a "−1" is assigned.

The resulting Q-matrix will always be of the form

$$Q = [\mathbf{1} \mid Q_1]$$

where $\mathbf{1}$ is the identity matrix, and in the above example, the Q-matrix is

$$
Q =
\begin{array}{c}
 \\
1 \\
2 \\
4
\end{array}
\begin{array}{ccc|ccc}
1 & 2 & 4 & 3 & 5 & 6 \\
\left[\begin{array}{ccc} 1 & 0 & 0 \end{array}\right. & & & 1 & 0 & -1 \\
0 & 1 & 0 & -1 & -1 & 0 \\
0 & 0 & 1 & 0 & 1 & 1
\end{array}
$$

9.7.3   The B-Matrix

The *basic loop matrix*, i.e., the B-matrix, describes the basic loops of the graph. The construction is similar to that of the Q-matrix. We will use the same graph as in Example 9.7 to illustrate the procedure.

Example 9.8: Construction of B-matrix — Referring again to the graph of figure 9.9 (b), we construct the B-matrix as follows.

1. Each column corresponds to a branch, and each row corresponds to a basic loop. The columns are arranged such that tree branches are on the left and co-tree branches on the right, i.e., 1, 2, 4, 3, 5, 6, as shown below.

2. For each basic loop (row), we put either a "1", "−1" or "0" in each column to indicate whether the branch corresponding to that column belongs to the loop. (Note that there is exactly one co-tree branch in a basic loop.)

3. For branches of the loop, either "1" or "−1" is assigned. In order to distinguish the direction of the branches, "1" is always assigned to the co-tree branch, and likewise to other tree branches in the loop that have the same direction as the co-tree branch, and otherwise a "−1" is assigned.

The resulting B-matrix will always be of the form

$$B = [B_1 \mid \mathbf{1}]$$

where $\mathbf{1}$ is the identity matrix, and in the above example, the B-matrix is

$$
B \quad = \quad
\begin{matrix}
 & \begin{matrix} 1 & 2 & 4 & 3 & 5 & 6 \end{matrix} \\
\begin{matrix} 3 \\ 5 \\ 6 \end{matrix} &
\left[\begin{array}{ccc|ccc}
-1 & 1 & 0 & 1 & 0 & 0 \\
0 & 1 & -1 & 0 & 1 & 0 \\
1 & 0 & -1 & 0 & 0 & 1
\end{array} \right]
\end{matrix}
$$

Remarks – It can be shown that the Q-matrix and B-matrix are not independent and that one can be derived from the other. Essentially, if $Q = [\mathbf{1} \mid Q_1]$, then

$$B = \left[-Q_1^T \mid \mathbf{1} \right]$$

i.e., $B_1 = -Q_1^T$ or $Q_1 = -B_1^T$. Thus, once we have found the Q-matrix, we get the B-matrix automatically, and vice versa.

9.8 Kirchhoff's Laws in Matrix Form

Let V_k and I_k be the voltage across and current in a branch, and let $\bar{V}$ be the vector of branch voltages and $\bar{I}$ be the vector of branch currents. In particular we define $\bar{V}$ and $\bar{I}$ as

$$\bar{V} = \begin{bmatrix} \bar{V}_t \\ \bar{V}_l \end{bmatrix} \quad \text{and} \quad \bar{I} = \begin{bmatrix} \bar{I}_t \\ \bar{I}_l \end{bmatrix}$$

where $\bar{V}_t$ and $\bar{V}_l$ are the vectors containing the tree and co-tree branch voltages respectively, and, $\bar{I}_t$ and $\bar{I}_l$ are the vectors containing the tree and co-tree branch currents respectively. The independent Kirchhoff's law equations are given by

KCL:	$Q\bar{I} = 0$
KVL:	$B\bar{V} = 0$

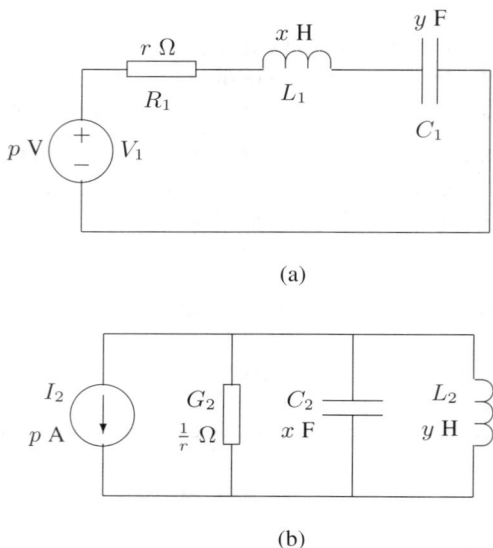

(a)

(b)

Figure 9.10: Dual circuits. (a) Series circuit; (b) parallel circuit

9.9   Duality

"Duality" is an important concept in circuit theory that relates two circuits of apparently different topologies, but having similar properties when certain quantities and components are interchanged between the two circuits. The concept of duality is only valid for *planar* circuits in which no branches pass over or under any other branches.

9.9.1   A Glimpse of Duality

We will now get a quick glimpse of the concept, before probing into details, by considering the pair of dual circuits shown in figure 9.10. Figure 9.10 (a) is a series LRC circuit and figure 9.10 (b) is a parallel LRC circuit. In particular we have chosen the component values in such a way that

1. the value of the capacitor in the series circuit is exactly equal to that of the inductor in the parallel circuit;
2. the value of the inductor in the series circuit is exactly equal to that of the capacitor in the parallel circuit;
3. the value of the resistor in the series circuit is exactly equal to that of the conductor (reciprocal of resistor) in the parallel circuit;
4. the value of the voltage source in the series circuit is exactly equal to that of the current source in the parallel circuit.

It can be shown that the current through the capacitor branch in the series circuit exactly equals the voltage across the inductor branch in the parallel circuit, and that

The series circuit	The parallel circuit
Current in V_1 (A)	Voltage in I_2 (V)
Current in R_1 (A)	Voltage in G_2 (V)
Voltage in R_1 (V)	Current in G_2 (A)
Current in C_1 (A)	Voltage in L_2 (V)
Voltage in C_1 (V)	Current in L_2 (A)
Current in L_1 (A)	Voltage in C_2 (V)
Voltage in L_1 (V)	Current in C_2 (A)

Table 9.1: Duality of series and parallel resonant circuits

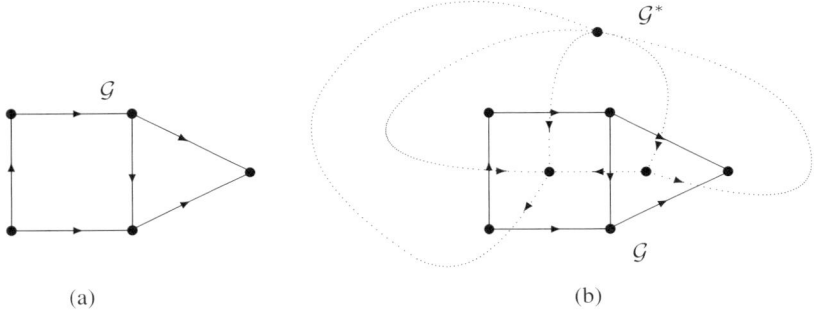

Figure 9.11: (a) A di-graph $\mathcal{G}$; and (b) its dual $\mathcal{G}^*$

voltage across the inductor branch in the series circuit exactly equals the current through the capacitor branch in the parallel branch, etc.

We can in fact summarize this amazing correspondence between the two dual circuits in Table 9.1 in which items listed in the left column are exactly equal to items in the right column of the same row.

We immediately appreciate that once we have solved the series circuit, we know everything about the parallel circuit, and vice versa. In fact, this is exactly what duality is about — if we know a circuit, we know its dual immediately. This could be a very powerful tool for circuit analysis if we could identify the dual of a given circuit.

9.9.2   Construction of a Dual Di-Graph

The first step to find the dual of a given circuit is to find its dual graph. In order to facilitate the mapping of currents into voltages, or vice versa, which involves polarities and directions, the di-graph (graph in which branches are directional) is considered. We will illustrate the procedure with an example.

Example 9.9: Construction of the dual di-graph — Suppose we are given a di-graph $\mathcal{G}$, as shown in figure 9.11 (a). The procedure for finding the dual di-graph $\mathcal{G}^*$ is as follows.

1. Introduce a node to each window of $\mathcal{G}$.

2. Add one node to the outer region as well.

3. The nodes created will form the nodes of the dual di-graph $\mathcal{G}^*$.

4. Create as many branches as possible which join up the nodes of $\mathcal{G}^*$, and in doing so make sure that each branch in $\mathcal{G}$ is cut exactly once.

5. The newly created branches form the branches of $\mathcal{G}^*$.

6. Check that the number of branches in $\mathcal{G}^*$ equals that in $\mathcal{G}$, and the number of nodes in $\mathcal{G}^*$ equals the number of windows in $\mathcal{G}$ plus one.

7. To find the direction of each branch in $\mathcal{G}^*$, we consider the pair of cutting branches (one from $\mathcal{G}^*$ and one from $\mathcal{G}$). We "imagine" rotating the branch of $\mathcal{G}$ clockwisely until it overlaps with its cutting partner from $\mathcal{G}^*$. Then we assign the direction for that particular branch of $\mathcal{G}^*$ to coincide with that of the rotated (imagined) branch of $\mathcal{G}$.

8. Direction for each branch can be found by repeating the above procedure. The resulting dual di-graph is shown in figure 9.11 (b).

9.9.3   The Dual of a Circuit

For a given planar circuit, we can derive the dual circuit by first obtaining the dual di-graph of the corresponding circuit di-graph. Then, the circuit elements are replaced by their dual counterparts, namely, inductors by capacitors, capacitors by inductors, resistors by conductors, current sources by voltage sources, voltage sources by current sources, open switches by closed switches, and closed switches by open switches. A complete dual translation table is shown in Table 9.2.

Example 9.10: Derivation of the dual circuit — Let us now go back to the earlier example of the series and parallel LRC circuits. Suppose we are given the series circuit. We can follow the above procedure to find its dual di-graph. After the dual di-graph is found, we place circuit elements back on the di-graph to form the dual circuit for the series LRC circuit. The rules for placing elements are simple. Basically, as shown in figure 9.12, we put (i) a *current source* on a branch of the dual circuit which cuts a *voltage source* of the original circuit; (ii) a *capacitor* on a branch of the dual circuit which cuts an *inductor* of the original circuit; (iii) an *inductor* on a branch of the dual circuit which cuts a *capacitor* of the original circuit; (iv) a *conductor* on a branch of the dual circuit which cuts a *resistor* of the original circuit.

In this example, we say that the following are dual elements:

1. capacitor (F) and inductor (H).

2. resistor (Ω) and conductor (S).

3. voltage source (V) and current source (A).

Quantity or Element	Dual
Current (A)	Voltage (V)
Voltage (V)	Current (A)
Resistance (Ω)	Conductance (S)
Conductance (S)	Resistance (Ω)
Reactance (Ω)	Susceptance (S)
Susceptance (S)	Reactance (Ω)
Impedance (Ω)	Admittance (S)
Admittance (S)	Impedance (Ω)
Capacitance (F)	Inductance (H)
Inductance (H)	Capacitance (F)
Independent current source (A)	Independent voltage source (V)
Independent voltage source (V)	Independent current source (A)
Dependent current source (A)	Dependent voltage source (V)
Dependent voltage source (V)	Dependent current source (A)
Open switch	Closed switch
Closed switch	Open switch
Loop	Cutset
Cutset	Loop
Branch	Node
Node	Branch
Tree	Co-tree
Co-tree	Tree

Table 9.2: Dual elements

9.9.4 Properties of the Dual Circuit

The very first property of the dual circuit is *uniqueness,* i.e., given a planar circuit, its dual is unique. In other words, we can find only one dual circuit for any given planar circuit.

The second property of the dual circuit is current–voltage correspondence. If we follow strictly the sign convention defined in figure 9.5, then values of voltages across branches of a circuit are equal to values of currents through the corresponding branches of its dual. For example, if the capacitor voltage of the series LRC circuit is 6V, then the inductor current of the parallel LRC circuit is exactly 6A. The direction of the current and the polarity of the voltage are given by the di-graph, using the sign convention of figure 9.5.

The third property of the dual circuit is the resemblance of the dynamic response. A circuit and its dual have the same set of eigenvalues (time constant in the case of a first-order circuit) and eigenvectors. Thus, the series and parallel LRC circuits shown in figure 9.12 (with component values chosen exactly as indicated in the figure) will have exactly the same set of resonant frequency, damping factor, Q-factor, etc.

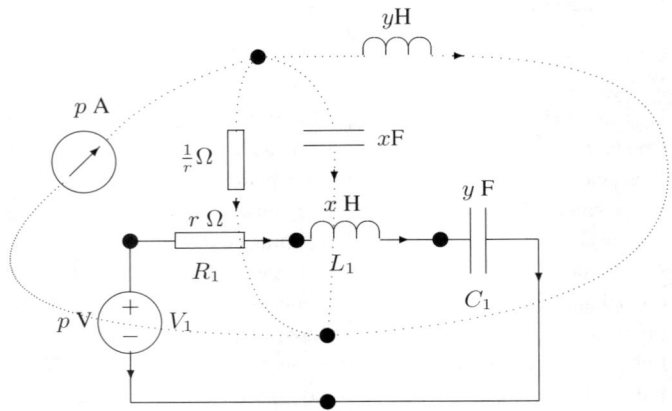

Figure 9.12: Duality of the series and parallel LRC circuits

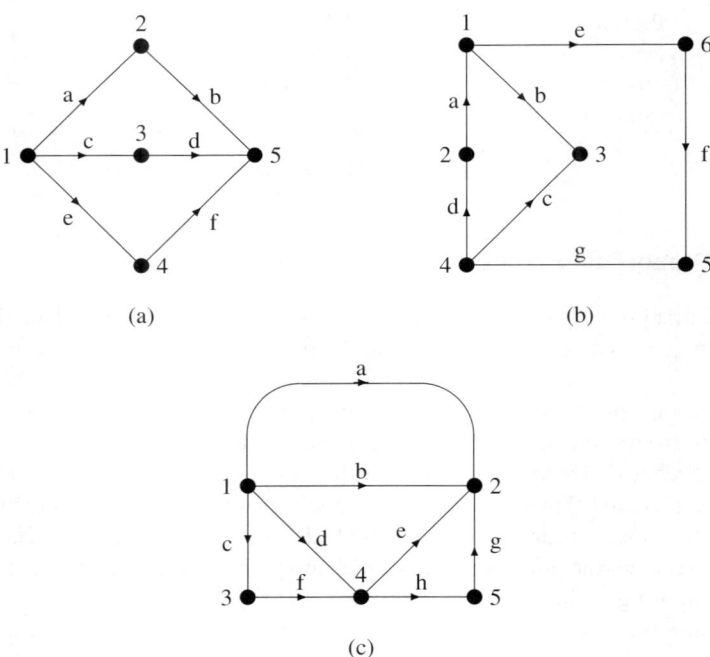

Figure 9.13: Circuits for Problems 1, 3, 5 and 8

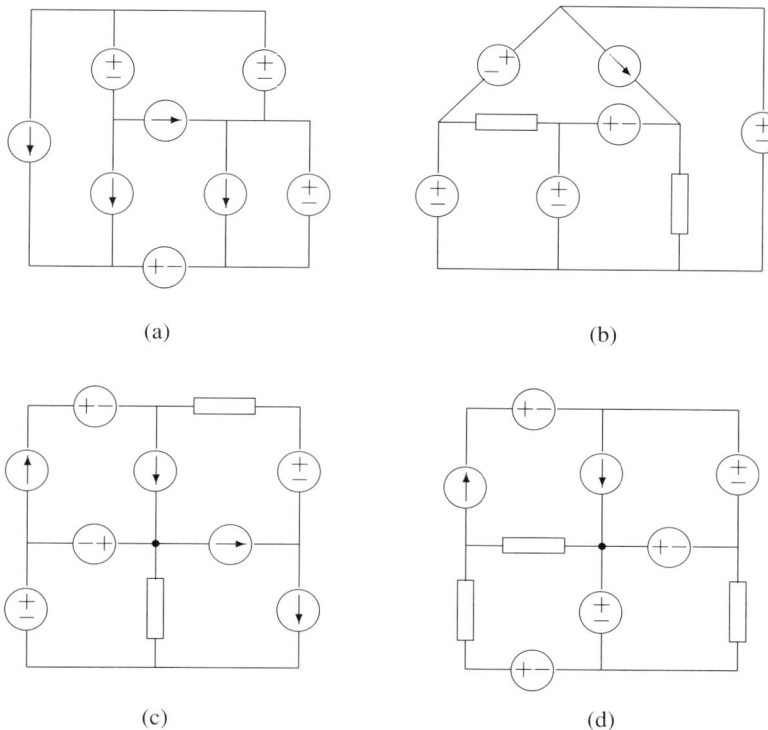

(a)   (b)

(c)   (d)

Figure 9.14: Circuits for Problem 2

9.10 Problems

1. For each of the graphs shown in figure 9.13, write down any three cutsets
   and three loops, in terms of a set of branches. Write down the corresponding
   Kirchhoff's law equation for each cutset and loop you have chosen. Use the
   sign convention defined in figure 9.5.

2. A circuit is unsolvable if it violates KCL or KVL. State which of the circuits
   shown in figure 9.14 is/are unsolvable. Explain your answers.

3. Referring to the circuits of figure 9.13, calculate the number of tree branches in
   each case. Choose one particular tree, and write a set of independent KVL and
   KCL equations in each case.

4. In the circuit of figure 9.15, calculate I_x, I_y and I_z, using a suitably chosen
   cutset for each unknown. There is no need to solve simultaneous equations if
   the right cutset is selected. Similarly, find V_x and V_y, using a suitably chosen
   loop for each unknown. Assuming all resistances are 1Ω, solve the circuit
   completely, i.e., find all voltages and currents. (Hint: Use Ohm's law to find the
   current in a resistor if its voltage is known, and vice versa. Choose appropriate

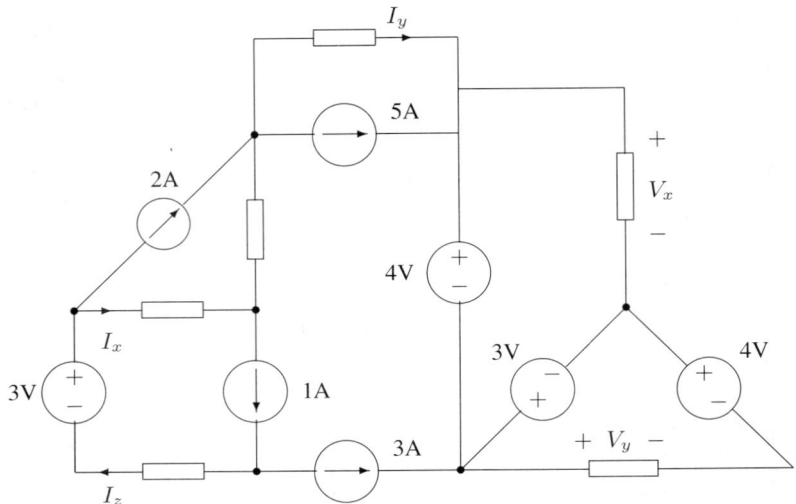

Figure 9.15: Circuits for Problem 4

cutsets and loops to find other currents and voltages. For this particular problem, you should be able to find every unknown voltage and current without solving simultaneous equations.)

5. Based on your answers to Problem 3, find the Q-matrix and B-matrix for each of the graphs shown in figure 9.13. Verify the dependence of the Q- and B-matrices.

6. For the circuit shown in figure 9.16 (a), find the Q-matrix and B-matrix based on the tree $\{E_0, E_4, R_1\}$. Hence, find a set of independent KVL equations and a set of independent KCL equations for the circuit.

7. For the same circuit of figure 9.16 (a), assume that all resistances are 1Ω, $E_0 = 5E_4 = 10$V, and $I_3 = 2I_6 = 2$A. Systematically solve this circuit using the independent Kirchhoff's law equations found in Problem 6 and also Ohm's law for the resistances.

8. Derive the dual di-graph for each of the di-graphs shown in figure 9.13.

9. Derive the dual for the circuit of figure 9.16 (b), and state its properties in relation to the original circuit.

10. Verify that a series circuit of a capacitor C, a resistor R_C and a voltage source V_i, is the dual of a parallel circuit of an inductor L, a resistor R_L and a current source I_i. Suppose you have found that the capacitor voltage in the series circuit is given by

$$v_C(t) = V_i(1 - e^{\frac{-t}{CR_C}}) \text{ V}$$

Use duality to derive the expression for the inductor current in the parallel circuit. Also, without working on the parallel circuit, find the voltage across the current source in the parallel circuit.

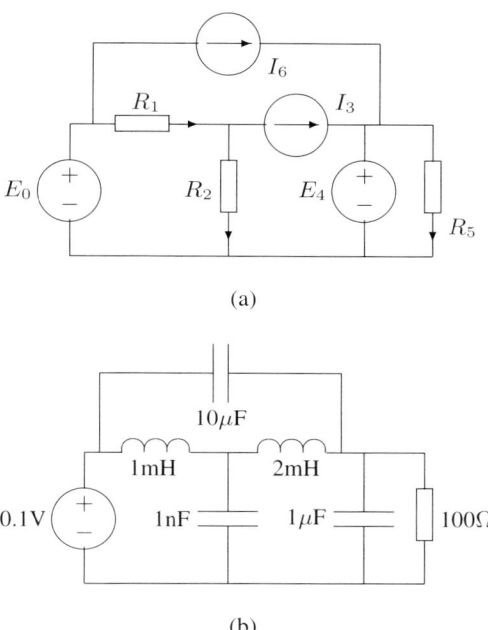

(a)

(b)

Figure 9.16: Circuits for Problems 6, 7 and 9

Chapter 10

Topological Approach to Resistive Circuit Analysis

In this chapter we introduce two systematic and economical methods for analyzing resistive circuits based on the concepts of basic loops and basic cutsets discussed in Chapter 9. The main feature of the methods to be introduced here is that the topology of the circuit is taken into account in the process of formulating the Kirchhoff's law equations. We recall that in the conventional mesh and nodal methods, Kirchhoff's law equations are always written for all meshes (or window loops) and all nodes except the reference node, regardless of the circuit topology. Under certain conditions, as we will see in this chapter, the resulting number of Kirchhoff's law equations from mesh or nodal analysis is more than what is actually needed. We will begin with a brief review of the mesh and nodal approaches, and in particular examine the application of the mesh and nodal methods to circuits containing independent current or voltage sources, where redundant Kirchhoff's law equations are set up resulting in inefficient solutions. In the process of setting up a minimal set of Kirchhoff's law equations, the graph theoretic concepts of trees, basic loops and basic cutsets play a crucial role. We will describe two systematic methods for deriving a minimal set of Kirchhoff's voltage or current law equations for any arbitrary circuit. These methods, known as the *loop-current* method and the *cutset-voltage* method, take advantage of the presence of independent voltage and current sources, with the help of graph theory, to minimize the size of the resulting linear system of equations.

10.1   Mesh and Nodal Methods Revisited

Let us start with a quick glimpse of the causes for inefficient solutions to a given circuit analysis problem. In particular we will find out why sometimes the mesh or nodal method would unnecessarily enlarge the size of the problem.

Consider a circuit containing n independent current sources. Suppose we wish

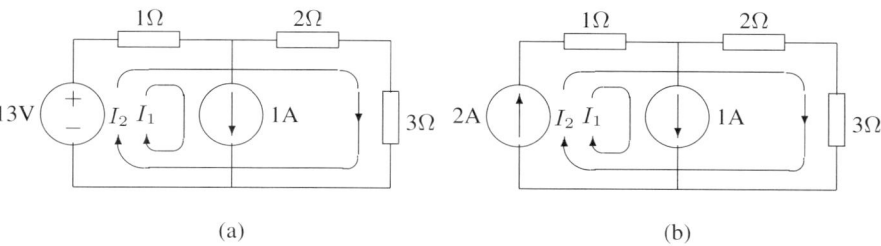

Figure 10.1: (a) Circuit containing a current source; (b) circuit with as many current sources as meshes

to analyze this circuit using the mesh method. We first assign the mesh currents as unknown variables. Then, we remove all the current sources temporarily and define the so-called "super-meshes", if any, as described in Chapter 3. If there are m meshes in the original circuit, we will naturally set up $m - n$ KVL equations (for the remaining meshes and super-meshes), and attempt to solve them for all the m unknowns. Moreover, in the original circuit, we have an equation relating each current source and two unknown mesh currents. Usually, the value of a current source is the sum or difference of two unknown mesh currents. Thus, we have n extra equations. Together with the previous $m - n$ KVL equations, we have a total of m equations to solve for the m unknown mesh currents. However, if we are allowed to set up KVL equations for loops other than the window loops, we can actually eliminate some KVL equations, thus making the problem easier to solve. An example will clarify this.

Example 10.1: Eliminating redundancy in mesh analysis — By definition, a mesh is a window loop, i.e., a loop containing no inner loops. In mesh analysis we set up KVL equations for the meshes and attempt to solve the equations. In this example, we will relax the restriction so that KVL equations can be written for any loop, and we will try to set up KVL equations more sensibly so that fewer equations need to be solved. Consider the circuit of figure 10.1 (a). It has two meshes and hence two KVL equations are expected from the mesh method. Let us instead write the KVL equations for the two loops marked in the figure. (One of them is not quite a mesh by definition.) Obviously, one equation is trivial, i.e., $I_1 = 1$A. This equation is not needed. In fact, we need to set up only one equation, which is

$$\text{Loop 2:} \quad (1 + I_2) + 2I_2 + 3I_2 = 13$$

Solving this equation gives $I_2 = (13 - 1)/6 = 2$A.

This example clearly shows that setting up KVL equations for a "sensible" choice of loops can more efficiently solve a circuit that contains independent current sources. In particular we conjecture that *the presence of one independent current source will eliminate one equation.* So, more current sources will make the circuit even easier to solve. Of course, the number of independent current sources cannot exceed the number

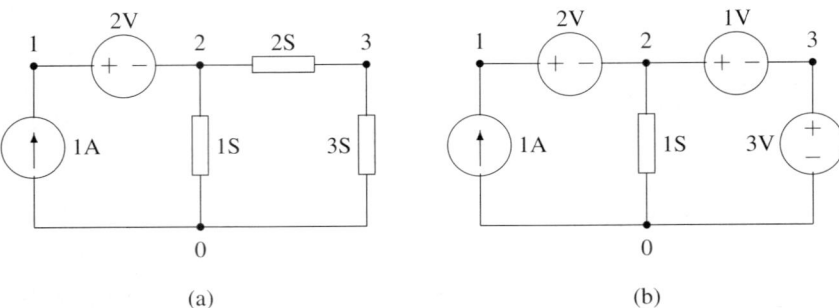

Figure 10.2: (a) Circuit containing a voltage source; (b) circuit with as many voltage sources as the number of nodes minus 1

of meshes.[1] An extreme case will reinforce this preliminary thought. Consider the circuit shown in figure 10.1 (b) which contains two meshes and two independent current sources. A moment's reflection will convince us that this circuit can be solved by inspection and no equations are required in the process. We conjecture that *if the number of independent current sources is equal to the number of meshes, we need not set up any KVL equation!*

Example 10.2: Eliminating redundancy in nodal analysis — A similar reasoning can be applied to the nodal analysis. In the circuit of figure 10.2 (a), nodal analysis will give three KCL equations, each corresponding to a node except the reference. The usual unknowns are V_1, V_2 and V_3, which are node voltages with respect to the reference. However, the presence of the voltage source can actually eliminate one equation if we set up KCL equations for a "sensible" choice of cutsets. Specifically let us choose only V_2 and V_3 as the unknown voltages. The two equations needed are

KCL for the cutset containing 1A, 1S and 2S: $\qquad V_2 + 2(V_2 - V_3) = 1$

KCL for the cutset containing 2S and 3S: $\qquad 2(V_3 - V_2) + 3V_3 = 0$

Here, V_1 is not needed because it is always equal to $V_2 + 2V$. Thus, we see that the presence of independent voltage sources actually makes the nodal problem easier to solve. We conjecture again that *the presence of one independent voltage source will eliminate one KCL equation.*

Let us consider the extreme case where there are as many independent voltage sources as the number of nodes minus 1.[2] As shown in figure 10.2 (b), the circuit can be solved by inspection, and no nodal equations are ever needed! Thus, we may conjecture that *if there are as many independent voltage sources as the number of nodes minus 1, no KCL equations need be set up.*

---

[1] When the number of independent current sources exceeds the number of meshes, the circuit is not solvable since KCL would have been violated.

[2] When the number of independent voltage sources equals or exceeds the number of nodes, the circuit is not solvable since KVL would have been violated.

The above conjectures concerning the effects of the presence of independent sources on the complexity of mesh and nodal methods are indeed well grounded, as will become apparent in the light of graph theory. In the following sections, we will attempt to develop some "smart" procedures for setting up a minimal set of (independent) Kirchhoff's law equations, on the basis of graph theory.

10.2 Standard Tree

A *standard tree* contains all independent voltage source branches and a maximum number of resistance branches. In practice, we select branches for inclusion in the standard tree in the order described as follows.

1. Select all independent voltage source branches;
2. Select as many resistance branches as possible, without forming a loop.

The tree constructed in this way is not unique, which means that it is possible to have more than one standard tree for a given circuit. Nevertheless, regardless of the choice of a particular standard tree, the total number of tree branches is always equal to the number of nodes minus one. Ignoring dependent sources for the time being, the number of resistance branches in a standard tree, t_r, is given by

$$t_r = (n - 1) - n_{VS}$$

where n is the number of nodes and n_{VS} is the number of independent voltage sources.

Besides being an important preparatory step for systematic circuit analysis, the process of choosing a standard tree is useful in identifying ill-posed circuits. Below are some important arguments.

1. If we could not include all independent voltage sources in the tree, then some independent voltage sources must be forming a loop. This violates KVL, and the circuit is not solvable. Thus, *if the circuit is solvable, it is always possible to include all the independent voltage sources in the tree.*

2. If we could include an independent current source in the tree after all resistances are taken, then that current source must be forming a cutset with other independent current sources because the co-tree would have only current sources in that case. This violates KCL, and the circuit is not solvable. Thus, *if the circuit is solvable, it is always possible to exclude all the independent current sources from the tree. In other words, all the independent current sources must be in the co-tree.*

Example 10.3: Solvability of circuits — Consider a circuit having 10 nodes and 15 branches. We can easily calculate the number of tree branches as $10 - 1 = 9$, and the number of co-tree branches as $15 - 9 = 6$. If there are more than 9 independent voltage sources, the circuit is not solvable because there must be a loop of independent voltage sources, thus violating KVL. Likewise, if there are more than 6 independent current sources, the circuit is not solvable because there must be a cutset of independent current sources, thus violating KCL.

10.3 Outline of Systematic Analysis

10.3.1 Objective

Before we derive the formal analysis procedures, we must clarify our objectives so that we will not set up and solve equations "blindly".

Fact — To solve a circuit completely, we need only to know *either* of the following:

1. All co-tree branch currents;
2. All tree branch voltages.

Verification — In the first case, if we know the values of all co-tree branch currents, all tree branch currents are automatically known from basic cutset KCL equations. (Remember that a basic cutset contains one tree branch and other co-tree branches.) Once we know the tree branch currents, we can derive the tree branch voltages from the Ohm's law equations. Finally, the remaining unknown co-tree branch voltages can be found from the basic loop KVL equations since all tree branch voltages are already known. Thus, the circuit can be completely solved once all co-tree branch currents are known.

In the second case, if we know the values of all tree branch voltages, all co-tree branch voltages are automatically known from basic loop KVL equations. (Remember that a basic loop contains one co-tree branch and other tree branches.) Once we know the co-tree branch voltages, we can derive the tree branch currents from the Ohm's law equations. Finally, the remaining unknown tree branch currents can be found from basic cutset KCL equations since all co-tree branch currents are already known. Thus, the circuit can be completely solved once all tree branch voltages are known.

10.3.2 Choice of Methods

The next question is: Given a circuit, should we try to find the currents of all co-tree branches, or the voltages of all tree branches? Naturally, we would use whichever method is easier. To decide on which method is easier, we can compare the number of unknown variables to be found in the two cases. Suppose there are n_{VS} voltage sources, n_{CS} current sources, n nodes and b branches. The number of tree branches, t, is

$$t = n - 1$$

and the number of co-tree branches, l, is

$$l = b - (n - 1)$$

Thus, the number of unknown tree branch voltages is the same as the number of tree branches minus the number of voltage sources, since the values of the voltage sources are known already.

$$\text{Number of unknown tree branch voltages} = n - 1 - n_{VS}$$

Similarly, the number of unknown co-tree branch currents is given by

$$\text{Number of unknown co-tree branch currents} = b - (n - 1) - n_{CS}$$

Therefore, if $n - 1 - n_{VS} < b - (n - 1) - n_{CS}$, we prefer to solve for the unknown tree branch voltages, otherwise, we should try to find the unknown co-tree branch currents instead.

Example 10.4: Choosing the right unknowns — Suppose a circuit has 10 nodes, 15 branches, 3 independent voltage sources and 4 independent current sources. Thus, the number of tree branches is $10 - 1 = 9$, and the number of co-tree branches is $15 - 9 = 6$. To solve this circuit, we have to find either all unknown tree branch voltages or all unknown co-tree branch currents. Using the above formulae, we have $9 - 3$, i.e., 6, unknown tree branch voltages, and $6 - 4$, i.e., 2, unknown co-tree branch currents. Clearly, it is more economical to find the unknown co-tree branch currents in this particular example.

10.4   The General Loop (Loop-Current) Analysis

For any given resistive circuit, we begin with selecting a standard tree according to the procedure outlined in the foregoing section. This tree always contains all the independent voltage sources and as many resistances as possible.

In the general loop method or loop-current method, the unknowns to be found are the $b - (n - 1) - n_{CS}$ co-tree branch currents. Since each co-tree branch forms one basic loop with other tree branches, we have $b - (n-1) - n_{CS}$ basic loops. We assume that the co-tree branch current circulates round the associated basic loop. Thus, the current in any branch is the sum or difference of two or more basic loop currents. In other words, any branch current can be expressed in terms of the unknowns.

For each of the basic loops, except the one associated with a current source, we write down the KVL equation. Altogether we have $b - (n - 1) - n_{CS}$ KVL equations to solve for the $b - (n - 1) - n_{CS}$ unknowns.

Example 10.5: Illustration of the loop-current method — Consider the circuit shown in figure 10.3. The tree branches are drawn in solid line, whereas the co-tree branches are shown in dotted line. The three basic loops are

1. Basic loop 1: R_1—R_4—V_o
2. Basic loop 2: R_2—R_4—R_3
3. Basic loop 3: I_6—R_5—V_o—R_3

We let I_1, I_2 and I_6 be the currents circulating round basic loops 1, 2 and 3 respectively. The current in a co-tree branch is simply the current in the corresponding basic loop. Current in any other branch is either the sum or difference of I_1, I_2 and I_6. For example, current in R_4 is $I_1 + I_2$ downwards, current in R_3 is $I_6 - I_2$ downwards, current in R_5 is I_6, etc.

We may now write down the KVL equations for the basic loops. Note that basic loop 3 can be ignored since we are given that $I_6 = 7A$. In fact we can put $I_6 = 7$ directly into all KVL equations.

$$\text{Basic loop 1:} \quad I_1 R_1 - V_o + (I_1 + I_2)R_4 = 0$$
$$\text{Basic loop 2:} \quad I_2 R_2 + (I_2 - 7)R_3 + (I_1 + I_2)R_4 = 0$$

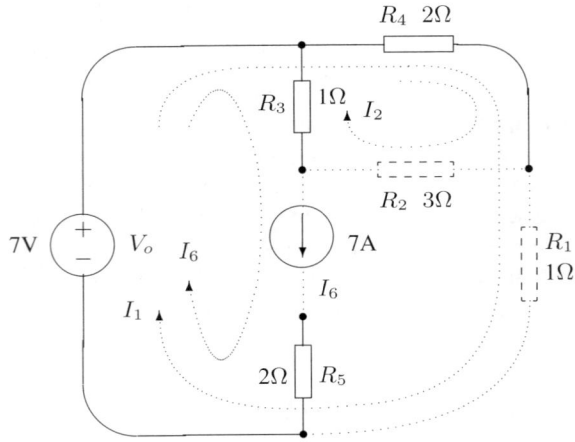

Figure 10.3: Illustration of the general loop method. Tree: solid line; co-tree: dotted line

$$\Rightarrow \quad \begin{cases} 3I_1 + 2I_2 &= 7 \\ 2I_1 + 6I_2 &= 7 \end{cases}$$

Solving this system of equation gives

$$I_1 = 2\text{A} \quad \text{and} \quad I_2 = 0.5\text{A}$$

Remarks — Although the circuit has three meshes, the general loop method requires only two equations (instead of three). The reason is that we have selected the loops in the right way so as to make good use of the current source value. As we can see in the above example, we choose loop 3 to lie exactly on the current source. Hence, loop 3 does not need to be solved.

10.5   The General Cutset (Cutset-Voltage) Method

The starting point for the general cutset method or cutset-voltage method is again the choice of a standard tree. The unknowns are the $n - 1 - n_{VS}$ tree branch resistance voltages. Our objective is to set up $n - 1 - n_{VS}$ KCL equations according to the $n - 1 - n_{VS}$ basic cutsets. To achieve this, we must express all involving variables in terms of the $n - 1 - n_{VS}$ unknown tree-branch resistance voltages.

Since a basic cutset has only one tree branch, the basic cutset KCL equation typically takes the form:

$$I_t + \sum_{\text{co-tree}} I_l = 0$$

where I_t is a tree branch resistance voltage and the summation term represents the other co-tree branch currents. We take the following procedure to express this equation in terms of the unknown voltages, i.e., tree branch resistance voltages.

1. Substitute V_t / R_t for I_t.

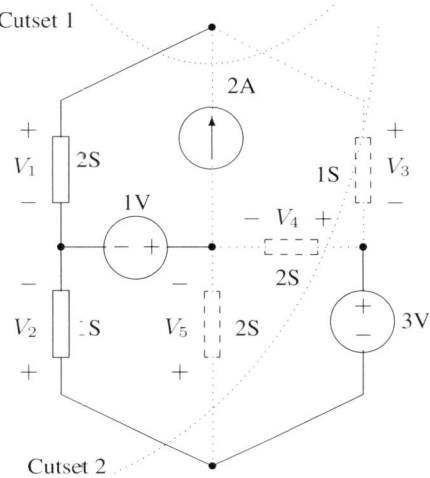

Figure 10.4: Illustration of the general cutset method. Tree: solid line; co-tree: dotted line

2. The other co-tree branch currents are either resistance currents or current source currents. For the resistance current, replace it by V_l/R_l, where V_l and R_l are the voltage and resistance of that co-tree branch.

3. Replace each co-tree branch voltage by tree branch voltages using the appropriate basic loop KVL equation.

Thus, any basic cutset KCL equation is always expressed in terms of the unknown tree branch resistance voltages. We will illustrate the procedure by an example.

Example 10.6: Illustration of cutset-voltage method — The circuit shown in figure 10.4 is to be solved using the cutset-voltage method. There are four tree branches, of which two correspond to independent voltage sources. Thus, the unknowns are V_1 and V_2. We consider only the basic cutsets corresponding to R_1 and R_2, while we ignore the basic cutsets corresponding to the voltage sources. The basic cutset KCL equations are

$$\text{Basic cutset 1:} \quad 2V_1 - 2 + V_3 = 0$$
$$\text{Basic cutset 2:} \quad V_2 + 2V_5 + 2V_4 - V_3 = 0$$

We must now express these equations in terms of V_1 and V_2. From some basic loop KVL equations, we have $V_3 = V_1 - V_2 - 1$, $V_4 = V_2 + 3 - 1$, and $V_5 = V_2 - 1$. Thus, the above basic cutset KCL equations become

$$\text{Basic cutset 1:} \quad 3V_1 - V_2 = 3$$
$$\text{Basic cutset 2:} \quad -V_1 + 6V_2 = -6$$

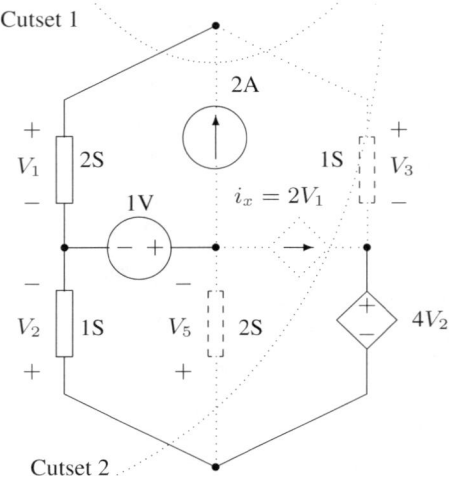

Figure 10.5: Circuit containing dependent sources. Tree: solid line; co-tree: dotted line

Solving these two equations gives

$$V_1 = \frac{15}{17}\text{V} \quad \text{and} \quad V_2 = -\frac{6}{17}\text{V}$$

10.6   Circuits Containing Dependent Sources

The analysis of circuits that contain dependent sources follows almost exactly the procedure discussed previously. Essentially we put as many dependent voltage sources as possible in the tree. If the general cutset method is preferred, we set up KCL equations for the basic cutsets, ignoring those cutsets that correspond to dependent or independent voltage sources. Thus, as before, exactly $n - 1 - n_{VS}$ KCL equations are set up. (Here, n_{VS} includes independent and dependent voltage sources.) However, these equations will contain some "extra" variables (not necessarily the tree-branch voltages) due to the presence of dependent sources. We must then write down all the constraint equations from the dependent sources and use them to eliminate the "extra" variables. The result is again a system of $n - 1 - n_{VS}$ equations with $n - 1 - n_{VS}$ unknowns.

Example 10.7: Circuit with dependent sources — The circuit shown in figure 10.5 is similar to the previous example, but contains a dependent voltage source and a dependent current source in lieu of a fixed voltage source and a resistor. The KCL equations corresponding to the two basic cutsets are

Basic cutset 1:	$2V_1 - 2 + (V_1 - V_2 - 4V_2) = 0$
Basic cutset 2:	$V_2 + 2(V_2 - 1) - i_x - (V_1 - V_2 - 4V_2) = 0$

Since $i_x = 2V_1$, we have

$$\text{Basic cutset 1:} \qquad 3V_1 - 5V_2 = 2$$
$$\text{Basic cutset 2:} \qquad -3V_1 + 8V_2 = 2$$

Solving these two equations give

$$V_1 = \frac{26}{9}\text{V} \quad \text{and} \quad V_2 = \frac{4}{3}\text{V}$$

In order to solve the circuit completely, we need to know all tree branch voltages. Thus, we need to know, in addition to V_1 and V_2, the voltage of the dependent voltage source. This is simply given by $4V_2$ which is 16/3 V.

10.7 Problems

1. If a circuit has 10 nodes, 20 branches, 3 independent voltage sources and 2 independent current sources, how many tree branches does the circuit have? If you are to solve the circuit using either the loop-current method or the cutset-voltage method, how would you make the choice?

2. State which of the following topologies are not solvable: (i) 5 nodes, 10 branches, 5 independent voltage sources; (ii) 7 nodes, 12 branches, 7 independent current sources; (iii) 4 nodes, 7 branches, 4 independent current sources; (iv) 6 nodes, 9 branches, 1 independent voltage source, 3 independent current sources; (v) 5 nodes, 9 branches, 3 independent voltage sources, 2 independent current sources.

3. For each solvable topology in Problem 2, find the minimum number of equations to be set up, and state whether the loop-current or the cutset-voltage method is preferred.

4. For each of the circuits shown in figure 10.6, select a standard tree and state whether the cutset-voltage method or the loop-current method would require fewer Kirchhoff's law equations to be set up. Hence, use the simpler method to solve the circuit completely, i.e., find the voltage and current in each branch of the circuit. (Note: You must find either the set of tree branch voltages or the set of co-tree branch currents first, and then proceed to find all other voltages and currents.)

5. Consider the circuit of figure 10.7 (a), which contains a dependent source. State whether the loop-current method or the cutset-voltage method will give fewer equations to be solved. Use the simpler method to solve the circuit. Note that a dependent source is not the same as an independent source, but is rather a trans-resistance or trans-conductance function. You may treat the dependent source initially like other independent sources, and later on introduce the constraint equation that governs the dependent source. You should have enough equations to solve the unknowns.

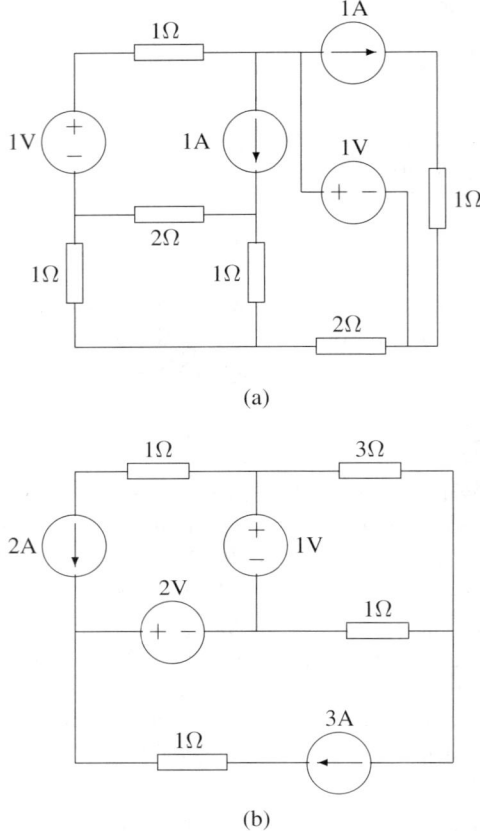

(a)

(b)

Figure 10.6: Circuits for Problem 4

6. The circuit shown in figure 10.7 (b) is a small-signal representation of an amplifier. Verify that the voltage gain is given by

$$\frac{v_o}{v_s} = \frac{r_\pi}{r_\pi + r_b} \times \frac{g_m r_L r_o}{r_L + r_r + r_o}$$

where v_o is the voltage across the load resistance r_L.

7. For the circuit shown in figure 10.8 (a), choose a value for g such that $v_x = 5$V. What is the power supplied/absorbed by the controlled current source?

8. Calculate the equivalent input resistance R_{in} for the circuit of figure 10.8 (b). (Hint: Open-circuit the 0.25A source, put a 1A current source on the input terminals, and find the ratio of input voltage to input current.)

9. Derive the Thévenin equivalent circuit for the circuit of figure 10.8 (b) by finding the open-circuit voltage and short-circuit current. Compare the Thévenin resistance with the one found in Problem 8.

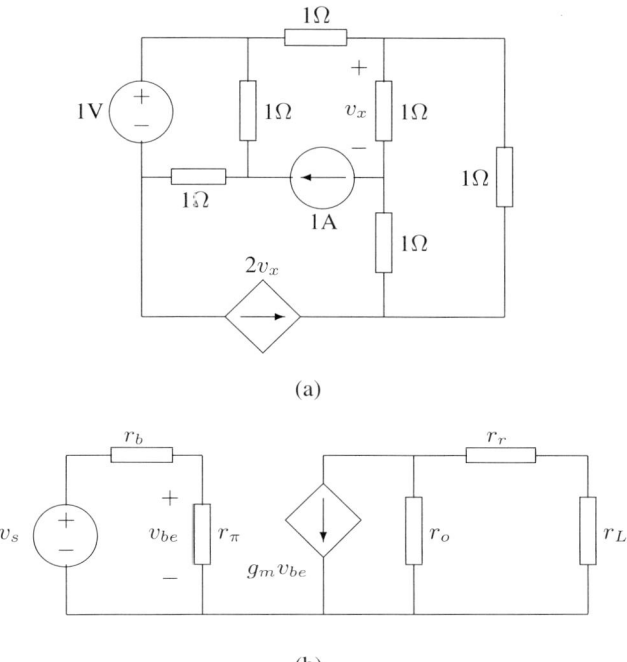

(a)

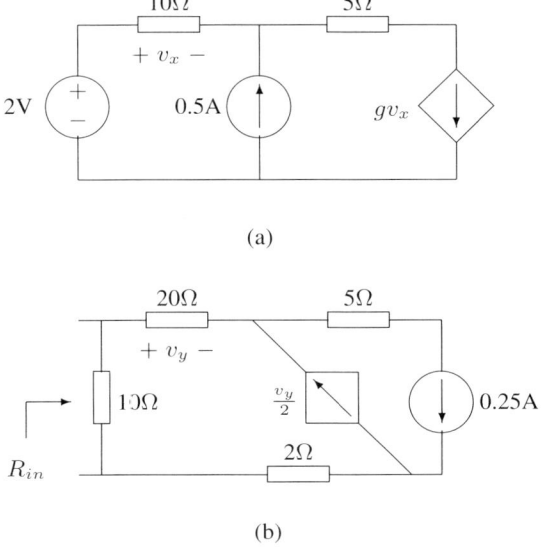

(b)

Figure 10.7: Circuits for Problems 5 and 6

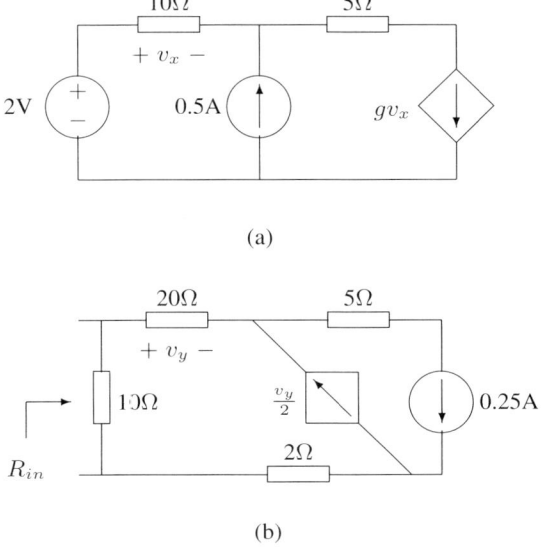

(a)

(b)

Figure 10.8: Circuits for Problems 7, 8 and 9

10. For the circuit of figure 10.9, find all branch currents. State whether the general-loop method or the general-cutset method is preferable, and explain why.

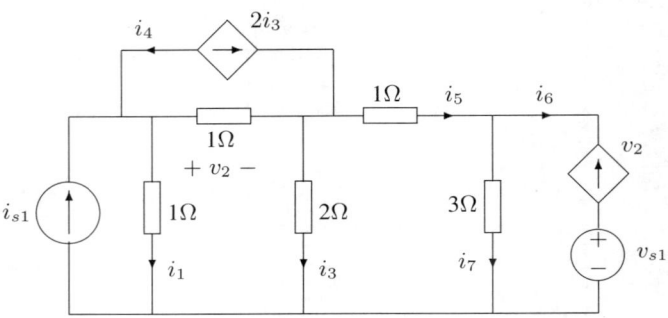

Figure 10.9: Circuit for Problem 10

Chapter 11

Topological Approach to Dynamic Circuit Analysis

A dynamic circuit in general consists of at least one element whose constitutive relation involves a time derivative. Thus, any circuit that has a capacitor or an inductor is a dynamic circuit. Two approaches are possible for the analysis of dynamic circuits. Either we treat everything in the time domain starting from the describing differential equation, or we resort to the algebraic equation in the complex frequency domain. In this chapter we will show both techniques and demonstrate their interchangeability. In particular, similar to what we did in the previous chapter for resistive circuits, we will make use of graph theory to develop a systematic method for deriving the describing equations which may take the form of either differential equations or algebraic equations.

11.1 Independent Energy Storage Elements

There are two types of energy storage element in electric circuits, namely the capacitor and the inductor. A first-order circuit has one storage element, and a second-order circuit has two storage elements. However, it is not generally true that an nth-order circuit has n storage elements, because some storage elements may be dependent on one another. Thus, it is necessary to know the number of independent storage elements in a circuit.

Shown in figure 11.1 (a) is a series connection of two capacitors and a voltage source. The voltages of C_1 and C_2 are related by $v_{C1} + v_{C2} = E$. They are therefore dependent, and the circuit has only one independent storage element. We can choose either C_1 or C_2 as the independent storage element. The circuit is thus first-order. Another example is shown in figure 11.1 (b) where two inductors are in parallel with a current source. In this case, the current of L_1 and that of L_2 are dependent since $i_{L1} + i_{L2} = I$. The circuit has only one storage element, and is thus first-order.

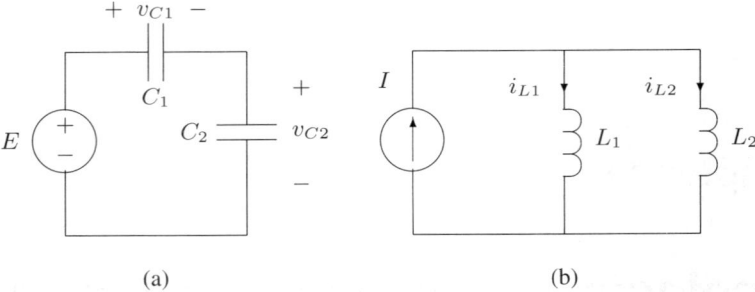

Figure 11.1: First-order circuits with more than one storage elements

The above two examples are particular cases of a more general result which gives the total number of independent storage elements. Essentially, if a loop consists of exclusively capacitors and voltage sources, one of the capacitors is disqualified from being an independent storage element. Similarly, if a cutset consists of exclusively inductors and current sources, one of the inductors is disqualified from being an independent storage element. We will give a more precise formula for calculating the number of independent storage elements, based on a standard tree, in a later section.

11.2 State Equation

For any linear dynamic circuit having n independent storage elements and driven by m independent sources, we can write an equation of the form

$$\dot{\bar{x}} = A\bar{x} + B\bar{u}$$

where $\bar{x}$ is a vector of dimension n, $\bar{u}$ is the driving signal vector of dimension m, A is an nth-order square matrix, and B is an $m \times n$ matrix.

The vector $\bar{x}$ is called the *state vector* and is composed of all independent capacitor voltages and inductor currents. The vector $\bar{u}$ is called the *input vector* or the *driving source vector*, and is composed of all independent driving sources. We will use an example to clarify our notations.

In the circuit of figure 11.2, we can easily write down the following two equations:

$$C\frac{dv_c}{dt} = i_L - i_R = i_L - \frac{v_c}{R} \quad \Rightarrow \quad \frac{dv_c}{dt} = \frac{-v_c}{CR} + \frac{i_L}{C}$$

and

$$L\frac{di_L}{dt} = u - v_c \quad \Rightarrow \quad \frac{di_L}{dt} = \frac{-v_c}{L} + \frac{u}{L}$$

Hence, we have

$$\frac{d}{dt}\begin{bmatrix} v_c \\ i_L \end{bmatrix} = \begin{bmatrix} -\frac{1}{CR} & \frac{1}{C} \\ -\frac{1}{L} & 0 \end{bmatrix}\begin{bmatrix} v_c \\ i_L \end{bmatrix} + \begin{bmatrix} 0 \\ \frac{1}{L} \end{bmatrix} u$$

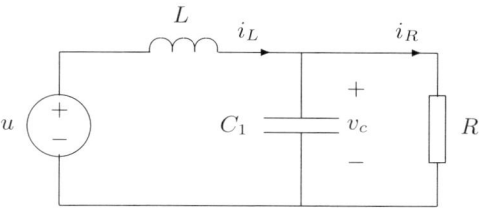

Figure 11.2: Circuit with two state variables

which is of the form $\dot{\bar{x}} = A\bar{x} + B\bar{u}$. In this case, the state vector is

$$\bar{x} = \begin{bmatrix} v_c \\ i_L \end{bmatrix}$$

and the source vector is one-dimensional or scalar. The A and B matrices can also be recognized from the above equation.

The above matrix equation, consisting of n first-order differential equations and involving n variables, is called the *normal form state equation* for the circuit. The n variables in the state vector are called the *state variables* of the circuit.

11.3 Time-Domain Descriptions

The above circuit can also be represented by a differential equation involving one variable, either v_c or i_L. By eliminating i_L, for instance, we get

$$\frac{d^2}{dt^2}v_c + \frac{1}{CR}\frac{d}{dt}v_c + \frac{v_c}{LC} = \frac{u}{LC}$$

This is a second-order ODE, and involves only v_c. Thus, we have two ways of describing a linear dynamic circuit:

1. Normal form state equation: $\dot{\bar{x}} = A\bar{x} + B\bar{u}$.
2. nth-order ODE in one variable: $\dfrac{d^n}{dt^n}x + a_1\dfrac{d^{n-1}}{dt^{n-1}}x + \cdots + a_n x = f(u)$.

The order of the circuit, in the first case, is reflected by the dimension of $\bar{x}$, and in the second case, is the same as the order of the ODE. Both representations yield the same solution. In this chapter we will focus on the normal form state equation since it allows systematic derivation based on the circuit graph, and can be applied to any given dynamic circuit.

11.4 Systematic Analysis of Dynamic Circuits

The purpose here is to derive the state equation for any given dynamic circuit. The procedure involves

1. choosing a state vector, i.e., independent state variables;
2. finding the matrices A and B.

11.4.1   Standard Tree and Choice of State Vector

As discussed earlier in the chapter, not all capacitor voltages and inductor currents in a circuit are independent. In particular the problem lies in capacitor-voltage-source loops and inductor-current-source cutsets. If we can detect these loops and cutsets, we are able to eliminate the dependent capacitor voltages and inductor currents. To this end, we make use of a tree of the circuit graph. An obvious construction is to base the choice of tree on the following order of priorities:

1. All independent voltage sources.
2. Maximum number of capacitors.
3. Maximum number of resistors.
4. Inductors to complete the tree.

A tree that is formed using the above order of choice is called a *standard tree*. Note that a standard tree will always contain all independent voltage sources, and capacitors which form no loops with themselves and/or voltage sources. Conversely, the co-tree will always have all current sources, and inductors which form no cutsets with themselves and/or current sources.

By properly numbering the branches on the graph, it is always possible to write down the Q-matrix and the B-matrix in the following forms:

$$
\begin{aligned}
Q &= [\, \mathbf{1} \mid Q_1 \,] \\
B &= [\, B_1 \mid \mathbf{1} \,]
\end{aligned}
$$

where $\mathbf{1}$ is the unit matrix. Also, from Chapter 9, we know that $B_1 = -Q_1{}^T$.

Once a standard tree is chosen, the choice of a state vector is straightforward. Since the tree has all the independent capacitors and the co-tree has all the independent inductors, *the state vector is composed of all tree-branch capacitor voltages and co-tree branch inductor currents.* Since the order of a dynamic circuit is the total number of state variables, i.e., the dimension of the state vector, we have

$$
\begin{aligned}
\text{Order of circuit} &= \text{dimension of state vector} \\
&= \text{number of tree capacitors} + \text{number of co-tree inductors}
\end{aligned}
$$

Example 11.1: Choice of state variables — Suppose we want to choose a set of state variables for a dynamic circuit containing 10 nodes, 4 independent voltage sources, 7 capacitors, 3 resistors, 3 inductors and 2 independent current sources. The number of tree branches is $10 - 1 = 9$. According to the order of priority stated above, we take all voltage sources and any 5 capacitors into the tree. The remaining 2 capacitors, all resistors, inductors and current sources will form the co-tree. Thus, the state variables are the 5 capacitor voltages in the tree and the 3 inductor currents in the co-tree. The size of the state vector is $5 + 3 = 8$, which is also the order of the circuit.

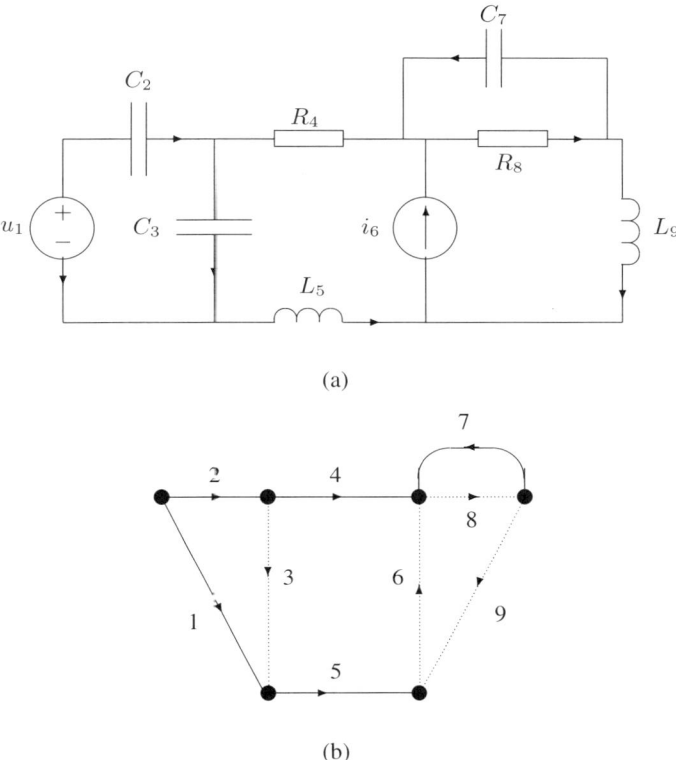

(a)

(b)

Figure 11.3: (a) Dynamic circuit; (b) di-graph showing standard tree

From this example, we see that we are able to tell the order of the circuit and hence the relative complexity of the solution without knowing the details of the circuit. All we need to know are the number of nodes (which actually gives the number of tree branches) and the number of each type of element. In the next example, we will go a step further to derive the B-matrix and Q-matrix for a dynamic circuit.

Example 11.2: Obtaining the Q-matrix and B-matrix for a dynamic circuit – the first step to systematic analysis — We will use the circuit of figure 11.3 (a) to illustrate the procedure for obtaining the Q-matrix and B-matrix. Before we proceed, let us clarify the sign convention. Throughout our discussion we will assign the branch direction as the current direction. The branch voltage is such that current goes into the positive pole of the branch element and leaves from its negative pole.

An arbitrary direction is first assigned to each branch, as denoted by the arrow direction in figure 11.3. The current direction and voltage polarity are then understood. Referring to figure 11.3 (a), for example, the current through the voltage source u_1 is assumed downward, and u_1 is the voltage of the branch, with positive on the top and negative at the bottom.

To find a standard tree, we follow the order of priorities outlined previously. We first take u_1, then C_7, and either C_2 or C_3. Next we take R_4, and finally either L_5 or L_9 to complete the tree. Suppose we choose u_1, C_2, C_7, R_4 and L_5. The di-graph showing our choice of standard tree is shown in figure 11.3 (b), where the tree branches are drawn with solid lines and the co-tree branches with dotted lines. Based on this tree, we choose the state variables as v_2, v_7 and i_9, i.e., the state vector as

$$\bar{x} = \begin{bmatrix} v_2 \\ v_7 \\ i_9 \end{bmatrix}$$

Also, the Q- and B-matrices corresponding to this standard tree are:

$$
Q = \begin{array}{c} \\ 1 \\ 2 \\ 7 \\ 4 \\ 5 \end{array}
\begin{array}{ccccc} 1 & 2 & 7 & 4 & 5 \end{array}
\begin{array}{cccc} 3 & 8 & 9 & 6 \end{array}
\left[
\begin{array}{ccccc|cccc}
1 & 0 & 0 & 0 & 0 & 1 & 0 & 1 & -1 \\
0 & 1 & 0 & 0 & 0 & -1 & 0 & -1 & 1 \\
0 & 0 & 1 & 0 & 0 & 0 & -1 & 1 & 0 \\
0 & 0 & 0 & 1 & 0 & 0 & 0 & -1 & 1 \\
0 & 0 & 0 & 0 & 1 & 0 & 0 & 1 & -1
\end{array}
\right]
$$

$$
B = \begin{array}{c} \\ 3 \\ 8 \\ 9 \\ 6 \end{array}
\begin{array}{ccccc} 1 & 2 & 7 & 4 & 5 \end{array}
\begin{array}{cccc} 3 & 8 & 9 & 6 \end{array}
\left[
\begin{array}{ccccc|cccc}
-1 & 1 & 0 & 0 & 0 & 1 & 0 & 0 & 0 \\
0 & 0 & 1 & 0 & 0 & 0 & 1 & 0 & 0 \\
-1 & 1 & -1 & 1 & -1 & 0 & 0 & 1 & 0 \\
1 & -1 & 0 & -1 & 1 & 0 & 0 & 0 & 1
\end{array}
\right]
$$

11.4.2 Derivation of State Equations

The purpose is to find a first-order differential equation for each of the state variables. Since the state variables are either capacitor voltages or inductor currents, we will discuss the derivation procedure separately for these two cases. The circuit of figure 11.3 (a) will serve as an illustrative example.

Example 11.3: Illustration of the procedure for state equation derivation — In the previous example, we have chosen v_2, v_7 and i_9 as the state variables for the circuit of figure 11.3 (a). Our job now is to get a differential equation for each of these state variables. Our specific aim is to put the differential equations in the normal form, i.e., each differential equation must involve only the state variables, the input sources and the time derivative of one state variable.

Case 1: the state variable is capacitor voltage — In this case, we always begin with the basic cutset KCL equation involving the tree capacitor concerned. We may get this equation by inspection or from the Q-matrix. For C_2 in the above circuit, the basic cutset KCL equation is

$$i_2 = i_3 + i_9 - i_6$$

Now we "transform" this equation into a voltage equation by the following procedure:

1. Retain all current source(s) and state variable current(s), i.e., i_6 and i_9 to remain untouched.
2. Put $i_c = C\frac{dv_c}{dt}$ for all capacitor current(s). Then, express the co-tree capacitor voltage(s) in terms of the state variable(s) and/or current source(s).

$$i_2 = C_2\frac{dv_2}{dt}$$

$$i_3 = C_3\frac{dv_3}{dt} = C_3\frac{d}{dt}(u-1-v_2)$$

3. Put $i_R = \frac{v_R}{R}$ for all resistor current(s). Then, express all variables in terms of the state variable(s) and the voltage source(s).

After steps 1, 2 and 3 are done, the original KCL equation becomes

$$C_2\frac{dv_2}{dt} = C_3\frac{d}{dt}(u_1 - v_2) + i_9 - i_6$$

$$\Rightarrow \quad \frac{dv_2}{dt} = \frac{i_9}{C_2+C_3} - \frac{i_6}{C_2+C_3} + \frac{C_3}{C_2+C_3}\frac{du_1}{dt} \tag{V2}$$

We have completed the derivation of the first-order differential equation involving dv_2/dt.

Now the same procedure can be repeated for C_7 to obtain the first-order differential equation involving dv_7/dt. The result is

$$\frac{dv_7}{dt} = -\frac{v_7}{C_7 R_8} - \frac{i_9}{C_7} \tag{V7}$$

Case 2: the state variable is inductor current — In this case, we always begin with the basic loop KVL equation involving the co-tree inductor concerned. We may get this equation by inspection or from the B-matrix. For L_9 in the above circuit, the basic loop KVL equation is

$$v_9 = u_1 - v_2 + v_7 - v_4 + v_5$$

Now we "transform" this equation into a current equation by the following procedure:

1. Retain all voltage source(s) and state variable voltage(s), i.e., v_2 v_7 and u_1 to remain untouched.
2. Put $v_L = L\frac{di_L}{dt}$ for all inductor voltage(s). Then, express the co-tree inductor current(s) in terms of the state variable(s) and/or voltage source(s).

$$v_5 = L_5\frac{di_5}{dt} = L_5\frac{d}{dt}(i_6 - i_9)$$

3. Put $v_R = i_R R$ for all resistor voltage(s). Then, express all variables in terms of the state variable(s) and the current source(s).

$$v_4 = R_4 i_4 = R_4(i_9 - i_6)$$

After steps 1, 2 and 3 are done, the original KVL equation becomes

$$L_9 \frac{di_9}{dt} = u_1 - v_2 + v_7 - R_4(i_9 - i_6) - L_5 \frac{d}{dt}(i_6 - i_9)$$

$$\Rightarrow \quad \frac{di_9}{dt} = \frac{1}{L_5 + L_9}\left(-v_2 + v_7 - R_4 i_9 + u_1 + R_4 i_6 + L_5 \frac{di_6}{dt}\right) \tag{I9}$$

We have now obtained all three differential equations for the circuit of figure 11.3 (a). Indeed, equations (V2), (V7) and (I9) are the required state equations. The final step is to put them in the normal form, i.e., $\dot{\bar{x}} = A\bar{x} + B\bar{u}$.

$$\frac{d}{dt}\begin{bmatrix} v_2 \\ v_7 \\ i_9 \end{bmatrix} = \begin{bmatrix} 0 & 0 & \frac{1}{C_2 + C_3} \\ 0 & -\frac{1}{R_8 C_7} & -\frac{1}{C_7} \\ -\frac{1}{L_5 + L_9} & \frac{1}{L_5 + L_9} & -\frac{R_4}{L_5 + L_9} \end{bmatrix}\begin{bmatrix} v_2 \\ v_7 \\ i_9 \end{bmatrix}$$

$$+ \begin{bmatrix} \frac{C_3}{C_2 + C_3} & 0 \\ 0 & 0 \\ 0 & \frac{L_5}{L_5 + L_9} \end{bmatrix}\begin{bmatrix} \frac{du_1}{dt} \\ \frac{di_6}{dt} \end{bmatrix}$$

$$+ \begin{bmatrix} 0 & -\frac{1}{C_2 + C_3} \\ 0 & 0 \\ \frac{1}{L_5 + L_9} & \frac{R_4}{L_5 + L_9} \end{bmatrix}\begin{bmatrix} u_1 \\ i_6 \end{bmatrix}$$

Remarks — In the above example, we observe that in equations (V2), (V7) and (I9), the LHS involves the first derivative of only one state variable. Hence, the three equations directly give

$$\frac{d\bar{x}}{dt} = A\bar{x} + B\bar{u} + D\dot{\bar{x}}$$

Two points are worth noting.

1. The RHS has derivative of $\bar{u}$, which seems to differ from the normal form state equation. This should not, however, be regarded as a different form of equation since it can be reduced to the form $\dot{\bar{x}} = X\bar{x} + B'\bar{u}'$, by letting $B' = [B\ D]$ and $\bar{u}' = \begin{bmatrix} \bar{u} \\ \dot{\bar{u}} \end{bmatrix}$.

2. In some cases, the LHS may have first derivatives of more than one state variable, for example,

$$C_2 \frac{dv_2}{dt} + C_7 \frac{dv_7}{dt} = \cdots$$

This leads to a state equation of the form $P\dot{\bar{x}} = Q\bar{x} + R\bar{u}$, where P is not a unit matrix or multiple of a unit matrix. In such cases, we can obtain the standard normal form equation by left-multiplying both sides by P^{-1}, i.e.,

$$\dot{\bar{x}} = P^{-1}Q\bar{x} + P^{-1}R\bar{u} = A\bar{x} + B\bar{u}$$

11.5 Derivation of Transfer Functions

It is possible to obtain some particular transfer functions in the frequency domain directly from the normal form state equation. Referring to the above example, an algebraic matrix equation in the complex frequency domain can be obtained by replacing d/dt by s (based on the same reasoning as explained in Chapter 8), $\bar{x}(t)$ by $\bar{X}(s)$, and $\bar{u}(t)$ by $\bar{U}(s)$.

$$\begin{bmatrix} sV_2(s) \\ sV_7(s) \\ sI_9(s) \end{bmatrix} = A \begin{bmatrix} V_2(s) \\ V_7(s) \\ I_9(s) \end{bmatrix} + B \begin{bmatrix} sU_1(s) \\ sI_6(s) \end{bmatrix} + D \begin{bmatrix} U_1(s) \\ I_6(s) \end{bmatrix}$$

Hence, we have

$$\begin{bmatrix} V_2(s) \\ V_7(s) \\ I_9(s) \end{bmatrix} = [s\mathbf{1} - A]^{-1} \left\{ B \begin{bmatrix} sU_1(s) \\ sI_6(s) \end{bmatrix} + D \begin{bmatrix} U_1(s) \\ I_6(s) \end{bmatrix} \right\}$$

Example 11.4: Derivation of transfer function — Assuming that in the circuit of the previous example, all capacitors are 1F, all inductors are 1H and all resistances are 1Ω, we have

$$\begin{bmatrix} V_2(s) \\ V_7(s) \\ I_9(s) \end{bmatrix} = \frac{\begin{bmatrix} 1 + 1.5s + s^2 & 0.25 & 0.5(1+s^2) \\ 0.5 & 0.5 + 0.5s + s^2 & -s \\ -0.5(1+s) & 0.5s & s(1+s) \end{bmatrix}}{0.25 + 1.25s + 1.5s^2 + s^3}$$
$$\begin{bmatrix} 0.5s & -0.5 \\ 0 & 0 \\ 0.5 & 0.5(1+s) \end{bmatrix} \begin{bmatrix} U_1(s) \\ I_6(s) \end{bmatrix}$$

from which we may obtain the transfer functions V_2/U_1, V_2/I_6, V_7/U_1, V_7/I_6, I_9/U_1 and I_9/I_6. For example,

$$\frac{V_2}{U_1} = \frac{1 + 2s + 4s^2 + 2s^3}{1 + 5s + 6s^2 + 4s^3}$$
$$\frac{V_7}{U_1} = \frac{-s}{1 + 5s + 6s^2 + 4s^3}$$
$$\frac{I_9}{U_1} = \frac{2s(1+s)}{1 + 5s + 6s^2 + 4s^3} \quad \text{etc.}$$

It is then possible to obtain the frequency response for any of the transfer functions using the Bode technique or otherwise.

Remarks — In the foregoing discussion of transfer functions, no knowledge is required about the types or forms of the driving sources. Neither is the initial condition relevant for deriving the transfer functions. However, as mentioned in the previous chapter, the frequency response subtly tells us about the type of transient or natural response. This amazing linkage is provided by the characteristic equation. Essentially, we observe that all the transfer functions above share the same denominator $1 + 5s + 6s^2 + 4s^3$, and hence the same characteristic equation. This also shows that the natural response is everywhere the same for the same given circuit.

11.6 Characteristic Equation

In the derivation of the transfer functions outlined in the previous section, we see that the common denominator $D(s)$ comes into play when we take the inverse of $[s\mathbf{1} - A]$, and $D(s)$ is exactly the determinant $|s\mathbf{1} - A|$. Thus, the characteristic equation is given by

$$|s\mathbf{1} - A| = 0$$

Example 11.5: Derivation of characteristic equation — Once we have derived the state equation, we get the characteristic equation right away. For the circuit studied in the previous examples, for example, all we need to do to obtain the characteristic equation is to evaluate the determinant $|s\mathbf{1} - A|$. The result is

$$\begin{vmatrix} s & 0 & -0.5 \\ 0 & s+1 & 1 \\ 0.5 & -0.5 & s+0.5 \end{vmatrix} = 0$$

$$\Rightarrow \quad 1 + 5s + 6s^2 + 4s^3 = 0$$

11.7 Solution by Laplace Transform

If the driving functions and the initial values of $\bar{x}$ are given, then the complete time-domain solution can be obtained. To this end, any method that can solve the state equation can be invoked. In the following we will briefly introduce the Laplace Transform method, which is very popular in electrical engineering. However, we will leave the rigorous pursuit of the mathematics to an engineering mathematics course. A discussion is found in Appendix C of this book.

The method involves transforming the state equation to an algebraic matrix equation via the Laplace Transformation, solving the algebraic equation, then inverse-transforming the answer back to the time domain. The rules for transforming the state equation to the Laplace algebraic form are as follows:

1. For the LHS, we transform $\dot{\bar{x}}$ to $(s\bar{X}(s) - \bar{x}(0^-))$. The same is applied to any first derivative of time, i.e., $\dot{f}(t)$ is transformed to $(sF(s) - f(0^-))$.

2. For the RHS, we simply write $\bar{x}$ as $\bar{X}(s)$ which is the unknown to be found. The transformation of $\bar{u}$ depends on what $\bar{u}$ is. Usually we transform each source function individually to obtain $\bar{U}(s)$. The Laplace Transform of some common functions are listed in Table 11.1.

Now, all state variables and the sources are functions of s, and we have obtained an algebraic matrix equation of the form

$$s\bar{X}(s) = A\bar{X}(s) + B\bar{U}(s) + \bar{x}(0^-)$$

Note that the presence of initial condition $\bar{x}(0^-)$ is crucial to the determination of the complete transient response. Solving the above Laplace equation for $\bar{X}(s)$ yields

$$\bar{X}(s) = [s\mathbf{1} - A]^{-1}[B\bar{U}(s) + \bar{x}(0^-)]$$

which will give all the state variables as functions of s as

$$\bar{X}(s) = \begin{bmatrix} X_1(s) \\ X_2(s) \\ \vdots \\ X_n(s) \end{bmatrix}$$

The final step is to perform an inverse Laplace Transform on every state variable to take it back to the time domain.

$$\bar{x}(t) = \mathcal{L}^{-1}\left\{\bar{X}(s)\right\}$$

This inverse operation may be performed with the help of a Laplace Transform table. Table 11.1 contains some samples.

Example 11.6: Laplace Transform method for solving state equations — Suppose a circuit is described by the following state equation, and the driving sources are a unit step $u_s(t)$ and a unit impulse $\delta(t)$.[1]

$$\dot{\bar{x}} = \begin{bmatrix} -1 & 0 \\ -1 & -2 \end{bmatrix} \bar{x} + \begin{bmatrix} 1 & 0 \\ 0 & 1 \end{bmatrix} \begin{bmatrix} u_s(t) \\ \delta(t) \end{bmatrix}$$

Assume that initially $x_1(0^-) = 10$ and $x_2(0^-) = 0$. The Laplace Transform equation is thus given by

$$s\bar{X}(s) - \begin{bmatrix} 10 \\ 0 \end{bmatrix} = \begin{bmatrix} -1 & 0 \\ -1 & -2 \end{bmatrix} \bar{X}(s) + \begin{bmatrix} 1 & 0 \\ 0 & 1 \end{bmatrix} \begin{bmatrix} \frac{1}{s} \\ 1 \end{bmatrix}$$

---

[1] The unit step function $u_s(t)$ takes the value zero for $t < 0$ and 1 for $t > 0$. The unit impulse function $\delta(t)$ has value zero everywhere except at $t = 0$ where it is undefined. The area under the unit impulse from $-\infty$ to $+\infty$ is one, i.e., $\int_{-\infty}^{+\infty} \delta(t)dt = 1$. The essential material relating to the step and impulse functions is usually covered in an engineering math course.

$f(t)$	$F(s)$
$\delta(t)$	1
$u_s(t)$	$\frac{1}{s}$
$t\,u_s(t)$	$\frac{1}{s^2}$
$t^n u_s(t)$	$\frac{n!}{s^{n+1}}$
$e^{-at} u_s(t)$	$\frac{1}{s+a}$
$te^{-at} u_s(t)$	$\frac{1}{(s+a)^2}$
$t^n e^{-at} u_s(t)$	$\frac{n!}{(s+a)^{n+1}}$
$\sin \omega t\; u_s(t)$	$\frac{\omega}{s^2+\omega^2}$
$\cos \omega t\; u_s(t)$	$\frac{s}{s^2+\omega^2}$
$e^{-at} \sin \omega t\; u_s(t)$	$\frac{\omega}{(s+a)^2+\omega^2}$
$e^{-at} \cos \omega t\; u_s(t)$	$\frac{s+a}{(s+a)^2+\omega^2}$
$kf(t)u_s(t)$	$kF(s)$
$e^{-at} f(t)u_s(t)$	$F(s+a)$

Table 11.1: Laplace Transform pairs

Rearranging this equation gives

$$
\begin{aligned}
\bar{X}(s) &= \begin{bmatrix} s+1 & 0 \\ 1 & s+2 \end{bmatrix}^{-1} \left\{ \begin{bmatrix} 1 & 0 \\ 0 & 1 \end{bmatrix} \begin{bmatrix} \frac{1}{s} \\ 1 \end{bmatrix} + \begin{bmatrix} 10 \\ 0 \end{bmatrix} \right\} \\
&= \frac{1}{(s+1)(s+2)} \begin{bmatrix} s+2 & -1 \\ 0 & s+1 \end{bmatrix} \begin{bmatrix} \frac{1}{s}+10 \\ 1 \end{bmatrix} \\
&= \begin{bmatrix} \dfrac{10s+1}{s(s+1)} - \dfrac{1}{(s+1)(s+2)} \\[3mm] \dfrac{1}{s+2} \end{bmatrix}
\end{aligned}
$$

In order to facilitate the inverse transformation, we take the partial fraction expansion of each element in $\bar{X}(s)$:

$$
\bar{X}(s) = \begin{bmatrix} \dfrac{1}{s} + \dfrac{10}{s+1} - \dfrac{1}{s+2} \\[3mm] \dfrac{1}{s+2} \end{bmatrix}
$$

With the help of Table 11.1, we obtain the time-domain solution of $\bar{x}(t)$ as

$$
\bar{x}(t) = \begin{bmatrix} x_1(t) \\ x_2(t) \end{bmatrix} = \begin{bmatrix} \left(1 + 10e^{-t} - e^{-2t} \right) u_s(t) \\ e^{-2t} u_s(t) \end{bmatrix}
$$

Remarks — The above Laplace Transform approach gives the complete solution, i.e., both transient and steady-state responses. In the above example, we see that in the steady state both x_1 and x_2 stay at 1. An interesting observation is made here about the jump of $x_2(t)$ at $t = 0$, since $x_2(0^-) = 0$ and $x_2(0^+) = 1$. Physically, we may regard such an effect as being resulted from the $\delta(t)$ driving function. Suppose x_2 is a capacitor voltage. The state equation says that its derivative partly equals an impulse at $t = 0$. This is like injecting a fixed amount of charge to the capacitor momentarily at $t = 0$. Hence, its voltage jumps up at $t = 0$ by the amount determined by the value of its capacitance. On the other hand, from the state equation we see that $\delta(t)$ does not affect x_1. Hence, we do not observe any discontinuity in its value. This example clearly demonstrates that the Laplace Transform method is a powerful approach which consistently accounts for all possible situations.

11.8 Problems

1. Calculate the size of the state vector and hence the order of the circuit that contains 8 nodes, 2 independent voltage sources, 2 capacitors, 1 resistors and 4 inductors.

2. A circuit exhibits a typical second-order response in all experiments of the circuit. An inspection of the circuit reveals that there are 3 capacitors and 2 inductors in the circuit. What can you say about the circuit topology?

3. State the number of independent storage elements in the circuit of figure 11.4, and choose a state vector for the circuit. Is this state vector unique for this circuit? What is the order of the circuit?

4. Write down the Q- and B-matrices for the circuit of figure 11.4, based on your choice of standard tree in Problem 3.

5. For the circuit of figure 11.5, choose a standard tree, write down the Q-matrix and the B-matrix, derive the normal form state equation, and obtain the characteristic equation. Deduce the type of transient or natural response.

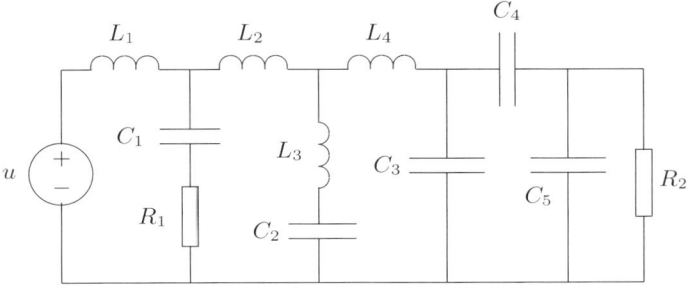

Figure 11.4: Circuit for Problems 3 and 4

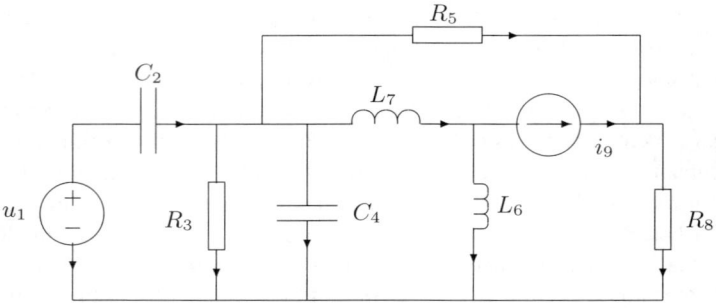

Figure 11.5: Circuit for Problems 5 to 7

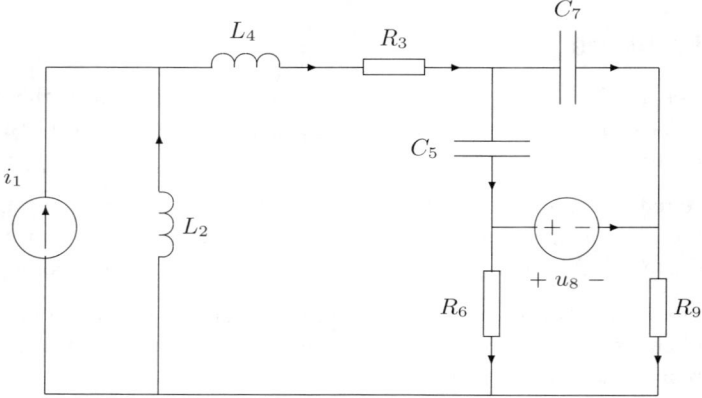

Figure 11.6: Circuit for Problem 8

6. For the circuit of figure 11.5, obtain the transfer functions V_2/U_1 and I_7/U_1 in the s-domain. Then, find the poles and zeros for these two transfer functions, given that all capacitors are 0.5F, all inductors are 0.5H, all resistors are 1Ω.

7. For the circuit figure 11.5, assuming zero initial condition, find $v_2(t)$ and $i_7(t)$ given that all capacitors are 0.5F, all inductors are 0.5H, all resistors are 1Ω, $u_1 = 10u_s(t)$, and $i_9 = 5u_s(t)$, where $u_s(t)$ is the unit step function.

8. For the circuit of figure 11.6, derive the state equation in the form $\dot{x} = Ax + Bu + C\dot{u}$. Choose a standard tree that gives the state vector as $[v_5\ i_4]^T$.

9. Consider the following state equation which describes a dynamic system.

$$\frac{d\bar{x}}{dt} = \begin{bmatrix} 0 & -1 \\ 1 & -3 \end{bmatrix} \bar{x} + \begin{bmatrix} 1 \\ 0 \end{bmatrix} f(t)$$

What is the natural response of the system? What is the characteristic equation? Derive the transfer functions $X_1(s)/F(s)$ and $X_2(s)/F(s)$. Now, given that

$\bar{x}(0^-) = [10\ 5]^T$ and $f(t) = \sin 5t$, solve the equation completely. Give your answer in the time domain.

10. A dynamic circuit is described by the following state equation:

$$\frac{d}{dt}\begin{bmatrix} v_1 \\ i_2 \end{bmatrix} = \begin{bmatrix} 1 & -1 \\ 4 & -3 \end{bmatrix}\begin{bmatrix} v_1 \\ i_2 \end{bmatrix} + \begin{bmatrix} \delta(t) \\ u(t) \end{bmatrix}$$

Find the characteristic equation. Solve the equation with initial condition given by $[1\ 0]^T$. (The Laplace transform method is usually treated in depth in an engineering mathematics text. Appendix C summarizes the essential procedure.)

Chapter 12

Passive Two-Port Networks

In the previous chapters, we have studied methods for solving general circuit problems. Some problems, however, possess certain properties that may allow the solution to be obtained more rapidly with somewhat less general methods. In this chapter we consider some special methods for solving circuits that are constructed of interconnected sub-circuits with a relatively small number of connections. The kinds of sub-circuits to be considered have a pair of terminals to which current and voltage can be applied, as shown in figure 12.1. Such sub-circuits are called *two-port* elements. Figure 12.2 shows three two-port elements $\mathcal{N}_1, \mathcal{N}_2$ and $\mathcal{N}_3$ interconnected together. We may regard a two-port element as an extended one-port element which we have been dealing with previously.

12.1 Characterization of Passive Two-Port Elements

A passive one-port element has a pair of terminals to which current or voltage can be applied. Since the one-port is associated with only one current and one voltage, a single relationship is needed to describe it. Essentially, the current and voltage are related by the constitutive relation of the one-port element. For instance, a resistor is a one-port component defined by $V = IR$. Assuming that linearity holds, any one-port can be represented by either an impedance or an admittance.

12.1.1 Immittance Matrices [Z] and [Y]

A passive two-port element has two pairs of terminals. Characterizing it would certainly require a more complicated notation. For brevity we denote the column vector of the port voltages by $\bar{v}$, and that of the port currents by $\bar{i}$. The voltage polarity and current direction are as defined in figure 12.1. We will refer to the left-side port as

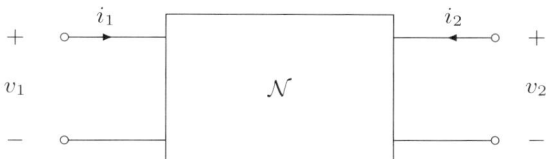

Figure 12.1: Passive two-port

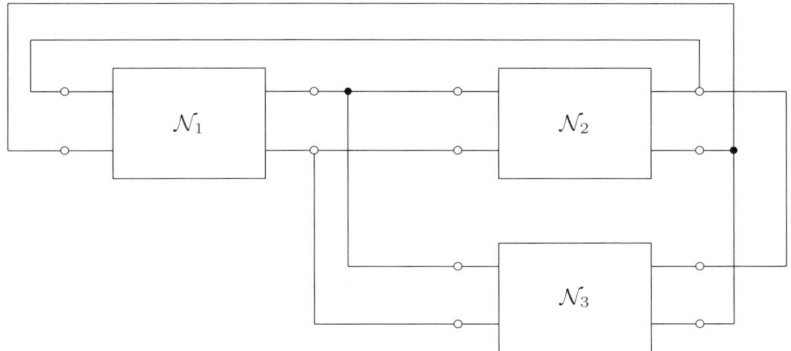

Figure 12.2: Circuits viewed as interconnection of two-port elements

port-1, and the right-side port as port-2.

$$\bar{v} = \left[\begin{array}{c} v_1 \\ v_2 \end{array} \right] \quad \text{and} \quad \bar{i} = \left[\begin{array}{c} i_1 \\ i_2 \end{array} \right]$$

Two obvious ways of characterizing a two-port, as inspired by the one-port case, are to define the two-port by four impedances or admittances:

$$\left[\begin{array}{c} v_1 \\ v_2 \end{array} \right] = \left[\begin{array}{cc} z_{11} & z_{12} \\ z_{21} & z_{22} \end{array} \right] \left[\begin{array}{c} i_1 \\ i_2 \end{array} \right] \qquad \text{(Z)}$$

$$\left[\begin{array}{c} i_1 \\ i_2 \end{array} \right] = \left[\begin{array}{cc} y_{11} & y_{12} \\ y_{21} & y_{22} \end{array} \right] \left[\begin{array}{c} v_1 \\ v_2 \end{array} \right] \qquad \text{(Y)}$$

or we may simply write

$$\bar{v} = [Z]\bar{i} \quad \text{and} \quad \bar{i} = [Y]\bar{v}$$

where matrix $[Z]$ and matrix $[Y]$ are called the *impedance matrix* and the *admittance matrix* respectively. The elements z_{11}, z_{12}, z_{21} and z_{22} are called *z-parameters,* and the elements y_{11}, y_{12}, y_{21} and y_{22} are called *y-parameters.*

Now suppose port-2 is terminated with an impedance z_L, as shown in figure 12.3. Then, the current and voltage of port-2 are constrained by

$$v_2 = -z_L i_2$$

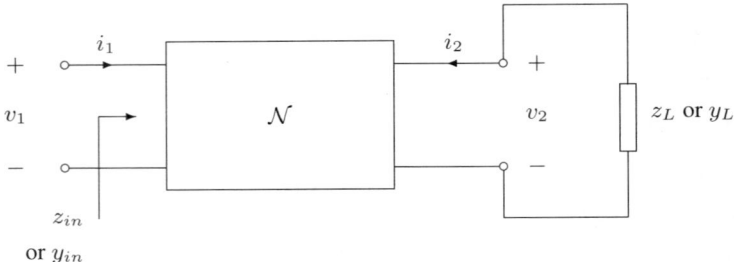

Figure 12.3: Two-port with load impedance/admittance

Putting this constraint into the port-2 voltage expression gives

$$-z_L i_2 = z_{21} i_1 + z_{22} i_2 \quad \Rightarrow \quad \frac{i_2}{i_1} = \frac{-z_{21}}{z_{22} + z_L}$$

Also, the voltage of port-1 is given by

$$v_1 = z_{11} i_1 + z_{12} i_2 \quad \Rightarrow \quad \frac{v_1}{i_1} = z_{11} + z_{12} \frac{i_2}{i_1}$$

Hence, the input impedance observed from port-1 is given by

$$z_{in} = \frac{v_1}{i_1}$$
$$= z_{11} - \frac{z_{12} z_{21}}{z_L + z_{22}}$$

Similarly, we can show that if the terminating admittance at port-2 is y_L, the input admittance observed from port-1 is given by

$$y_{in} = y_{11} - \frac{y_{12} y_{21}}{y_L + y_{22}}$$

Remarks — Existence of the z-parameters and y-parameters is generally not guaranteed. In other words, there are two-ports for which z-parameters do not exist, and there are two-ports for which y-parameters do not exist. There are also two-ports for which both z-parameters and y-parameters do not exist. Some examples are shown in figure 12.4. When they exist, we can prove that $z_{12} = z_{21}$ and $y_{12} = y_{21}$, i.e., the $[Z]$ and $[Y]$ matrices are symmetric.

12.1.2 Hybrid Matrices [*H*] and [*G*]

We may also characterize a two-port by expressing v_1 and i_2 in terms of v_2 and i_1. The parameters used to establish the necessary relations are called *h-parameters,* namely h_{11}, h_{12}, h_{21} and h_{22}.

$$\begin{bmatrix} v_1 \\ i_2 \end{bmatrix} = \begin{bmatrix} h_{11} & h_{12} \\ h_{21} & h_{22} \end{bmatrix} \begin{bmatrix} i_1 \\ v_2 \end{bmatrix} \quad \text{or} \quad \begin{bmatrix} v_1 \\ i_2 \end{bmatrix} = [H] \begin{bmatrix} i_1 \\ v_2 \end{bmatrix} \tag{H}$$

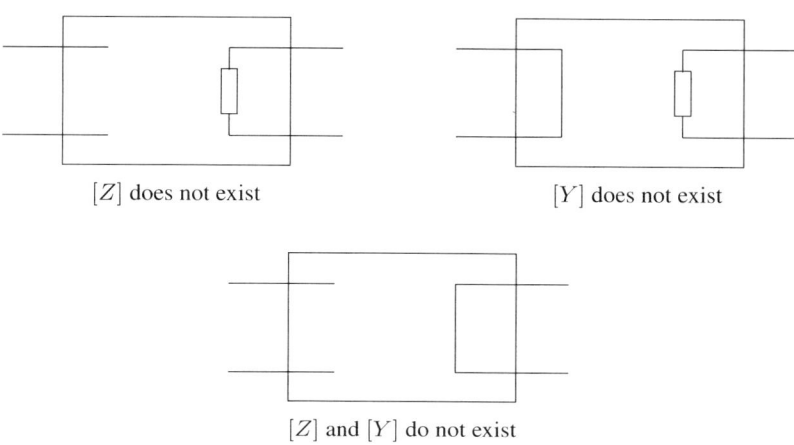

Figure 12.4: Examples of two-ports having no [Z] or [Y]

If we interchange rows and columns and suffices 1 and 2 in the above equations, we can obtain another set of equations which equally characterizes the two-port.

$$\begin{bmatrix} i_1 \\ v_2 \end{bmatrix} = \begin{bmatrix} g_{11} & g_{12} \\ g_{21} & g_{22} \end{bmatrix} \begin{bmatrix} v_1 \\ i_2 \end{bmatrix} \quad \text{or} \quad \begin{bmatrix} i_1 \\ v_2 \end{bmatrix} = [G] \begin{bmatrix} v_1 \\ i_2 \end{bmatrix} \tag{G}$$

where the g_{ij}s are called *g-parameters,* and $[G]$ is called the G-matrix.

Remarks — Unlike the case of immittance matrices, existence of a hybrid matrix (either $[H]$ or $[G]$ or both) is always guaranteed. Furthermore, we can prove that $h_{12} = -h_{21}$ and $g_{12} = -g_{21}$ for passive two-ports.

12.2 Determination of Two-Port Parameters

12.2.1 *z*-Parameters

Referring to equation (Z), we can determine z_{11} by measuring the input impedance at port-1 with i_2 set to zero. Likewise, we can find all parameters as

$$z_{11} = \left. \frac{v_1}{i_1} \right|_{i_2=0} \quad \text{(open-circuit port-2)}$$

$$z_{12} = \left. \frac{v_1}{i_2} \right|_{i_1=0} \quad \text{(open-circuit port-1)}$$

$$z_{21} = \left. \frac{v_2}{i_1} \right|_{i_2=0} \quad \text{(open-circuit port-2)}$$

$$z_{22} = \left. \frac{v_2}{i_2} \right|_{i_1=0} \quad \text{(open-circuit port-1)}$$

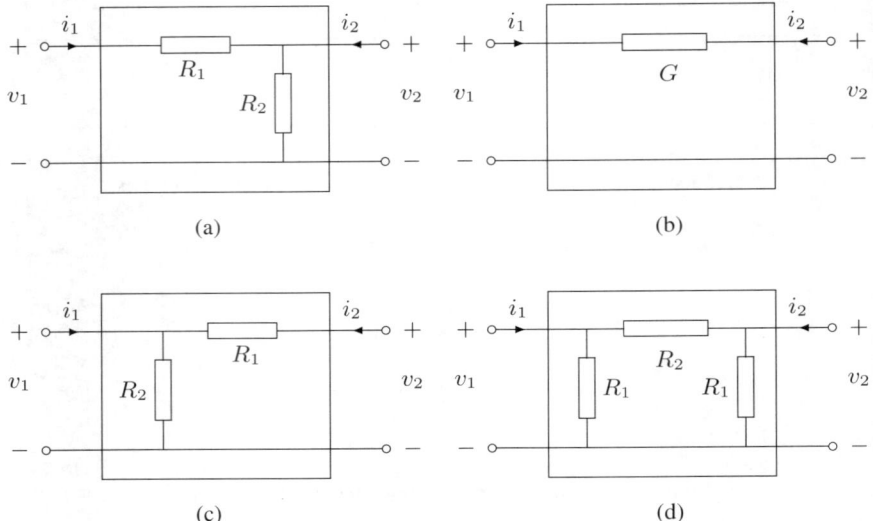

Figure 12.5: Examples of passive two-ports

Example 12.1: Determination of z-parameters — Consider the ladder circuit shown in figure 12.5 (a). We can represent this two-port using z-parameters. Open-circuiting port-2, we have $v_1 = (R_1 + R_2)i_1$. Thus, we obtain z_{11} as

$$z_{11} = \left.\frac{v_1}{i_1}\right|_{i_2=0} = R_1 + R_2$$

With port-2 open-circuited, v_2 is simply equal to $i_1 R_2$. Hence, we have

$$z_{21} = \left.\frac{v_2}{i_1}\right|_{i_2=0} = R_2$$

Now if we open-circuit port-1, no current flows in R_1, and hence $v_1 = v_2$. Both z_{12} and z_{22} are thus given by

$$z_{12} = z_{22} = \left.\frac{v_2}{i_2}\right|_{i_1=0} = R_2$$

We may write the $[Z]$ matrix as

$$[Z] = \left[\begin{array}{cc} R_1 + R_2 & R_2 \\ R_2 & R_2 \end{array} \right]$$

12.2.2 *y*-Parameters

In a similar fashion, we can find the *y*-parameters by suitably short-circuiting port-1 or port-2 and calculating the appropriate admittances (see equation (Y)).

$$y_{11} = \left.\frac{i_1}{v_1}\right|_{v_2=0} \quad \text{(short-circuit port-2)}$$

$$y_{12} = \left.\frac{i_1}{v_2}\right|_{v_1=0} \quad \text{(short-circuit port-1)}$$

$$y_{21} = \left.\frac{i_2}{v_1}\right|_{v_2=0} \quad \text{(short-circuit port-2)}$$

$$y_{22} = \left.\frac{i_2}{v_2}\right|_{v_1=0} \quad \text{(short-circuit port-1)}$$

Example 12.2: Determination of y-parameters — Consider the two-port shown in figure 12.5 (b). Short-circuiting port-2 gives $i_1 = -i_2$, and hence

$$y_{11} = G \quad \text{and} \quad y_{12} = -G$$

Short-circuiting port-1, similarly, gives

$$y_{21} = -G \quad \text{and} \quad y_{22} = G$$

The $[Y]$ matrix for the two-port of figure 12.5 (b) is

$$[Y] = \begin{bmatrix} G & -G \\ -G & G \end{bmatrix}$$

Remarks — If we attempt to find $[Z]$ for the two-port of figure 12.5 (b), we will end up with some undefined parameters. In this case, we say that $[Z]$ does not exist.

12.2.3 *h*-Parameters

From equation (H), we can write the following formulae for determining the *h*-parameters:

$$h_{11} = \left.\frac{v_1}{i_1}\right|_{v_2=0} \quad \text{(short-circuit port-2)}$$

$$h_{12} = \left.\frac{v_1}{v_2}\right|_{i_1=0} \quad \text{(open-circuit port-1)}$$

$$h_{21} = \left.\frac{i_2}{i_1}\right|_{v_2=0} \quad \text{(short-circuit port-2)}$$

$$h_{22} = \left.\frac{i_2}{v_2}\right|_{i_1=0} \quad \text{(open-circuit port-1)}$$

Example 12.3: Determination of h-parameters — We will find the h-parameters for the two-port of figure 12.5 (c). Firstly, we short-circuit port-2 and find h_{11} as

$$h_{11} = R_1 \| R_2 = \frac{R_1 R_2}{R_1 + R_2}$$

When port-2 is short-circuited, i_2 equals the current in R_1 and is given by the current divider formula $-i_2 = i_1 \frac{R_2}{R_1 + R_2}$. Thus, we have

$$h_{21} = \frac{i_2}{i_1}\Big|_{v_2=0} = \frac{-R_2}{R_1 + R_2}$$

Now open-circuit port-1. The ratio v_1/v_2 can be obtained by a voltage divider as

$$h_{12} = \frac{v_1}{v_2}\Big|_{i_1=0} = \frac{R_2}{R_1 + R_2}$$

Finally, with port-1 open-circuited, i_2 flows into R_1 and R_2.

$$h_{22} = \frac{i_2}{v_2}\Big|_{i_1=0} = \frac{1}{R_1 + R_2}$$

12.2.4 *g*-Parameters

From equation (G), we can write the following formulae for determining the g-parameters:

$$g_{11} = \frac{i_1}{v_1}\Big|_{i_2=0} \qquad \text{(open-circuit port-2)}$$

$$g_{12} = \frac{i_1}{i_2}\Big|_{v_1=0} \qquad \text{(short-circuit port-1)}$$

$$g_{21} = \frac{v_2}{v_1}\Big|_{i_2=0} \qquad \text{(open-circuit port-2)}$$

$$g_{22} = \frac{v_2}{i_2}\Big|_{v_1=0} \qquad \text{(short-circuit port-1)}$$

Example 12.4: Determination of g-parameters — Using the above formulae, we obtain the g-parameters for the circuit of figure 12.5 (d) by suitably open-circuiting port-2 or short-circuiting port-1. With port-2 open-circuited, the circuit observed from port-1 is equivalent to R_1 in parallel with $R_1 + R_2$. With port-1 short-circuited, the circuit observed from port-2 is equivalent to R_1 in parallel with R_2. By observing the input admittance and output impedance, and using the appropriate voltage divider or current divider formulae, we obtain the g-parameters as

$$g_{11} = \frac{1}{R_1 \| (R_1 + R_2)}, \quad g_{12} = \frac{-R_1}{R_1 + R_2}, \quad g_{21} = \frac{R_1}{R_1 + R_2}, \quad g_{22} = R_1 \| R_2$$

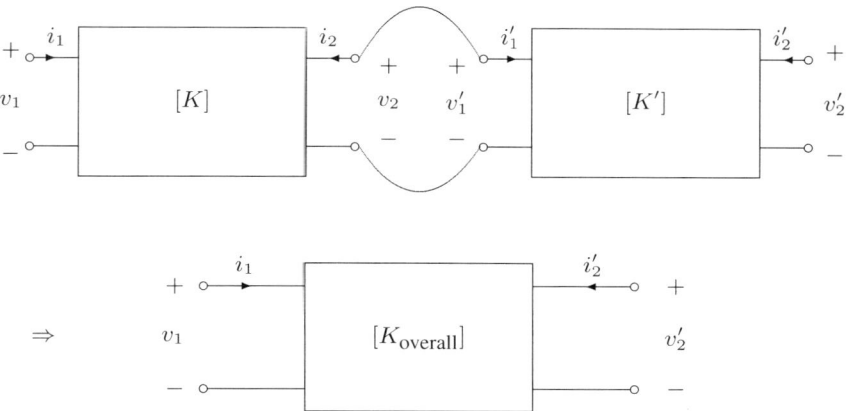

Figure 12.6: Chain matrix for cascaded networks

12.3 Chain Matrix

Another important matrix is the chain matrix. It relates the voltage and current of port-1 with those of port-2. With the same voltage polarity and current direction as defined in figure 12.1, we define the the the chain matrix as

$$
\begin{bmatrix} v_1 \\ i_1 \end{bmatrix} = \begin{bmatrix} k_{11} & k_{12} \\ k_{21} & k_{22} \end{bmatrix} \begin{bmatrix} v_2 \\ -i_2 \end{bmatrix} \quad \text{or} \quad \begin{bmatrix} v_1 \\ i_1 \end{bmatrix} = [K] \begin{bmatrix} v_2 \\ -i_2 \end{bmatrix} \tag{K}
$$

Note the negative sign prefixed to i_2. This negative sign is deliberately assigned here to facilitate the use of chain matrix in describing two-port networks connected in cascade.

Figure 12.6 shows two two-ports which are described by chain matrices $[K]$ and $[K']$. The equations relating the port voltages and currents are

$$
\begin{bmatrix} v_1 \\ i_1 \end{bmatrix} = [K] \begin{bmatrix} v_2 \\ -i_2 \end{bmatrix} \quad \text{and} \quad \begin{bmatrix} v_1' \\ i_1' \end{bmatrix} = [K'] \begin{bmatrix} v_2' \\ -i_2' \end{bmatrix}
$$

When the two two-ports are cascaded together with port-2 of the left network directly connected to port-1 of the right network, the following relations hold:

$$
v_2 = v_1' \quad \text{and} \quad i_2 = -i_1'
$$

Hence, we can derive the overall chain matrix of the cascaded network as

$$
\begin{bmatrix} v_1 \\ i_- \end{bmatrix} = [K] \begin{bmatrix} v_2 \\ -i_2 \end{bmatrix}
$$

$$
= [K] \begin{bmatrix} v_1' \\ i_1' \end{bmatrix}
$$

$$
= [K][K'] \begin{bmatrix} v_2' \\ -i_2' \end{bmatrix}
$$

The overall chain matrix is thus given by

$$[K_{\text{overall}}] = [K][K']$$

This result can be extended to the general case of a cascade connection of n two-port networks:

$$[K_{\text{overall}}] = [K_1][K_2] \cdots [K_n]$$

The determination of the chain matrix can be proceeded in a way similar to that of the other matrices. By open-circuiting port-2, we calculate k_{11} and k_{21} as

$$k_{11} = \left. \frac{v_1}{v_2} \right|_{i_2=0} \quad \text{(open-circuit port-2)}$$

$$k_{21} = \left. \frac{i_1}{v_2} \right|_{i_2=0} \quad \text{(open-circuit port-2)}$$

Likewise, short-circuiting port-2, we get

$$k_{12} = \left. -\frac{v_1}{i_2} \right|_{v_2=0} \quad \text{(short-circuit port-2)}$$

$$k_{22} = \left. -\frac{i_1}{i_2} \right|_{v_2=0} \quad \text{(short-circuit port-2)}$$

12.4   Connections of Two-ports

12.4.1   Series-series (Series) Connection

Figure 12.7 shows a series-series connection, or simply series connection, of two two-ports. Using z-parameters, the two two-ports are described as

$$\begin{bmatrix} v_1 \\ v_2 \end{bmatrix} = \begin{bmatrix} z_{11} & z_{12} \\ z_{21} & z_{22} \end{bmatrix} \begin{bmatrix} i_1 \\ i_2 \end{bmatrix}$$

$$\begin{bmatrix} v_1' \\ v_2' \end{bmatrix} = \begin{bmatrix} z_{11}' & z_{12}' \\ z_{21}' & z_{22}' \end{bmatrix} \begin{bmatrix} i_1' \\ i_2' \end{bmatrix}$$

For this series-series connection, we have

$$\begin{bmatrix} v_1^* \\ v_2^* \end{bmatrix} = \begin{bmatrix} v_1 \\ v_2 \end{bmatrix} + \begin{bmatrix} v_1' \\ v_2' \end{bmatrix}$$

and

$$\begin{bmatrix} i_1^* \\ i_2^* \end{bmatrix} = \begin{bmatrix} i_1 \\ i_2 \end{bmatrix} = \begin{bmatrix} i_1' \\ i_2' \end{bmatrix}$$

Hence, we have

$$\begin{bmatrix} v_1^* \\ v_2^* \end{bmatrix} = [[Z] + [Z']] \begin{bmatrix} i_1^* \\ i_2^* \end{bmatrix}$$

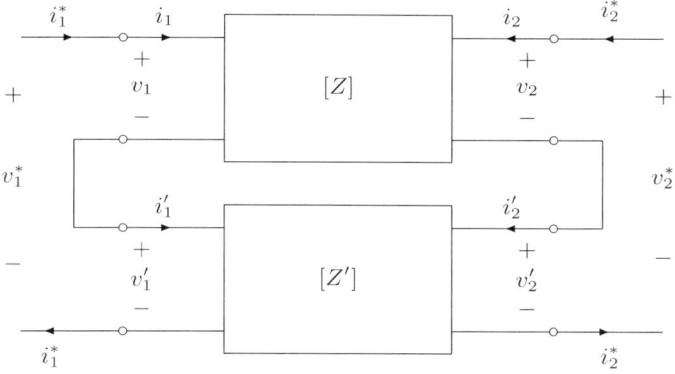

Figure 12.7: Series-series (series) connection

Thus, the overall impedance matrix is the sum of the two impedance matrices:

$$[Z^*] = [Z] + [Z']$$

It can be generalized by induction that the impedance matrix of the two-port that arises from a series connection of n two-ports is the sum of the n individual impedance matrices.

$$[Z^*] = [Z_1] + [Z_2] + \cdots + [Z_n]$$

Remarks — It must be borne in mind that the above result is valid only when $i_1 = i'_1$ and $i_2 = i'_2$. This in turns requires that the current going into the upper terminal of a port be equal to the current emerging from its lower terminal. Such a condition is called *port condition*. Although the port condition seems to be a trivial condition, it is not always satisfied in the case of interconnected two-port networks. Figure 12.8 shows two networks whose series-series connection results in violation of the port condition. Thus, in that case, $[Z^*] = [Z] + [Z']$ does not hold.

12.4.2 Shunt-shunt (Parallel) Connection

Figure 12.9 shows a shunt-shunt connection, or simply parallel connection, of two two-ports. In this case, we have $v_1 = v'_1$ and $v_2 = v'_2$. With the port condition satisfied, it can be shown that

$$[Y^*] = [Y] + [Y']$$

It can be generalized by induction that the admittance matrix of the two-port that arises from parallel connection of n two-ports is the sum of the n indivdual admittance matrices.

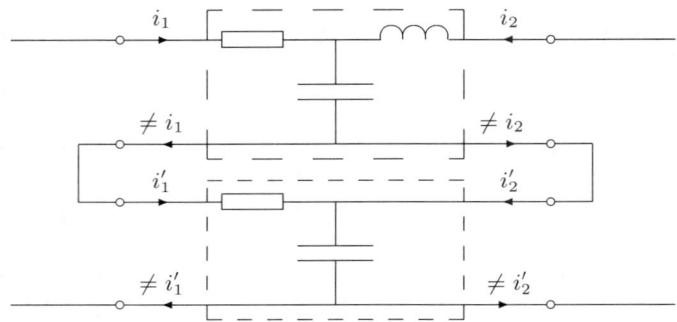

Figure 12.8: Violation of port condition

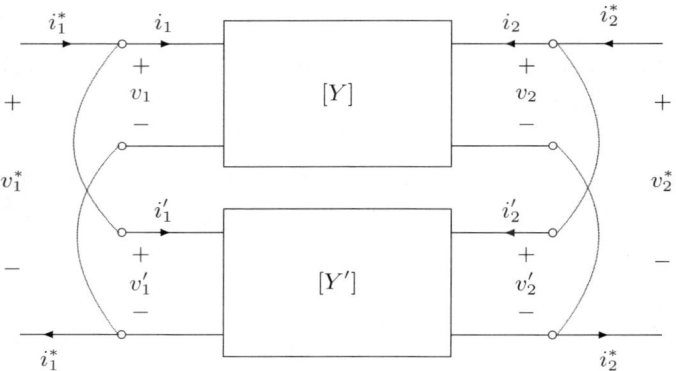

Figure 12.9: Shunt-shunt (parallel) connection

$$[Y^*] = [Y_1] + [Y_2] + \cdots + [Y_n]$$

12.4.3   Series-shunt Connection

Figure 12.10 shows a series-shunt connection of two two-ports. In this case, we have $i_1 = i_1'$ and $v_2 = v_2'$. Again, with the port condition satisfied, it can be shown that the overall H-matrix is the sum of the two individual H-matrices.

$$[H^*] = [H] + [H']$$

By induction, we have, for n two-ports connected in the series-shunt configuration,

$$[H^*] = [H_1] + [H_2] + \cdots + [H_n]$$

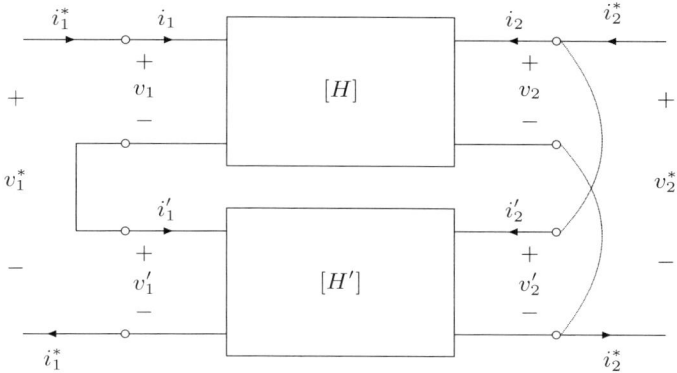

Figure 12.10: Series-shunt connection

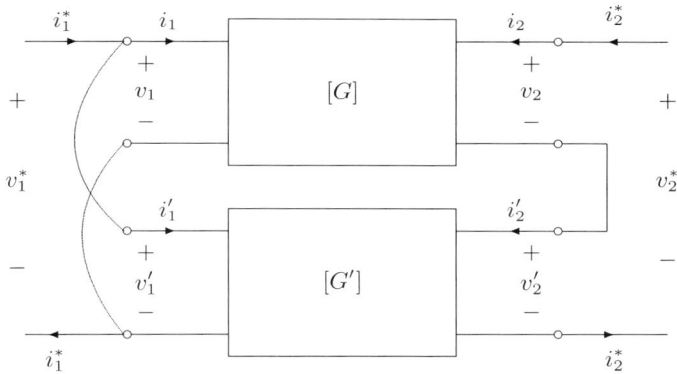

Figure 12.11: Shunt-series connection

12.4.4 Shunt-series Connection

Figure 12.11 shows a shunt-series connection of two two-ports. In this case, we have $v_1 = v_1'$ and $i_2 = i_2'$. Again, with the port condition satisfied, it can be shown that the overall G-matrix is the sum of the two individual G-matrices.

$$[G^*] = [G] + [G']$$

The general result for n two-ports connected in the shunt-series configuration is

$$[G^*] = [G_1] + [G_2] + \cdots + [G_n]$$

Example 12.5: Connection of two two-port networks — A passive two-port network has the following h-parameters:

$$h_{11} = 1\Omega, \quad h_{12} = 0.2, \quad h_{21} = -0.2, \quad \text{and} \quad h_{22} = 0.01\text{S}$$

This two-port is connected to the voltage-divider two-port shown in figure 12.5 (c), in a series-shunt configuration, and we wish to find the values of R_1 and R_2 in the voltage-divider network such that the overall network will have an h_{11} of 1.909Ω and an h_{22} of 0.1009S.

Clearly, since the h-parameters of the two networks add up to give the overall h-parameters, we can conclude that h_{11} and h_{22} of the voltage-divider network is $1.909 - 1 = 0.909\Omega$ and $0.1009 - 0.01 = 0.0909$S respectively. Also, from Example 12.3, we have

$$\frac{R_1 R_2}{R_1 + R_2} = 0.909 \quad \text{and} \quad \frac{1}{R_1 + R_2} = 0.0909$$

Solving these two equations gives $R_1 = 10\Omega$ and $R_2 = 1\Omega$, or $R_1 = 1\Omega$ and $R_2 = 10\Omega$. Note that interchanging R_1 and R_2 will not affect the overall h_{11} and h_{22}, but will alter h_{21} and h_{12}.

12.5   Circuit Models

Once the two-port parameters are found, a standard circuit model can be used to replace the original passive two-port network. The form of the circuit model depends on the type of parameters. Using z-parameters, the network is characterized by

$$v_1 = z_{11}i_1 + z_{12}i_2$$
$$v_2 = z_{12}i_1 + z_{22}i_2$$

From the first equation, we can model port-1 as a series combination of z_{11} and a current-controlled voltage source $z_{12}i_2$. Similarly, we can model port-2 as a series combination of z_{22} and a current-controlled voltage source $z_{21}i_1$. The z-parameter model is shown in figure 12.12 (a).

Using y-parameters, the network is characterized by

$$i_1 = y_{11}v_1 + y_{12}v_2$$
$$i_2 = y_{12}v_1 + y_{22}v_2$$

In this case, we can model port-1 as a parallel combination of y_{11} and a voltage-controlled current source $y_{12}v_2$. Similarly, we can model port-2 as a parallel combination of y_{22} and a voltage-controlled current source $y_{21}v_1$. The y-parameter model is shown in figure 12.12 (b).

In a likewise fashion, we can obtain standard models using h-parameters and g-parameters. These models are shown in figures 12.12 (c) and (d).

Example 12.6: Equivalent circuit model of two-port networks — The passive two-port circuit shown in figure 12.13 (a) is connected with other two-port networks in a series-shunt configuration. We therefore choose to represent it with h-parameters.

$$[H] = \begin{bmatrix} \frac{10}{7}\Omega & -\frac{2}{7} \\ \frac{2}{7} & \frac{1}{7}S \end{bmatrix}$$

The equivalent circuit for this two-port network is shown in figure 12.13 (b).

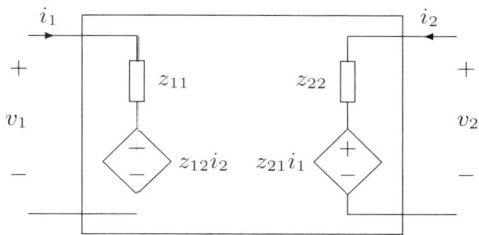

(a) z-parameter model

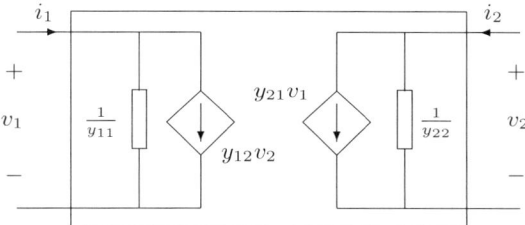

(b) y-parameter model

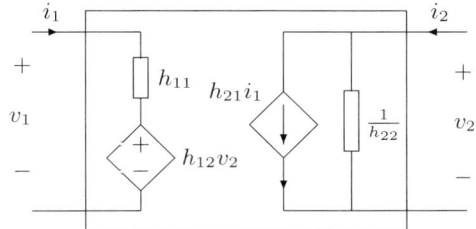

(c) h-parameter model

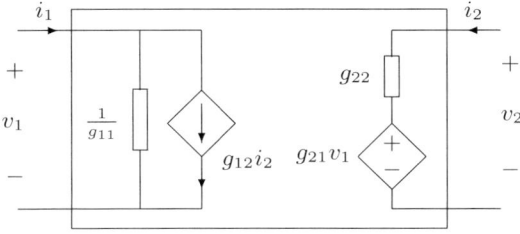

(d) g-parameter model

Figure 12.12: Circuit models

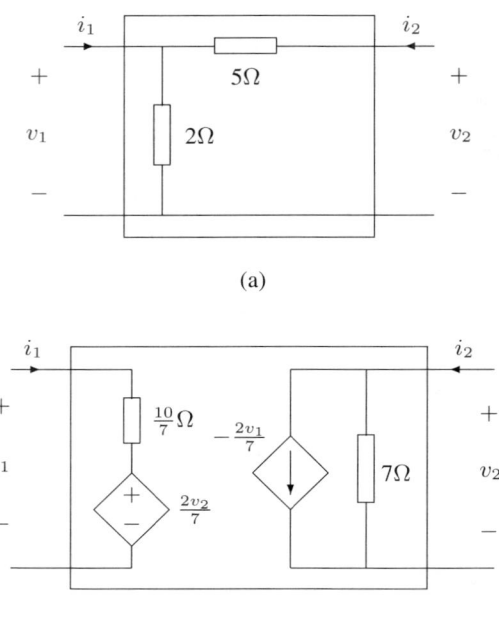

Figure 12.13: (a) Two-port circuit and (b) its equivalent model

12.6 Problems

1. Find the Y-matrix and Z-matrix for the two-port networks shown in figure 12.14. If, in each case, port-2 is terminated by an impedance z_L, determine the input impedance as seen from port-1.

2. Find the G-matrix and H-matrix for the two-port networks shown in figure 12.14.

3. Consider the symmetric Π network of figure 12.15. Show that the chain matrix is given by

$$[K] = \left[\begin{array}{cc} 1 + \frac{ZY}{2} & Z \\ Y + \frac{ZY^2}{4} & 1 + \frac{ZY}{2} \end{array} \right]$$

The maximum power transfer theorem suggests that each network matches the ones that precede and follow. Is there some value of Z_L for which the input impedance Z_{in} is equal to the terminating impedance Z_L? Find Z_L in terms of k_{12} and k_{21}. Show that this value of Z_L can also be found in terms of z-parameters and y-parameters:

$$Z_L = \sqrt{\frac{z_{11}}{y_{11}}}$$

4. Explain the meaning of port condition. Show that the port condition is not satisfied in each two-port in the series-series connection shown in figure 12.16. Find

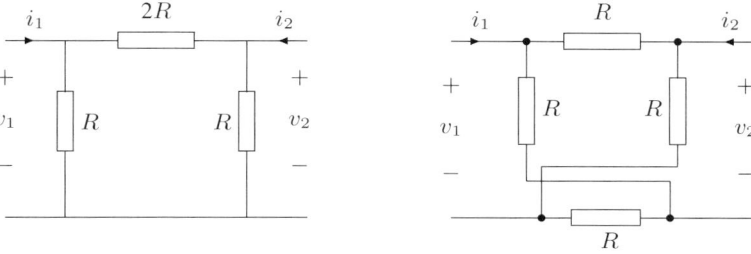

Figure 12.14: Circuits for Problems 1, 2, 7 to 10

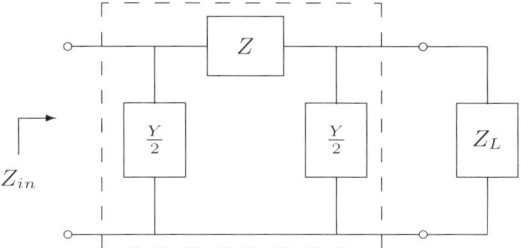

Figure 12.15: Circuit for Problem 3

the Z-matrix of the overall network arising from this series-series connection. Verify that

$$[Z^*] \neq [Z] + [Z']$$

where $[Z^*]$ is the overall Z-matrix, $[Z]$ and $[Z']$ are the Z-matrices for the upper and lower two-port respectively. In order to use the formula for series-series connection (i.e., adding up the z-parameters), port condition must be satisfied in each two-port. Suggest a way to force port condition. (Hint: consider ideal transformer.)

5. Determine the chain matrix of the two-ports of figure 12.17. Determine the chain matrix of the two-port which arises from a cascade connection of $[K_1]$ and $[K_2]$.

6. Find the overall Y-matrix for the network arising from a shunt-shunt connection of the two two-port networks shown in figure 12.14. Draw the equivalent circuit model.

7. Find the overall G-matrix for the network arising from a shunt-series connection of the two two-port networks shown in figure 12.14. Draw the equivalent circuit model.

8. Find the overall Z-matrix for the network arising from a series-series connection of the two two-port networks shown in figure 12.14. Draw the equivalent circuit model.

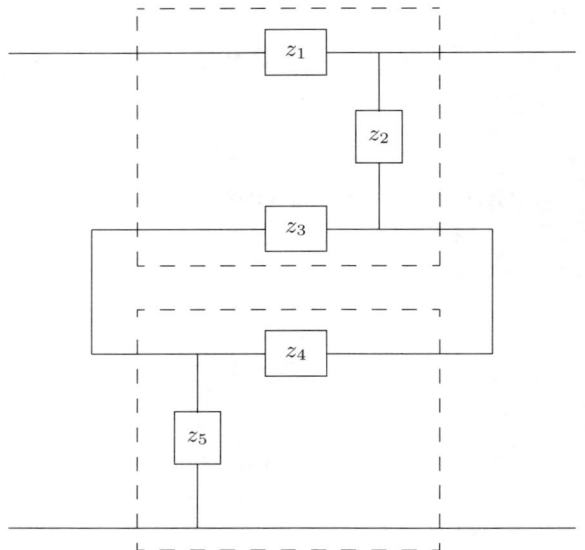

Figure 12.16: Circuit for Problem 4

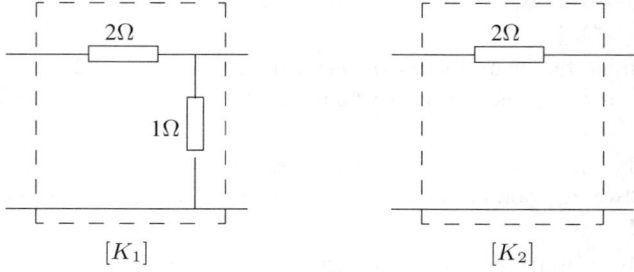

Figure 12.17: Circuit for Problem 5

9. Find the overall H-matrix for the network arising from a series-shunt connection of the two two-port networks shown in figure 12.14. Draw the equivalent circuit model.

10. Prove that the general reciprocity relation described in Chapter 2 (Section 2.2.2) holds for any two-port network that contains resistors, capacitors, inductors and ideal transformers. Note that the ideal transformer is itself a two-port defined by the relations

$$v_2 = nv_1 \quad \text{and} \quad i_2 = -\frac{1}{n}i_1$$

Appendix A

Glossary

Active power   Power measured in watts (W) which is actually used to do work or dissipated as heat. See also power and complex power.

Admittance   Ratio of current to voltage in an apparatus or an assembly of apparatus. Unit is siemens (S). It is in general a complex number. Its real part is the conductance and its imaginary part is the susceptance.

A-matrix   Matrix for specifying the connections between nodes and branches in a circuit. Its proper name is node incidence matrix.

Angular frequency of periodic function   Denoted by ω and defined as $2\pi/T$, where T is the repetition period of the function. Unit is rad/s.

Apparent power   Product of rms voltage and rms current in an apparatus or an assembly of apparatus. Unit is var, which is dimensionally equivalent to volt-ampere.

Asymptotic plot   Bode plot with straight line approximation for each portion of the plot. See also Bode plots.

Average value of periodic function   Integral average of the function over one period.

$$x_{av} = \frac{1}{T} \int_0^T x(t)\, dt$$

Basic cutset   Cutset that contains only one tree branch.

Basic cutset matrix   See Q-matrix.

Basic loop   Loop that contains only one co-tree branch.

Basic loop matrix   See B-matrix.

B-matrix   Matrix used to specify the basic loops, each of which contains only one co-tree branch for a circuit. Its proper name is basic loop matrix. See also basic loop.

Bode plots   Frequency response plots of magnitude versus frequency in log-log scale, and of phase shift versus frequency in semi-log scale. Usually the magnitude plot is also in semi-log scale with the magnitude expressed in dB.

Branch   Element in a graph that connects two nodes with no intermediate nodes.

Capacitance   Ratio of electric charge stored in a device to the voltage applied, i.e., $C = q/V$. Unit is farad (F). It may be regarded as a property of a device whose current is proportional to the rate of change of the voltage. The proportionality constant is also called capacitance.

Capacitor   Two-terminal device that has the property that the current is proportional to the rate of change of the voltage.

Characteristic equation   Algebraic equation obtained by replacing d/dt by s in a homogeneous differential equation that describes a circuit. Its roots give the eigenvalues of the circuit.

Circuit   An assembly of apparatus that is formed by interconnecting electrical components together.

Complex frequency   Complex number representation of frequency. $s = \sigma + j\omega$. Unit is rad/s.

Complex power   Complex number representation of power. Defined as dot product of voltage and the complex conjugate of current, all expressed in complex number form. $S = VI^* = P + jQ$. Its magnitude is the apparent power, its real part is the active or real power, and its imaginary part is the reactive power.

Conductance   Ratio of current to voltage in a resistor. It is equal to the reciprocal of resistance. Unit is siemens (S).

Corner frequency   Frequency where asymptotic lines in a Bode plot intersect.

Co-tree   The branches and nodes that are not in the tree. The union of the tree and the co-tree is the complete graph. See also tree.

Coupling coefficient   Measure of how tightly two inductors are coupled. Defined as ratio of mutual inductance to the square root of the product of self-inductances, i.e.,

$$\kappa = \frac{M}{\sqrt{L_1 L_2}}$$

Current   Rate of flow of electric charge. Defined as dq/dt. Unit is ampere (A), which is coulombs per second.

Cutset   A set of branches in a circuit graph which separates the graph into two disconnected parts.

Damping   Property of a circuit that ensures stability.

Damping factor ζ in the second-order characteristic equation $s^2 + 2\zeta\omega s + \omega^2 = 0$.

Decibel   Unit of the logarithmic gain multiplied by 20. In the case of power gain, it is the unit of the logarithmic gain multiplied by 10. The value in dB of a voltage gain V_o/V_i is $20 \log_{10} |V_o/V_i|$, and the value in dB of a power gain P_o/P_i is

$10 \log_{10} |P_o/P_i|$.

Delta connection   Connection style in which three devices are connected in series to form a delta.

Dependent source   Source whose intensity is a function of other voltages and currents in the circuit.

Di-graph   Circuit graph whose branches are marked with arrows indicating current direction and voltage polarity. See also graph.

Dirichlet's conditions   Conditions satisfying which a periodic function can be represented by an infinite Fourier series.

Dot convention   A symbolic convention in which dots are used to indicate polarity of voltage and direction of current in a group of coupled inductors. Each inductor is marked with a dot at one of its terminal, indicating that the terminal is positive with respect to the other terminal and that current enters the marked terminal.

Dual circuits   Two circuits having identical properties, when voltage and current are interchanged, R and G are interchanged, L and C are interchanged, etc., in one circuit.

Dynamic circuit   Circuit that contains inductor(s) and/or capacitor(s). The describing equation is a differential equation.

Eigenvalues   Roots of the characteristic equation. Denoted by λ. The natural response has a term proportional to $e^{\lambda t}$.

Energy storage element   Capacitor or inductor. Any apparatus that can store energy.

Equivalent circuit   Circuit, usually composed of a few elements connected in some simple way, which is equivalent to a complicated arrangement of elements.

Even function   Function with $x(t) = x(-t)$. Its waveform displays mirror symmetry about the y-axis.

First-order circuit   Circuit that can be described by a first-order differential equation. Its natural response is an exponential function of time.

Free-oscillating circuit   Circuit with all voltage sources short-circuited and all current sources open-circuited.

Free oscillation   Response of a circuit when the intensity of all sources is reduced to zero.

Fourier series   Series representation of periodic function. The standard form for a real periodic function with period $2\pi/\omega$ is

$$x(t) = \frac{a_0}{2} + \sum_{n=1}^{\infty} (a_n \cos n\omega t + b_n \sin n\omega t)$$

Frequency   Reciprocal of period, i.e., $f = 1/T$. Unit is hertz (Hz). See also period.

Frequency response   Relation between an input and an output, as frequency is varied. Usually frequency responses are described graphically by Bode plots.

Fundamental cutset   See basic cutset.

Fundamental loop   See basic loop.

Gibbs phenomenon   The phenomenon of oscillatory error manifested in Fourier series representation of a periodic function having a step jump in a period.

Graph   A set of nodes and branches defining the way in which elements are connected in a circuit.

Heavy damping   Property of a circuit with the damping factor exceeding 0.7071. See also damping factor.

Ideal transformer   Two-port element defined by the relation that the ratio of the input voltage to output voltage is equal to the ratio of the output current to input current, and this ratio is fixed.

Impedance   Ratio of voltage to current in an apparatus or an assembly of apparatus. Unit is ohm (Ω). It is in general a complex number. Its real part is the resistance and its imaginary part is the reactance.

Impulse function   A very short pulse defined by the mathematical function which is everywhere equal to 0 except at $t = 0$. Its integral from $-\infty$ to $+\infty$ equals 1.

$$\delta(t) = 0 \ \text{ for all } \ t \neq 0 \quad \text{and} \quad \int_{-\infty}^{\infty} \delta(t)dt = 1$$

Independent source   Source that is fixed or a function of time, independent of other circuit variables.

Inductance   Property of a device whose voltage is proportional to the rate of change of current. The proportionality constant is also called inductance, which is measured in henry (H).

Inductor   Two-terminal device that has the property that the voltage is proportional to the rate of change of current.

Instantaneous power   Product of the instantaneous voltage and the instantaneous current.

Inverse Laplace Transform   Mathematical transform to convert frequency domain functions to time domain functions.

Kirchhoff's laws   Basic laws governing the way in which electric charges flow in a circuit. There are two Kirchhoff's laws, one concerning currents and the other concerning voltages.

Laplace Transform   Mathematical transform to convert time domain functions to frequency domain functions.

Light damping   Property of a circuit with the damping factor less than 0.7071.

Linear element   Element to which superposition applies.

Link co-tree branch   See also co-tree.

Loop   A closed path in a circuit.

Maximum power transfer theorem   Maximum power is transfer from a source to a load when the load resistance is chosen to be equal to the source's internal resistance.

Mesh   Window loop in a circuit. A mesh has no inner loops.

Mesh current   Current circulating around a mesh in a circuit.

Mutual inductance   Property of two nearby inductors where the voltage of one inductor is proportional to the rate of change of current in the other inductor. The proportionality constant is also called mutual inductance. Unit is henry (H).

Natural frequencies   Eigenvalues. See also eigenvalues.

Node   Terminal common to two or more branches in a circuit.

Node incidence matrix   See A-matrix.

Node voltage   Voltage of a node with respect to the reference node or ground node.

Normal form state equation   State equation in the following form. See also state equation.

$$\frac{d\bar{x}}{dt} = A\bar{x} + B\bar{u}$$

Norton Theorem   Any linear circuit with two output terminals can be represented by a parallel combination of a resistor and a current source. The representation is equivalent from the point of view of the external apparatus connected to the output terminals.

Odd function   Function with $x(t) = -x(-t)$. Its waveform displays radial symmetry about the origin.

Ohm's law   The voltage across a resistor is proportional to the current through it. See also resistance.

Open circuit   Condition in which no current flows from one terminal to another regardless of the voltage applied across the two terminals.

Parseval's theorem   The power over a 1Ω resistor is equal to half of the sum of the square of the Fourier coefficients for the voltage across or the current in the resistor, i.e., $p = \frac{a_0^2}{4} + \frac{1}{2}\sum_{n=1}^{\infty}(a_1^2 + b_n^2)$. See also Fourier series.

Passive element   Element characterized by non-negative v-i product.

Period   Minimum period of time for which a periodic function must be defined in order to specify the function. It is also the period of repetition.

Periodic function   Function defined by the property that $x(t) = x(t + T)$, where T is called the period.

Phase shift   Phase angle difference between two periodic functions of the same frequency.

Phasor   Rotating vector under stroboscopic view representing a sinusoidal varying function. A phasor appears as vector or complex number on the Argand plane.

Phasor diagram   Diagram on which phasors are drawn to represent a number of sinusoidal varying functions of the same frequency. See also phasor.

Planar circuit   Circuit in which no branches cross over or under other branches.

Poles   Roots of characteristic equation. Points on the complex s-plane at which the magnitude of a transfer function is infinite.

Port condition   Condition that the current going into one terminal of a port equals that emerging from another terminal of the port in a two-port network.

Power   Rate of change of energy supplied by or work done in a device, i.e., dE/dt. Product of voltage and current. Unit is watt (W).

Power factor   Ratio of the active or real power to the apparent power, i.e., p.f = P/S. It is equal to the cosine of the phase shift between the voltage and the current for sinusoidally varying voltage and current, i.e., p.f. $= \cos \phi$.

Primary winding   Winding of a transformer. Usually the primary winding is connected to a power source.

Q-matrix   Matrix used to specify the basic cutsets, each of which contains only one tree branch for a circuit. Its proper name is basic cutset matrix. See also basic cutset.

Reactance   Ratio of voltage to current in a capacitor or inductor. Unit is ohm (Ω).

Reactive power   Product of rms voltage and rms current in a capacitor or inductor. Unit is var which is dimensionally equivalent to volt-ampere.

Reciprocity   Property of circuits that if a voltage in one branch produces a current in a second branch, the same voltage in the second branch will give the same current in the first branch.

Reference node   Node in a circuit selected as a reference (zero potential) from which other node voltages are measured.

Resistance   Property of a device whose voltage is proportional to the current. The proportionality constant is also called resistance which is measured in ohm (Ω).

Resistive circuits   Circuits having no transient response. See also dynamic circuit.

Resistivity   Ability to resist current flow of a material. For a piece of metal wire, resistance is proportional to the length and inversely proportional to the cross-sectional area. The proportionality constant is also called resistivity which is usually denoted by ρ, i.e., $\rho = RA/L$

Resistor   Two-terminal device having the property of resistance.

Root-mean-square value   Square root of the integral average of the square of a periodic function, i.e.,

$$x_{rms} = \sqrt{\frac{1}{T} \int_0^T x(t)^2 dt}$$

Secondary winding   Winding of a transformer. Usually the secondary winding is connected to a load.

Series connection   Connection style in which one end of an element is connected to one end of another element, and the connecting node is not connected to any third element.

Short circuit   Condition in which two terminals are connected together via zero impedance.

s-**plane**   Complex frequency plane.

Stability   State in which all variables in a circuit remain or will eventually remain close to a set of finite values or functions of time.

Star connection   Connection style in which three devices are connected to a common node forming a "star"-like structure.

State equation   Differential equation describing the dynamics of a circuit.

State variables   Variables independent of one another, which are used as variables in the construction of the describing state equations.

Steady state   State of a circuit after a sufficiently long period of time.

Step function   A function defined by a jump at $t = t_0$ from zero to a fixed value, i.e.,

$$x(t - t_0) = \begin{cases} 0 & \text{for } t \leq t_0 \\ A & \text{for } t > t_0 \end{cases}$$

When $A = 1$, $x(t - t_0)$ is called unit step function and is usually denoted by $u(t - t_0)$.

Super-mesh   One larger mesh formed from two meshes when the current source shared by the two meshes is removed (opened).

Super-node   One larger node formed from two nodes and a voltage source which connects the two nodes, with the voltage source removed (shorted).

Superposition   Characteristic of linear circuits. Response of a circuit due to a number of sources equals the sum of the individual responses due to the sources taken one at a time.

Susceptance   Ratio of current to voltage in a capacitor or inductor. Unit is siemens (S). See also admittance.

Thévenin theorem   Any linear circuit with two output terminals can be represented by a series combination of a resistor and a voltage source. The representation is equivalent from the point of view of the external apparatus connected to the output terminals.

Time constant   Time elapsed when a variable in a first-order circuit rises to 63 percent of its final value, or drops to 37 percent of its final value.

Topology   Way in which various elements are connected to form a circuit.

Transfer function   Ratio of a selected output to a selected input in the frequency domain.

Transformer   Two-port device formed by two windings wound around a common magnetic core and characterized by the property that the ratio of the voltage in the first port to that in the second port is equal to the ratio of the current in the second port to that in the first. This ratio is fixed and dependent on the physical turn ratio of the windings.

Transient response Response of dynamic circuits due to a stimulus or a switch activation.

Tree A maximal set of branches in a circuit graph which contains no loops.

Two-port network Networks having a pair of ports, each port being a pair of terminals.

Unit step function A function characterized by a jump at $t = t_0$ from 0 to 1, i.e.,

$$x(t - t_0) = \begin{cases} 0 & \text{for } t \leq t_0 \\ 1 & \text{for } t > t_0 \end{cases}$$

Voltage Potential difference between two points in a circuit. Current flows from a point of high potential to a point of low potential. Unit is volt (V).

Zeros Points in the complex frequency plane at which the magnitude of a transfer function is zero.

Appendix B

Linear Systems of Equations

A linear system of equations in general consists of n linear equations with n unknowns. The standard form is

$$
\begin{aligned}
a_{11}x_1 + a_{12}x_2 + \cdots + a_{1n}x_n &= c_1 \\
a_{21}x_1 + a_{22}x_2 + \cdots + a_{2n}x_n &= c_2 \\
&\cdots \\
a_{n1}x_1 + a_{n2}x_2 + \cdots + a_{nn}x_n &= c_n
\end{aligned}
$$

where $x_1, x_2, \ldots, x_n$ are the unknowns to be solved, and $c_1, c_2, \ldots, c_n$ are constants.

A number of techniques are available for solving a system of linear equations. For hand calculations, Cramer's rule is most widely used. For large systems of equations, however, Cramer's rule is unsuitable because of the difficulty associated with the calculation of determinants of high dimension. Instead, computer solutions are often employed for solving large systems of linear equations. In the formulation of computer algorithms for solving linear equations, the Gaussian elimination method and the LU factorization method are most often used.

1. Cramer's rule (for hand calculations);

2. Gaussian elimination (for hand calculations or computer implementations);

3. LU factorization (for computer implementations).

Determinants

Before we can use Cramer's rule, we must know how to evaluate determinants. We start from the second-order determinant, which is computed as follows.

$$
\begin{vmatrix} a_{11} & a_{12} \\ a_{21} & a_{22} \end{vmatrix} = a_{11}a_{22} - a_{21}a_{12}
$$

For a third-order determinant, the computation is as follows.

$$\begin{vmatrix} a_{11} & a_{12} & a_{13} \\ a_{21} & a_{22} & a_{23} \\ a_{31} & a_{32} & a_{33} \end{vmatrix} = a_{11} \begin{vmatrix} a_{22} & a_{23} \\ a_{32} & a_{33} \end{vmatrix} - a_{21} \begin{vmatrix} a_{12} & a_{13} \\ a_{32} & a_{33} \end{vmatrix} + a_{31} \begin{vmatrix} a_{12} & a_{13} \\ a_{22} & a_{23} \end{vmatrix}$$

In general, for a determinant of order n, we need to compute the *minors* which are determinants of order $n - 1$. Specifically, the minor M_{ij} of an nth-order determinant is formed by striking off the ith row and the jth column of the determinant. The nth order determinant in question is given by

$$\begin{vmatrix} a_{11} & a_{12} & \cdots & a_{1n} \\ a_{21} & a_{22} & \cdots & a_{2n} \\ \vdots & \vdots & \ddots & \vdots \\ a_{n1} & a_{n2} & \cdots & a_{nn} \end{vmatrix} = a_{11}M_{11} - a_{21}M_{21} + a_{31}M_{31} - \cdots + (-1)^{n+1}a_{n1}M_{n1}$$

Properties of Determinants

Some properties of determinants are worth noting. When used appropriately, these properties can lead to drastic simplification in the calculations of determinants.

1. If a_{ij} of a determinant is equal to a_{ji} of another determinant, then the two determinants are equal. In other words, interchanging rows and columns has no effect on the determinant value. For example,

$$\begin{vmatrix} 1 & 2 \\ 3 & 4 \end{vmatrix} = \begin{vmatrix} 1 & 3 \\ 2 & 4 \end{vmatrix}$$

2. If elements of one row or column are all multiplied by a factor k, the value of the determinant is multiplied by the same factor k. For example,

$$\begin{vmatrix} 1 & 2 \\ 3 & 4 \end{vmatrix} \times 2 = \begin{vmatrix} 1 \times 2 & 2 \times 2 \\ 3 & 4 \end{vmatrix} = \begin{vmatrix} 1 \times 2 & 2 \\ 3 \times 2 & 4 \end{vmatrix} \quad \text{etc.}$$

3. If the sign of all elements of one row or column is reversed, the determinant reverses its sign. For example,

$$-\begin{vmatrix} 1 & 2 \\ 3 & 4 \end{vmatrix} = \begin{vmatrix} -1 & -2 \\ 3 & 4 \end{vmatrix} = \begin{vmatrix} -1 & 2 \\ -3 & 4 \end{vmatrix} \quad \text{etc.}$$

4. If all elements of one row or column are zero, the determinant is equal to zero. For example,

$$\begin{vmatrix} 0 & 0 \\ 3 & 4 \end{vmatrix} = 0$$

5. If all elements of one row or column are k times the corresponding elements of another row or column, the determinant is equal to zero. For example,

$$\begin{vmatrix} 1 & 2 \\ 5 & 10 \end{vmatrix} = 0$$

6. If two rows or two columns are identical, the determinant is equal to zero. For example,

$$\begin{vmatrix} 1 & 2 \\ 1 & 2 \end{vmatrix} = 0$$

7. Interchanging two rows or two columns will reverse the sign of the determinant. For example,

$$\begin{vmatrix} 1 & 2 \\ 3 & 4 \end{vmatrix} = - \begin{vmatrix} 3 & 4 \\ 1 & 2 \end{vmatrix}$$

Cramer's Rule

Cramer's rule is useful for solving systems of equations of dimension less than or equal to 3, but is virtually useless for computer implementation. We will illustrate Cramer's rule for a system of three linear equations.

$$
\begin{aligned}
a_{11}x_1 + a_{12}x_2 + a_{13}x_3 &= c_1 \\
a_{21}x_1 + a_{22}x_2 + a_{23}x_3 &= c_2 \\
a_{31}x_1 + a_{32}x_2 + a_{33}x_3 &= c_3
\end{aligned}
$$

Case 1: Not all of c_1, c_2 and c_3 are zero.

In this case, we compute the determinant formed by the coefficients a_{ij}. We denote this determinant by Δ.

$$\Delta = \begin{vmatrix} a_{11} & a_{12} & a_{13} \\ a_{21} & a_{22} & a_{23} \\ a_{31} & a_{32} & a_{33} \end{vmatrix}$$

Then, we compute the following determinants which are formed by replacing one column in Δ by c_1, c_2 and c_3.

$$\Delta_1 = \begin{vmatrix} c_1 & a_{12} & a_{13} \\ c_2 & a_{22} & a_{23} \\ c_3 & a_{32} & a_{33} \end{vmatrix}$$

$$\Delta_2 = \begin{vmatrix} a_{11} & c_1 & a_{13} \\ a_{21} & c_2 & a_{23} \\ a_{31} & c_3 & a_{33} \end{vmatrix}$$

$$\Delta_3 = \begin{vmatrix} a_{11} & a_{12} & c_1 \\ a_{21} & a_{22} & c_2 \\ a_{31} & a_{32} & c_3 \end{vmatrix}$$

The solution is

$$
\begin{aligned}
x_1 &= \frac{\Delta_1}{\Delta} \\
x_2 &= \frac{\Delta_2}{\Delta} \\
x_3 &= \frac{\Delta_3}{\Delta}
\end{aligned}
$$

Case 2: All of c_1, c_2 and c_3 are zero.
In this case, the solution is trivial if $\Delta \neq 0$, i.e.,

$$x_1 = x_2 = x_3 = 0$$

Moreover, if $\Delta = 0$, then the system of equations is redundant. The solution is not unique. Specifically, we may write

$$a_{11}x_1 + a_{12}x_2 = -a_{13}x_3$$
$$a_{21}x_1 + a_{22}x_2 = -a_{23}x_3$$

The solution is

$$x_1 = \frac{\begin{vmatrix} -a_{13} & a_{12} \\ -a_{23} & a_{22} \end{vmatrix}}{\begin{vmatrix} a_{11} & a_{12} \\ a_{21} & a_{22} \end{vmatrix}} x_3$$

$$x_2 = \frac{\begin{vmatrix} a_{11} & -a_{13} \\ a_{21} & -a_{23} \end{vmatrix}}{\begin{vmatrix} a_{11} & a_{12} \\ a_{21} & a_{22} \end{vmatrix}} x_3$$

Note that any values of x_1, x_2 and x_3 bearing the ratios defined above will satisfy the system of equations.

Gaussian Elimination

The method of Gaussian elimination involves a number of steps which essentially perform subtraction or addition of a multiple of a row from or to another. The aim is to put the equations in the following form:

$$a'_{11}x_1 + a'_{12}x_2 + \cdots + a'_{1n}x_n = c'_1$$
$$a'_{12}x_2 + \cdots + a'_{1n}x_n = c'_1$$
$$\vdots \quad = \quad \vdots$$
$$a'_{nn}x_n = c'_n$$

Once the above form is reached, a back-subsitution process will give the unknowns, one by one, starting from x_n. Thus, the Gaussian elimination method involves two principal steps:

1. forward elimination
2. back-substitution

An example will clarify the procedure. Consider the following system of equations.

$$\begin{array}{rcrcrcr} 2x & + & y & + & z & = & 1 \\ 4x & + & y & & & = & -2 \\ -2x & + & 2y & + & z & = & 7 \end{array}$$

In the elimination process, we first multiply the first equation by 2 and subtract it from the second equation, and then we multiply the first equation by -1 and subtract it from the third equation. The result is

$$
\begin{array}{rcrcrcr}
2x & + & y & + & z & = & 1 \\
   & - & y & - & 2z & = & -4 \\
   &   & 3y & + & 2z & = & 8
\end{array}
$$

The coefficient 2 of x in the first equation is called the *pivot* of the first elimination stage. Next, we perform another stage of elimination. We multiply the second equation by -3 and subtract it from the third, resulting in

$$
\begin{array}{rcrcrcr}
2x & + & y & + & z & = & 1 \\
   & - & y & - & 2z & = & -4 \\
   &   &   & - & 4z & = & -4
\end{array}
$$

The coefficient -1 of y in the second equation is the *pivot* of the second elimination stage. Also, the coefficient -4 of z in the third equation is another pivot. In this third-order system, we see that in two stages of elimination we get the equation in the desired form for back-substitution.

Basically, the back-substitution begins with z, and proceeds in the reverse order. Specifically, from the third equation, we get

$$z = 1$$

Then, putting $z = 1$ in the second equation gives

$$y = 2$$

Finally, putting $z = 1$ and $y = 2$ in the first equation gives

$$x = -1$$

LU Factorization

The LU factorization method is highly suited for computer implementation. The basis of this method is the Gaussian elimination process. The LU factorization is simply a matrix description of the Gaussian process. The system of equations is first written as a matrix equation.

$$
\begin{pmatrix}
a_{11} & a_{12} & \cdots & a_{1n} \\
a_{21} & a_{22} & \cdots & a_{2n} \\
\vdots & \vdots & \ddots & \vdots \\
a_{n1} & a_{n2} & \cdots & a_{nn}
\end{pmatrix}
\begin{pmatrix}
x_1 \\
x_2 \\
\vdots \\
x_n
\end{pmatrix}
=
\begin{pmatrix}
c_1 \\
c_2 \\
\vdots \\
c_n
\end{pmatrix}
$$

or simply as

$$Ax = c$$

where A is the coefficient matrix and c is the constant vector. The algorithm involves factorization of the coefficient matrix A as a product of two triangle matrices, i.e.,

$$A = LU$$

where L and U are of the following forms:

$$L = \begin{pmatrix} 1 & 0 & \cdots & 0 \\ l_{21} & 1 & \cdots & 0 \\ \vdots & \vdots & \ddots & \vdots \\ l_{n1} & l_{n2} & \cdots & 1 \end{pmatrix}$$

and

$$U = \begin{pmatrix} u_{11} & u_{12} & \cdots & u_{1n} \\ 0 & u_{22} & \cdots & u_{2n} \\ \vdots & \vdots & \ddots & \vdots \\ 0 & 0 & \cdots & u_{nn} \end{pmatrix}$$

The computer algorithm is set to generate L through a series of elementary matrix operations, with each operation corresponding to an elimination process. Referring to the previous example, the matrix equation to be solved is

$$\begin{pmatrix} 2 & 1 & 1 \\ 4 & 1 & 0 \\ -2 & 2 & 1 \end{pmatrix} \begin{pmatrix} x \\ y \\ z \end{pmatrix} = \begin{pmatrix} 1 \\ -2 \\ 7 \end{pmatrix}$$

In the first elimination stage we multiply the first row by 2 and subtract it from the second row. This operation is equivalent to left-multiplying the coefficient matrix by the following elementary matrix:

$$E_{21} = \begin{pmatrix} 1 & 0 & 0 \\ -2 & 1 & 0 \\ 0 & 0 & 1 \end{pmatrix}$$

Also, in the first elimination stage we multiply the first row by -1 and subtract it from the third row. This operation is equivalent to left-multiplying the coefficient matrix by the following elementary matrix:

$$E_{31} = \begin{pmatrix} 1 & 0 & 0 \\ 0 & 1 & 0 \\ 1 & 0 & 1 \end{pmatrix}$$

In the second elimination stage we multiply the second row by -3 and subtract it from the third row. This operation is equivalent to left-multiplying the coefficient matrix by the following elementary matrix:

$$E_{32} = \begin{pmatrix} 1 & 0 & 0 \\ 0 & 1 & 0 \\ 0 & 3 & 1 \end{pmatrix}$$

The three operations are equivalent to left-multiplying the coefficient matrix by the following matrix:

$$E_{32}E_{31}E_{21} = \begin{pmatrix} 1 & 0 & 0 \\ -2 & 1 & 0 \\ -5 & 3 & 1 \end{pmatrix}$$

The result must be the system after the elimination process is completed, i.e.,

$$E_{32}E_{31}E_{21}Ax = \begin{pmatrix} 2 & 1 & 1 \\ 0 & -1 & -2 \\ 0 & 0 & -4 \end{pmatrix} \begin{pmatrix} x \\ y \\ z \end{pmatrix} = \begin{pmatrix} -1 \\ -4 \\ -4 \end{pmatrix} = E_{32}E_{31}E_{21}c$$

We note that the above coefficient matrix is an upper triangle. Let us denote it by U. Thus, we have

$$Ux = E_{32}E_{31}E_{21}c$$

or

$$E_{21}^{-1}E_{31}^{-1}E_{32}^{-1}Ux = c$$

Hence, we can write

$$A = LU$$

where L and U are the lower triangle and upper triangle given by

$$L = E_{21}^{-1}E_{31}^{-1}E_{32}^{-1} = \begin{pmatrix} 1 & 0 & 0 \\ 2 & 1 & 0 \\ -1 & -3 & 1 \end{pmatrix}$$

and

$$U = E_{32}E_{31}E_{21}A = \begin{pmatrix} 2 & 1 & 1 \\ 0 & -1 & -2 \\ 0 & 0 & -4 \end{pmatrix}$$

Clearly, the factorization process involves finding the elementary matrices that perform the forward elimination. In the three-dimensional case, we have $L = E_{21}^{-1}E_{31}^{-1}E_{32}^{-1}$ and $U = E_{32}E_{31}E_{21}A$. Now, recognizing that

$$Ax = LUx = c$$

we can get the solution easily by first finding c' in

$$Lc' = c$$

Then, with c' found, we can get x by solving

$$Ux = c'$$

A moment's reflection will convince us that the LU factorization method is, in essence, the Gaussian elimination method, with the forward elimination implemented by $Lc' = c$ and the back-substitution implemented by $Ux = c'$.

The Problem of Zero Pivots

Provided no pivots are zero, the afore-described Gaussian elimination process or LU factorization will work. However, if a zero pivot is found after several operations along the elimination process, the algorithm must stop prematurely. For example, consider the following system.

$$\begin{pmatrix} 0 & 1 \\ 3 & 4 \end{pmatrix} \begin{pmatrix} x \\ y \end{pmatrix} = \begin{pmatrix} 5 \\ 8 \end{pmatrix}$$

Obviously, no elimination step can be performed in the order defined in the foregoing sections. A simple fix for this case is to interchange the two rows such that the system becomes

$$\begin{pmatrix} 3 & 4 \\ 0 & 1 \end{pmatrix} \begin{pmatrix} x \\ y \end{pmatrix} = \begin{pmatrix} 8 \\ 5 \end{pmatrix}$$

This system is now ready for back-substitution.

In general, we can interchange a row with another row as soon as we hit a zero pivot. Provided the new row has a non-zero pivot, the elimination process can continue. The question is how to ensure that the exchange of rows can give non-zero pivots all the way through the entire elimination process.

Suppose we run into a zero pivot at the jth row. The general approach is to look at the jth column in the rows below the jth row. If some entry is non-zero, say in row m, then we exchange the jth row and the mth row. This will permit the elimination process to proceed. However, if all elements in the jth column below are zero, then the system is *singular,* and either no solution exists or infinitely many solutions exist. We will not pursue the singular case in this book.

Appendix C

Laplace Transforms

The Laplace Transform is an extremely useful tool in linear system analysis. In this appendix we summarize the important concepts and applications of the Laplace Transform. For a rigorous pursuit of the mathematics, however, readers should refer to an engineering mathematics text. One purpose of this appendix, among others, is to illustrate how the Laplace Transform can be used to solve differential equations systematically.

Laplace Transform

The Laplace Transform of a function $f(t)$ is defined by the following integral:

$$F(s) = \mathcal{L}\{f(t)\} = \int_0^\infty e^{-st} f(t) \, dt$$

This operation takes a time-domain function to a complex-frequency domain in which algebraic equations, rather than differential equations, are to be dealt with. For this reason, the mathematics in the transformed domain is much easier to handle.

The inverse operation is called the inverse Laplace Transform, which is given by

$$f(t) = \mathcal{L}^{-1}\{F(s)\} = \frac{1}{2\pi j} \int_{c-j\infty}^{c+j\infty} F(s) e^{st} \, dt$$

This inverse operation takes a complex-frequency-domain function back to the time domain.

It should be noted that the Laplace Transform integral is taken only for $t > 0$. Thus, the part of a function defined for $t < 0$ is not relevant. It is customary to multiply a function by a unit step function so that the resulting function has identically zero values for all $t < 0$. In other words, the Laplace transform of $f(t)$ is really the Laplace

Transform of $f(t)u_s(t)$, where $u_s(t)$ is defined by

$$u_s(t) = \begin{cases} 1 & \text{for } t \geq 0 \\ 0 & \text{for } t < 0 \end{cases}$$

Laplace Transforms of Common Functions

First of all, the unit step function is frequently used in the study of transient behaviour. For example, switching a 1V voltage source to a circuit at $t = 0$ can be regarded as applying a unit step function to the circuit. The Laplace Transform of the unit step function is

$$\mathcal{L}\{u_s(t)\} = \int_0^\infty e^{-st} u_s(t) \, dt = \frac{1}{s}$$

In first-order circuits, the solution of the form $e^{-\alpha t}$ frequently occurs. Let us find the Laplace Transform for this simple exponential function from first principles.

$$\mathcal{L}\{e^{-\alpha t}\} = \int_0^\infty e^{-st} e^{-\alpha t} dt = \frac{1}{s + \alpha}$$

Also found in first-order circuits is the solution of the form $(1 - e^{-\alpha t})$. The Laplace Transform of this function is

$$\mathcal{L}\{1 - e^{-\alpha t}\} = \int_0^\infty e^{-st}(1 - e^{-\alpha t}) dt = \frac{1}{s} - \frac{1}{s + \alpha} = \frac{\alpha}{s(s + \alpha)}$$

In second-order circuits, the solution of the form $e^{-\alpha t} \sin \omega t$ is common. From first principles, the Laplace Transform of this function is

$$\mathcal{L}\{e^{-\alpha t} \sin \omega t\} = \int_0^\infty e^{-st}(e^{-\alpha t} \sin \omega t) dt = \frac{\omega}{(s + \alpha)^2 + \omega^2}$$

Similarly, we have

$$\mathcal{L}\{e^{-\alpha t} \cos \omega t\} = \int_0^\infty e^{-st}(e^{-\alpha t} \cos \omega t) dt = \frac{s + \omega}{(s + \alpha)^2 + \omega^2}$$

For sinusoidally varying functions, we can use the above formulae to find their Laplace Transforms by setting α to zero, i.e.,

$$\mathcal{L}\{\sin \omega t\} = \frac{\omega}{s^2 + \omega^2}$$

and

$$\mathcal{L}\{\cos \omega t\} = \frac{s}{s^2 + \omega^2}$$

Laplace Transforms of Common Operations

In this section we derive some common operations in the complex-frequency domain corresponding to some common operations performed in the time domain.

1. We first consider the Laplace Transform of a function which is multiplied by a constant k, i.e., $k \times f(t)$.

$$\mathcal{L}\{kf(t)\} = \int_0^\infty e^{-st} kf(t)\, dt = k \int_0^\infty e^{-st} f(t)\, dt = k\mathcal{L}\{f(t)\}$$

Using this property, we can say that the Laplace Transform of a step function $ku_s(t)$ is given by k/s. Similarly, we have the Laplace Transform of $ke^{-\alpha t}$ equal to $k/(s+\alpha)$, etc.

2. Linearity remains an important property of the Laplace Transform. Essentially, the Laplace Transform of the sum of two functions, $f(t) + g(t)$, is the sum of the Laplace Transforms of the two functions, i.e.,

$$\mathcal{L}\{f(t) + g(t)\} = \mathcal{L}\{f(t)\} + \mathcal{L}\{g(t)\}$$

Using this property and the previous property, we can generalize that

$$\mathcal{L}\left\{ \sum_{n=1}^N k_n f_n(t) \right\} = \sum_{n=1}^N \{k_n \mathcal{L}\{f_n(t)\}\}$$

3. Next, we consider the Laplace Transform of a derivative of a function. Using the standard formula of integration by parts, we can write

$$\int_0^\infty f(t)d\left\{ \frac{-1}{s} e^{-st} \right\} = \left[-\frac{1}{s} f(t)e^{-st} \right]_0^\infty + \int_0^\infty \frac{1}{s} e^{-st} d\{f(t)\}$$

Thus, we have

$$\mathcal{L}\{f(t)\} = \frac{1}{s} f(0^+) + \frac{1}{s} \int_0^\infty f'(t)e^{-st} dt$$

Rearranging gives
$$\mathcal{L}\{f'(t)\} = s\mathcal{L}\{f(t)\} - f(0^+)$$

4. In a likewise manner, we can get the Laplace Transform of the second derivative of $f(t)$ as
$$\mathcal{L}\{f''(t)\} = s\mathcal{L}\{f'(t)\} - f'(0^+)$$

Substituting $\mathcal{L}\{f'(t)\}$ in the above expression gives

$$\mathcal{L}\{f''(t)\} = s^2\mathcal{L}\{f(t)\} - sf(0^+) - f'(0^+)$$

We can likewise derive the Laplace Transform of the nth derivative of $f(t)$ by induction.

5. Consider the Laplace Transform of the integral of $f(t)$. Let $g(t)$ be the indefinite integral of $f(t)$, i.e.,

$$g(t) = \int f(t) \, dt$$

Using the formula of integration by parts, we get

$$\int_0^\infty e^{-st} d\{g(t)\} = \left[e^{-st} g(t) \right]_0^\infty + \int_0^\infty e^{-st} sg(t) dt$$

Since $d\{g(t)\} = f(t)dt$. we have

$$\mathcal{L}\{f(t)\} = \left[e^{-st} g(t) \right]_0^\infty + \int_0^\infty e^{-st} sg(t) dt$$

which gives

$$\mathcal{L}\{g(t)\} = \frac{\mathcal{L}\{f(t)\}}{s} + \frac{g(0^+)}{s}$$

6. To find the Laplace Transform of the definite integral $\int_0^t f(t) dt$, we simply use the above property and the relation $\int_0^t f(t) dt = g(t) - g(0^+)$ to get

$$\mathcal{L}\left\{ \int_0^t f(t) dt \right\} = \frac{\mathcal{L}\{f(t)\}}{s}$$

Solution of First-Order Differential Equations

One important application of the Laplace Transform is in solving differential equations. The essential procedure involves converting the differential equation in question to the complex-frequency domain by the Laplace Transform. In the complex-frequency domain, we can put the unknown in explicit form by simple algebraic manipulations. Finally, we perform the inverse transform to obtain the time-domain solution. As an example, consider the following differential equation:

$$\frac{dx(t)}{dt} + bx(t) = f(t)$$

The Laplace Transform of this equation is

$$sX(s) - x(0^+) + bX(s) = F(s)$$

where $X(s)$ and $F(s)$ denote the Laplace Transforms of $x(t)$ and $f(t)$ respectively. We may now express $X(s)$ algebraically as

$$X(s) = \frac{F(s) + x(0^+)}{s + b}$$

$f(t)$	$F(s)$
1	$\dfrac{1}{s}$
$e^{-\alpha t}$	$\dfrac{1}{s+\alpha}$
$\dfrac{1}{\alpha}(1-e^{-\alpha t})$	$\dfrac{1}{s(s+\alpha)}$
$\dfrac{e^{-at}-e^{-bt}}{b-a}$	$\dfrac{a}{(s+a)(s+b)}$
$te^{-\alpha t}$	$\dfrac{1}{s+\alpha^2}$
$e^{-\alpha t}\sin\omega t$	$\dfrac{\omega}{(s+\alpha)^2+\omega^2}$
$e^{-\alpha t}\cos\omega t$	$\dfrac{s+\alpha}{(s+\alpha)^2+\omega^2}$
$\sin\omega t$	$\dfrac{\omega}{s^2+\omega^2}$
$\cos\omega t$	$\dfrac{s}{s^2+\omega^2}$
$\dfrac{-ae^{-at}+be^{-bt}}{-a+b}$	$\dfrac{s}{(s+a)(s+b)}$
t	$\dfrac{1}{s^2}$
$\delta(t)$	1

Table C.1: Laplace Transform pairs

Suppose the initial condition is $x(0^+) = x_0$. We then have

$$X(s) = \frac{F(s)+x_0}{s+b} = \frac{F(s)}{s+b} + \frac{x_0}{s+b}$$

Taking the inverse Laplace Transform, with the help of Tables C.1 and C.2, gives

$$x(t) = e^{-bt}\int_0^t e^{bt}f(t)dt + x_0 e^{-bt}$$

Note that the last equation requires integrating a function involving $f(t)$. In practice, we would never need to perform this integration because if we were given $f(t)$, we would have known $F(s)$. Thus, we can tackle $F(s)$ algebraically in the complex-frequency domain well before we get back to the time domain. For example, if $f(t) = e^{-ct}$, then $F(s) = 1/(s+c)$, and $X(s)$ is given by

$$X(s) = \frac{1}{(s+c)(s+b)} + \frac{x_0}{s+b}$$

Time-domain operation	Laplace Transform operation
$Af(t)$	$AF(s)$
$f_1(t) + f_2(t)$	$F_1(s) + F_2(s)$
$\dfrac{d}{dt}f(t)$	$sF(s) - f(0^+)$
$\dfrac{d^2}{dt^2}f(t)$	$s^2F(s) - sf(0^+) - f'(0^+)$
$\displaystyle\int f(t)\,dt$	$\dfrac{F(s)}{s} + \dfrac{1}{s}\displaystyle\int f(t)dt\Big|_{t=0^+}$
$\displaystyle\int_0^t f(t)dt$	$\dfrac{F(s)}{s}$
$e^{-\alpha t}f(t)$	$F(s + \alpha)$
$e^{-\alpha t}\displaystyle\int_0^t e^{\alpha t}f(t)dt$	$\dfrac{F(s)}{s + \alpha}$
$f(t - \alpha)u_s(t - \alpha)$	$e^{-s\alpha}F(s)$

Table C.2: Laplace Transform operations

Now we take the inverse, with the help of Table C.1, to yield the answer.

$$x(t) = \frac{-ce^{-ct} + be^{-bt}}{-c + b} + x_0e^{-bt}$$

Solution of State Equation in Normal Form

The procedure for solving state equations in normal form is similar to the above procedure for solving first-order differential equations. The only difference is that we are now dealing with matrices instead of numbers. The state equation in general is of the form:

$$\frac{d}{dt}\bar{x} = A\bar{x} + B\bar{u}$$

where A is an n-dim square matrix, $\bar{u}$ is an m-dim input vector, and B is an $n \times m$ matrix. Taking the Laplace Transform of this matrix state equation gives

$$s\bar{X}(s) - \bar{x}(0^+) = A\bar{X}(s) + B\bar{U}(s)$$

Rearranging, based on the rules of linear algebra, gives

$$\bar{X}(s) = [s\mathbf{1} - A]^{-1}[B\bar{U}(s) + \bar{x}(0^+)]$$

where $\mathbf{1}$ is the unit matrix, i.e.,

$$\mathbf{1} = \begin{pmatrix} 1 & 0 & \cdots & 0 \\ 0 & 1 & \cdots & 0 \\ \vdots & \vdots & \ddots & \vdots \\ 0 & 0 & \cdots & 1 \end{pmatrix}$$

Taking the inverse Laplace transform yields

$$\bar{x}(t) = \mathcal{L}^{-1}\left\{[s\mathbf{1} - A]^{-1}[B\bar{U}(s) + \bar{x}(0^+)]\right\}$$

which is the time-domain solution of the state equation.

Note that the general form of $\bar{X}(s)$ given above has a common denominator which is the determinant $|s\mathbf{1} - A|$. In other words, all elements in $\bar{X}(s)$ will take the following form:

$$X_1(s) = \frac{N_1(s)}{D(s)}$$

$$X_2(s) = \frac{N_2(s)}{D(s)}$$

$$X_3(s) = \frac{N_3(s)}{D(s)}$$

$$\vdots \qquad \vdots$$

where

$$D(s) = |s\mathbf{1} - A|$$

Thus, $D(s) = 0$ is the characteristic equation, solving which gives the eigenvalues or natural frequencies. Moreover, $D(s)$ is a polynomial in s, i.e.,

$$D(s) = a_n s^n + a_{n-1} s^{n-1} + a_{n-2} s^{n-2} + \cdots + a_1 s + a_0$$

Hence, $X_1(s)$, $X_2(s)$, etc., may take the following form:

$$X_i(s) = \frac{N_i(s)}{s^k(s + \alpha)(s + \beta)^2 \cdots (s^2 + \gamma s + \delta) \cdots}$$

Such a form is not readily used in the inverse transformation and must be expanded into partial fractions to facilitate table look-up. In Appendix D, we will describe the techniques for partial fraction expansion.

Examples of the Inverse Laplace Transform

We illustrate the procedure of the inverse transform with two examples. In the first example, suppose the solution of the state equation in normal form has generated the following frequency-domain solution for one of the state variables.

$$X_1(s) = \frac{\frac{60}{s} + 6}{\frac{s}{2} + 12}$$

First of all, we need to put it in the form of $N(s)/D(s)$, where $N(s)$ and $D(s)$ are polynomials in s.

$$X_1(s) = \frac{12(s+10)}{s(s+24)}$$

The next important step is partial fraction expansion. Here, we look for an expansion of the following form:

$$X_1(s) = \frac{A}{s} + \frac{B}{s+24}$$

The immediate problem now is to find A and B such that the partial fraction expansion is equivalent to the original expression. This can be done by expressing the partial fraction expression in the previous form and comparing coefficients, i.e.,

$$\frac{A(s+24) + sB}{s(s+24)} = \frac{12(10+s)}{s(s+24)}$$

Putting $s = -24$ gives $B = 7$, and putting $s = 0$ gives $A = 5$. Thus, the partial fraction expansion is

$$X_1(s) = \frac{5}{s} + \frac{7}{s+24}$$

Now, we can look up the table to get the time-domain solution.

$$x_1(t) = 5 + 7e^{-24t}$$

As another example, consider the solution of $X_2(s)$ in the complex-frequency domain given by

$$X_2(s) = \frac{10s+1}{(s^2 + 4s + 5)(s+1)}$$

The partial fraction expansion is

$$X_2(s) = \frac{As+B}{s^2 + 4s + 5} + \frac{C}{s+1}$$

where A, B and C can be found as follows. Adding up the two terms in the above expansion and comparing with the original function, we have

$$10s + 1 = (As+B)(s+1) + C(s^2 + 4s + 5)$$

Putting $s = -1$ gives

$$C = \frac{-9}{2} = -4.5$$

Putting $s = 0$ gives

$$B = \frac{1 + 5 \times 4.5}{1} = 23.5$$

Finally, putting $s = 1$ gives

$$11 = 2(A + B) + 10C$$

which can be solved to give

$$A = 4.5$$

Hence, we can write

$$X_2(s) = \frac{4.5s + 23.5}{s^2 + 4s + 5} - \frac{4.5}{s + 1}$$

Now, the inverse Laplace Transform of the second term is straightforward, but that of the first term does not seem to appear anywhere in the table. Recognizing that the inverse transform of $\omega/(s^2 + \omega^2)$ is $\sin \omega t$, and that of $s/(s^2 + \omega^2)$ is $\cos \omega t$, we may try rearranging the first term as follows.

$$
\begin{aligned}
X_2(s) &= \frac{14.5 + 4.5(s + 2)}{(s + 2)^2 + 1} - \frac{4.5}{s + 1} \\
&= \frac{14.5}{(s + 2)^2 + 1} + \frac{4.5(s + 2)}{(s + 2)^2 + 1} - \frac{4.5}{s + 1}
\end{aligned}
$$

From Table C.2, we know that when $F(s)$ is replaced by $F(s+a)$, the inverse Laplace Transform is multiplied by a factor of e^{-at}. Thus, we have

$$x_2(t) = 14.5e^{-2t} \sin t + 4.5e^{-2t} \cos t - 4.5e^{-t}$$

Initial Value Theorem and Final Value Theorem

Two useful results can be obtained by considering the limiting cases of s tending towards infinity and zero in the Laplace Transform of the time derivative of $f(t)$. We recall that

$$\mathcal{L}\{f'(t)\} = sF(s) - f(0^+)$$

As $s \to \infty$, the Laplace Transform integral vanishes, i.e.,

$$\lim_{s \to \infty} \mathcal{L}\{f'(t)\} = 0$$

Thus, we have

$$f(0^+) = \lim_{s \to \infty} sF(s)$$

This result is known as the *initial value theorem*.

Furthermore, the formula for the Laplace Transform of $f'(t)$, as s tends to 0, is

$$\int_0^\infty f'(t)dt = \lim_{s \to 0} sF(s) - f(0^+)$$

Since $\int_0^\infty f'(t)dt = \lim_{t \to \infty} f(t) - f(0^+)$, we have

$$\lim_{t \to \infty} f(t) = \lim_{s \to 0} sF(s)$$

This result is known as the *final value theorem*.

Appendix D

Partial Fractions

Consider a function of the form $N(s)/D(s)$ where $N(s)$ and $D(s)$ are polynomials in s. We assume that the degree of $N(s)$ is less than that of $D(s)$, and that $D(s)$ has been factorized as $s^k(s+a)^m \cdots (s^2+bs+c)^p \cdots$. The problem is to express $N(s)/D(s)$ as

$$\frac{N(s)}{D(s)} = \frac{A}{s} + \frac{B}{s^2} + \cdots + \frac{C}{s+a} + \cdots + \frac{D}{(s+a)^2} + \cdots + \frac{Es+F}{s^2+bs+c} + \cdots$$

In general, the procedure of partial fraction expansion involves guessing the right forms of fractions and finding the unknown coefficients in the numerators of the individual fractions.

"Guessing" the Right Form of an Expansion

We will start with the simplest form of $D(s)$ which consists of distinct linear factors, i.e., the characteristic equation has distinct real roots.

$$\frac{N(s)}{D(s)} = \frac{N(s)}{(s+a)(s+b)(s+c)}$$

where $a \neq b \neq c$. In this case, we should look for an expansion of the form

$$\frac{N(s)}{(s+a)(s+b)(s+c)} = \frac{A}{s+a} + \frac{B}{s+b} + \frac{C}{s+c}$$

This also applies to the case when either a, b or c is zero, i.e.,

$$\frac{N(s)}{s(s+b)(s+c)} = \frac{A}{s} + \frac{B}{s+b} + \frac{C}{s+c}$$

The second case to be considered here corresponds to the presence of double linear factors in $D(s)$, i.e., the characteristic equation has double roots. In this case, we should look for a term $A/(s+a)$ as well as $B/(s+a)^2$, e.g.,

$$\frac{N(s)}{(s+a)^2(s+c)} = \frac{A}{s+a} + \frac{B}{(s+a)^2} + \frac{C}{s+c}$$

This again applies to the case where a is zero, i.e.,

$$\frac{N(s)}{s^2(s+c)} = \frac{A}{s} + \frac{B}{s^2} + \frac{C}{s+c}$$

The third case to be considered here corresponds to the presence of the factor $(s^2 + a^2)$ in $D(s)$. In this case we should look for expansion of the form $(As+B)/(s^2+a^2)$, e.g.,

$$\frac{N(s)}{(s^2+a^2)(s+b)(s+c)} = \frac{As+B}{s^2+a^2} + \frac{C}{s+b} + \frac{D}{s+c}$$

We must stress here that the degree of $N(s)$ is assumed to be smaller than that of $D(s)$. If the degree of $N(s)$ is larger than that of $D(s)$, we need to perform a polynomial division before we can start the expansion process. For example, consider the function

$$F(s) = \frac{2s^2 + 3s}{(s+2)(s+3)}$$

We should not attempt to expand it as $\frac{A}{s+2} + \frac{B}{s+3}$ since the numerator is of the same degree as the denominator. We must first perform a division to extract the constant term, i.e.,

$$F(s) = 2 - \frac{7s + 12}{(s+2)(s+3)}$$

Now we can proceed to find its partial fraction expansion. In this example, the expansion is

$$F(s) = 2 - \left[\frac{A}{s+2} + \frac{B}{s+3} \right]$$

Finding the Unknown Coefficients

After we have made the right choice of the form of the expansion, we need to calculate all the unknown coefficients that appear in the numerators. Basically we find all these coefficients by comparing the partial fraction expansion with the original expression. To do this, we substitute some suitable value for s so that some unknowns disappear during comparison.

We consider first the simple expansion of the form

$$\frac{N(s)}{(s+a)(s+b)(s+c)} = \frac{A}{s+a} + \frac{B}{s+b} + \frac{C}{s+c}$$

Assume that the degree of $N(s)$ is less than three. Our job now is to find A, B and C. The equation for comparison of coefficients is given by

$$N(s) = A(s + b)(s + c) + B(s + a)(s + c) + C(s + a)(s + b)$$

If we put $s = -a$, then the B term and the C term will vanish, giving

$$A = \frac{N(-a)}{(b - a)(c - a)}$$

Likewise, putting $s = -b$, we have

$$B = \frac{N(-b)}{(a - b)(c - b)}$$

Finally, putting $s = -c$, we have

$$C = \frac{N(-c)}{(a - c)(b - c)}$$

We see that in each substitution process, all coefficients but one disappear, making it very easy to find the coefficients. This procedure usually works, except for the case of double linear factors, which we will examine now. Consider the expansion of the form

$$\frac{N(s)}{(s + a)^2(s + b)} = \frac{A}{s + a} + \frac{B}{(s + a)^2} + \frac{C}{s + c}$$

We put the right side in the form of $N(s)/D(s)$, and compare the numerator with that of the original function. This gives

$$N(s) = A(s + a)(s + b) + B(s + b) + C(s + a)^2$$

Let us apply the substitution process as before. First, putting $s = -b$, we get

$$C = \frac{N(-b)}{(a - b)^2}$$

Next, we put $s = -a$ to get

$$B = \frac{N(-a)}{b - a}$$

However, we cannot find any suitable substitution such that the B term and the C term disappear simultaneously. Of course, we can put any value and use the above expressions for B and C to solve for A. Let us make an easy substitution with $s = 0$. This gives

$$A = \frac{N(0) - bB - a^2 C}{ab}$$

Finally, we consider the function of the form

$$\frac{N(s)}{D(s)} = \frac{As + B}{s^2 + bs + a^2} + \frac{C}{s + c}$$

The coefficients can be found again by comparison. We put the right side in the form of $N(s)/D(s)$, and compare the numerator with that of the original function.

$$N(s) = (As + B)(s + c) + C(s^2 + bs + a^2)$$

As usual, we put $s = -c$ to make the first term disappear. This gives

$$C = \frac{N(-c)}{c^2 - bc + a^2}$$

Then, we put $s = 0$ to make the A term vanish. Since we have found C, we can get B as

$$B = \frac{N(0) - a^2 C}{c}$$

Finally, we put any value for s to find A.

Examples

Let us illustrate the procedure with a few examples.

Example D.1 — First consider the following function:

$$F(s) = \frac{1}{s(s^2 + 4s + 3)}$$

Factorizing the denominator gives

$$F(s) = \frac{1}{s(s + 3)(s + 1)}$$

We look for an expansion of the form

$$F(s) = \frac{A}{s} + \frac{B}{s + 3} + \frac{C}{s + 1}$$

where the coefficients A, B and C are found by comparison, i.e.,

$$1 = A(s + 3)(s + 1) + Bs(s + 1) + Cs(s + 3)$$

Putting $s = 0$ will make the B term and the C term disappear. Thus, we have

$$A = \frac{1}{3}$$

Likewise, putting $s = -3$ gives

$$B = \frac{1}{6}$$

Finally, putting $s = -1$ gives

$$C = \frac{-1}{2}$$

Hence, the partial fraction expansion of $F(s)$ is given by

$$F(s) = \frac{1}{3s} + \frac{1}{6(s+3)} - \frac{1}{2(s+1)}$$

Example D.2 — As another example, consider the function with double factors in the denominator.

$$G(s) = \frac{2}{s(s+1)^2}$$

The correct form of the partial fraction expansion for $G(s)$ is

$$G(s) = \frac{A}{s} + \frac{B}{s+1} + \frac{C}{(s+1)^2}$$

The equation for comparing coefficients is written as

$$2 = A(s+1)^2 + Bs(s+1) + Cs$$

Putting $s = 0$ will make the B term and the C term disappear, giving

$$A = 2$$

Likewise, putting $s = -1$ will get rid of the A term and the B term. Thus, we have

$$C = -2$$

Now, no specific substitution will make A and C disappear. We put any value for s, say 1, to get

$$2 = 4A + 2B + C$$

which can be solved to give

$$B = -2$$

Hence, the partial fraction expansion for $G(s)$ is

$$G(s) = \frac{2}{s} - \frac{2}{s+1} - \frac{2}{(s+1)^2}$$

Example D.3 — As a final example, we consider the following function:

$$H(s) = \frac{2s+1}{(s^2+4)(s+1)}$$

The correct form of the expansion is

$$H(s) = \frac{As+B}{s^2+4} + \frac{C}{s+1}$$

The equation for comparison of coefficients is

$$2s+1 = (As+B)(s+1) + C(s^2+4)$$

We put $s = -1$ to get rid of the A and B terms, giving

$$C = -\frac{1}{5}$$

Putting $s = 0$ gives

$$1 = B + 4C$$

which can be solved to give

$$B = \frac{9}{5}$$

Finally, we put any value for s, say 1, to get

$$3 = 2(A + B) + 5C$$

Solving this equation gives

$$A = \frac{1}{5}$$

Hence, the partial fraction expansion of $H(s)$ is

$$H(s) = \frac{s + 9}{5(s^2 + 4)} - \frac{1}{5(s + 1)}$$

General Approach for *D(s)* Containing Distinct Linear Factors

The method can be generalized to a simple form of expression for each unknown coefficient in the partial fraction expansion if the denominator of the given function contains distinct linear factors.

$$\frac{N(s)}{D(s)} = \frac{A_1}{s + a_1} + \frac{A_2}{s + a_2} + \frac{A_3}{s + a_3} + \cdots$$

Provided all a_i are distinct, we can find the coefficients using the following formula:

$$A_i = \frac{N(-a_i)}{D'(-a_i)}$$

where $D'(s)$ denotes $dD(s)/ds$. For example, consider the function

$$\frac{3s + 1}{s(s + 1)(s + 2)} = \frac{A}{s} + \frac{B}{s + 1} + \frac{C}{s + 2}$$

Here, $D(s) = s^3 + 3s^2 + 2s$ and hence $D'(s) = 3s^2 + 6s + 2$. The coefficients are given by

$$A = \left.\frac{3s + 1}{3s^2 + 6s + 2}\right|_{s=0} = \frac{1}{2}$$

$$B = \left.\frac{3s + 1}{3s^2 + 6s + 2}\right|_{s=-1} = \frac{-2}{-1} = 2$$

$$C = \left.\frac{3s + 1}{3s^2 + 6s + 2}\right|_{s=-2} = \frac{-5}{2}$$

Hence, the partial fraction expansion is

$$\frac{N(s)}{D(s)} = \frac{1}{2s} + \frac{2}{s+1} - \frac{5}{2(s+2)}$$

Note that this method is not applicable to functions whose $D(s)$ contains repeated linear factors, e.g., $D(s) = (s+a)^2(s+b)$.

Appendix E

Complex Numbers

Complex numbers arise from solving polynomial equations having no real solutions. For example, the equation $x^2 + 2 = 0$ has no real solutions. However, if $x = \sqrt{-2} = \sqrt{2}\sqrt{-1}$ is acceptable, then we may solve other polynomial equations having no real solutions by introducing a new "imaginary" number $\sqrt{-1}$. Let us denote this number by j, i.e.,

$$j = \sqrt{-1}$$

Thus, quadratic equations of the form $as^2 + bs + c = 0$ will have solutions even if $b^2 - 4ac < 0$. Specifically, the solutions are

$$s = -\frac{b}{2a} \pm \frac{j\sqrt{4ac - b^2}}{2a}$$

We refer to this kind of solution as complex solutions, and to numbers of the form $p + jq$ as *complex numbers*.

Real Part, Imaginary Part and Conjugate Pairs

A complex number z has a real part and an imaginary part, denoted here by p and jq respectively, where both p and q are real numbers. It should be noted that complex solutions of a polynomial equation always come in pairs which are of the form $p \pm jq$. We call $p \pm jq$ a *conjugate pair*. It is also customary to call $p + jq$ the complex conjugate of $p - jq$, and vice versa. For brevity we denote the complex conjugate of a complex number z by z^*.

Representations of Complex Numbers

Complex numbers are often represented by points on the Argand plane. The horizontal axis is the real number axis, and the vertical axis is the imaginary number axis. Thus, $1 + j1$ is represented by the point $(1, 1)$ on the Argand plane.

We refer to the above representation of the form $p + jq$ as *rectangular form*. Clearly, we may also represent a complex number in polar form, i.e.,

$$z = r\angle\theta$$

For example, the number $1 + j1$ can be represented as $\sqrt{2}\angle 45°$. The translation between the rectangular form and the polar form is given by

$$r = \sqrt{p^2 + q^2} \quad \text{and} \quad \theta = \arctan\left(\frac{q}{p}\right)$$

Conversely, we have

$$p = r\cos\theta \quad \text{and} \quad q = r\sin\theta$$

Of course, we need to take care of the signs of p and q, and assign the correct value of θ according to the quadrant to which θ belongs. For example, $-1 - j1$ should be written as $\sqrt{2}\angle -225°$, not $\sqrt{2}\angle 45°$.

Basic Operations of Complex Numbers

When we add up two complex numbers, we add up the real parts and the imaginary parts separately. Letting $z_1 = p_1 + jq_1$ and $z_2 = p_2 + jq_2$, we have

$$z_1 + z_2 = (p_1 + p_2) + j(q_1 + q_2)$$

We perform subtraction of two complex numbers in a likewise manner.

$$z_1 - z_2 = (p_1 - p_2) + j(q_1 - q_2)$$

The graphical illustrations of addition and subtraction on the Argand plane are shown in figure E.1.

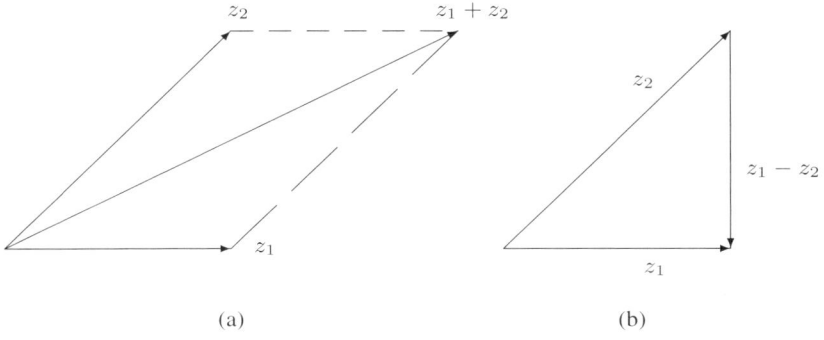

(a)   (b)

Figure E.1: (a) Addition; and (b) subtraction on the Argand plane

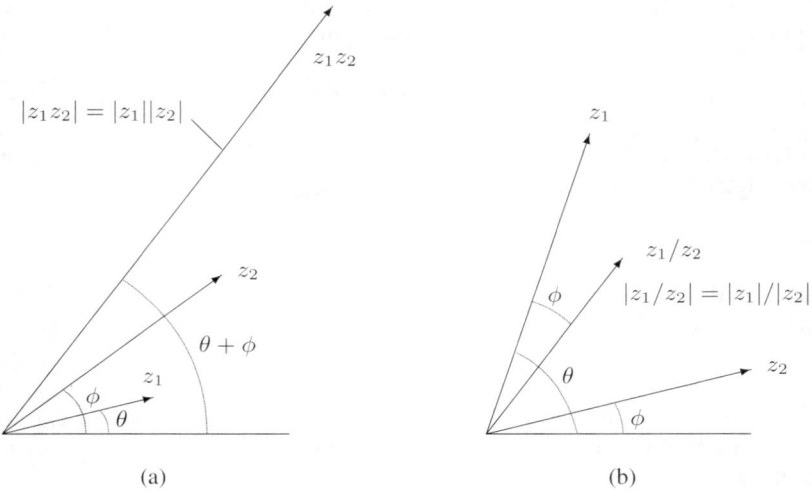

Figure E.2: (a) Multiplication; and (b) division on the Argand plane

The multiplication of two complex numbers in rectangular form proceeds as follows, with j^2 being taken as -1.

$$
\begin{aligned}
z_1 z_2 &= (p_1 + jq_1)(p_2 + jq_2) \\
&= (p_1 p_2 + j^2 q_1 q_2) + j(p_1 q_2 + q_1 p_2) \\
&= (p_1 p_2 - q_1 q_2) + j(p_1 q_2 + q_1 p_2)
\end{aligned}
$$

In polar form, we write

$$
\begin{aligned}
z_1 z_2 &= r_1 \angle \theta_1 r_2 \angle \theta_2 \\
&= r_1 r_2 \angle (\theta_1 + \theta_2)
\end{aligned}
$$

The graphical illustration of multiplication on the Argand plane is shown in figure E.2 (a).

The division of two complex numbers in rectangular form proceeds as follows. Here, we need to multiply the dividend and the divisor by the complex conjugate of the divisor.

$$
\begin{aligned}
\frac{z_1}{z_2} &= \frac{z_1 z_2^*}{z_2 z_2^*} \\
&= \frac{(p_1 + jq_1)(p_2 - jq_2)}{(p_2 + jq_2)(p_2 - jq_2)} \\
&= \frac{(p_1 p_2 + q_1 q_2) + j(q_1 p_2 - p_1 q_2)}{p_2^2 + q_2^2}
\end{aligned}
$$

In polar form, we write

$$\frac{z_1}{z_2} = \frac{r_1 \angle \theta_1}{r_2 \angle \theta_2}$$

$$= \frac{r_1}{r_2} \angle (\theta_1 - \theta_2)$$

The graphical illustration of division on the Argand plane is shown in figure E.2 (b).

Clearly, it is more convenient to perform addition and subtraction in rectangular form, and to perform multiplication and division in polar form.

Finally, we note the following useful rules involving moduli and conjugates of complex numbers:

$$|z^*| = |z|$$

$$|z_1 z_2^*| = |z_1|.|z_2^*| = |z_1|.|z_2|$$

$$z.z^* = |z.z^*| = |z|^2$$

$$(z_1 + z_2)^* = z_1^* + z_2^*$$

$$(z_1.z_2)^* = z_1^*.z_2^*$$

$$\left(\frac{z_1}{z_2}\right)^* = \frac{z_1^*}{z_2^*}$$

Example E.1 — Suppose $z_1 = 3 + j4$ and $z_2 = 2 - j9$. Then, the sum of z_1 and z_2 is

$$z_1 + z_2 = (3 + 2) + j(4 - 9) = 5 - j5$$

The difference is

$$z_1 - z_2 = (3 - 2) + j(4 + 9) = 1 + j13$$

The product is

$$z_1 z_2 = (3 + j4)(2 - j9) = (6 + 36) + j(8 - 27) = 42 - j19$$

The quotient is

$$\frac{z_1}{z_2} = \frac{z_1 z_2^*}{z_2 z_2^*} = \frac{(3 + j4)(2 + j9)}{85} = \frac{-6}{17} + j\frac{7}{17}$$

Example E.2 — Multiplication and division are more easily done in polar form. Suppose $z_1 = 3 + j4$ and $z_2 = 2 - j9$. We first convert z_1 and z_2 into polar forms, i.e.,

$$z_1 = 5\angle 53.13° \quad \text{and} \quad z_2 = 9.22\angle -77.47°$$

Multiplication is performed by multiplying the moduli and adding up the arguments, i.e.,

$$z_1 z_2 = (5\angle 53.13°)(9.22\angle -77.47°) = 46.1\angle -24.34°$$

Division is performed by dividing the moduli and subtracting the arguments, i.e.,

$$\frac{z_1}{z_2} = \frac{5\angle 53.13°}{9.22\angle -77.47°} = 0.54\angle 130.6°$$

We can convert the answers back to rectangular form if we wish.

$$z_1 z_2 = 42 - j19 \quad \text{and} \quad \frac{z_1}{z_2} = -0.35 + j0.41$$

Appendix F

Common Mathematical Formulae

Roots of Quadratic Equations

The roots of the equation $as^2 + bs + c = 0$ are given by

$$s = \begin{cases} \dfrac{-b \pm \sqrt{b^2 - 4ac}}{2a} & \text{if } 4ac \leq b^2 \\[2ex] \dfrac{-b \pm j\sqrt{4ac - b^2}}{2a} & \text{if } 4ac > b^2 \end{cases}$$

When the magnitudes of the two terms b and $\sqrt{b^2 - 4ac}$ in the formula are close, precision in calculating the roots is generally hard to achieve using the above forms of expression. The following alternative forms are preferable.

$$s = \begin{cases} \dfrac{-2c}{b \pm \sqrt{b^2 - 4ac}} & \text{if } 4ac \leq b^2 \\[2ex] \dfrac{-2c}{b = j\sqrt{4ac - b^2}} & \text{if } 4ac > b^2 \end{cases}$$

If one the roots, say s_1, has been found precisely, we can use the following formula to find the other root, s_2.

$$s_2 = -s_1 - \frac{b}{a} \quad \text{or} \quad s_2 = \frac{c}{as_1}$$

Differentiation

By definition, the derivative of $f(t)$ with respect to t, or the rate of change of $f(t)$, is given by

$$\frac{df(t)}{dt} = \lim_{\delta t \to 0} \frac{f(t + \delta t) - f(t)}{\delta t}$$

The following are derivatives of some common functions.

$$\frac{dt^n}{dt} = nt^{n-1}$$

$$\frac{d(t^{-n})}{dt} = -nt^{-(n+1)}$$

$$\frac{de^t}{dt} = e^t$$

$$\frac{d\sin t}{dt} = \cos t$$

$$\frac{d\cos t}{dt} = -\sin t$$

$$\frac{d}{dt}af(t) = a\frac{d}{dt}f(t)$$

$$\frac{d}{dt}[f(t)+g(t)] = \frac{d}{dt}f(t) + \frac{d}{dt}g(t)$$

$$\frac{d}{dt}[f(t)g(t)] = g(t)\frac{d}{dt}f(t) + f(t)\frac{d}{dt}g(t)$$

$$\frac{d}{dt}[f(t)g(t)h(t)] = g(t)h(t)\frac{d}{dt}f(t) + f(t)h(t)\frac{d}{dt}g(t) + f(t)g(t)\frac{d}{dt}h(t)$$

$$\frac{d}{dt}\left[\frac{f(t)}{g(t)}\right] = \frac{g(t)\frac{d}{dt}f(t) - f(t)\frac{d}{dt}g(t)}{g(t)^2}$$

$$\frac{d}{dx}f(t) = \frac{d}{dt}f(t)\frac{dt}{dx}$$

L'Hospital's Rule

If $F(\omega) = 0$ and $G(\omega) = 0$, or if $F(\omega) = \infty$ and $G(\omega) = \infty$, then

$$\lim_{s\to\omega}\frac{F(s)}{G(s)} = \lim_{s\to\omega}\frac{F'(s)}{G'(s)}$$

where $F'(s)$ is $dF(s)/ds$ and $G'(s)$ is $dG(s)/ds$. Moreover, if $F'(\omega) = 0$ and $G'(\omega) = 0$, or if $F'(\omega) = \infty$ and $G'(\omega) = \infty$, then

$$\lim_{s\to\omega}\frac{F(s)}{G(s)} = \lim_{s\to\omega}\frac{F''(s)}{G''(s)}$$

Integration by Parts

The following are useful in evaluating Laplace Transform integrals.

$$\int f(t)d\{g(t)\} = f(t)g(t) - \int g(t)d\{f(t)\}$$

$$\int_a^b f(t)d\{g(t)\} = [f(t)g(t)]_a^b - \int_a^b g(t)d\{f(t)\}$$

Integrals

Integration is the inverse of differentiation. The problem is to find a function whose derivative is a given function. Thus, the integral of $f(t)$ is $g(t)$ if $g'(t) = f(t)$. The following are integrals of commonly encountered functions. They are used in finding average and root-mean-square values, and in evaluating Fourier coefficients and Laplace Transform integrals.

Indefinite Integrals

For brevity, we omit the integration constant in each case.

$$\int t \, dt = \frac{t^2}{2}$$

$$\int \frac{1}{t} \, dt = \ln|t|$$

$$\int t^n \, dt = \frac{t^{n+1}}{n+1}$$

$$\int \frac{1}{t^n} \, dt = -\frac{1}{(n-1)t^{n-1}}$$

$$\int \frac{1}{a+bt} \, dt = \frac{1}{b} \ln|a+bt|$$

$$\int t^{p/2} \, dt = \frac{2}{p+2} t^{(p+2)/2}$$

$$\int \sin t \, dt = -\cos t$$

$$\int \sin(a+bt) \, dt = -\frac{1}{b} \cos(a+bt)$$

$$\int \sin\left(\frac{t}{a}\right) \, dt = -a \cos\left(\frac{t}{a}\right)$$

$$\int t \sin t \, dt = \sin t - t \cos t$$

$$\int \sin^2 t \, dt = \frac{t}{2} - \frac{\sin 2t}{4}$$

$$\int \sin mt \cos nt \, dt = \frac{\sin(m-n)t}{2(m-n)} - \frac{(m+n)t}{2(m+n)}$$

$$\int \cos t \, dt = \sin t$$

$$\int \cos(a + bt)\ dt \;=\; \frac{1}{b}\sin(a + bt)$$

$$\int \cos\left(\frac{t}{a}\right)\ dt \;=\; a\sin\left(\frac{t}{a}\right)$$

$$\int t\cos t\ dt \;=\; \cos t + t\sin t$$

$$\int \cos^2 t\ dt \;=\; \frac{t}{2} + \frac{\sin 2t}{4}$$

$$\int e^{at} dt \;=\; \frac{1}{a}e^{at}$$

$$\int t e^{at} dt \;=\; e^{at}\left(\frac{t}{a} - \frac{1}{a^2}\right)$$

$$\int e^{at}\sin t\ dt \;=\; \frac{e^{at}}{a^2 + 1}(a\sin t - \cos t)$$

$$\int e^{at}\cos t\ dt \;=\; \frac{e^{at}}{a^2 + 1}(a\cos t + \sin t)$$

$$\int e^{at}\sin nt\ dt \;=\; \frac{e^{at}}{a^2 + n^2}(a\sin nt - n\cos nt)$$

$$\int e^{at}\cos nt\ dt \;=\; \frac{e^{at}}{a^2 + n^2}(a\cos nt + n\sin nt)$$

$$\int \tan t\ dt \;=\; -\ln|\cos t|$$

Definite Integrals

$$\int_0^\infty e^{-at} dt \;=\; \frac{1}{a}$$

$$\int_0^\infty t e^{-at} dt \;=\; \frac{1}{a^2}$$

$$\int_0^\infty t^2 e^{-at} dt \;=\; \frac{2}{a^3}$$

$$\int_0^\infty \sqrt{t}\,e^{-at} \;=\; \frac{\sqrt{\pi}}{2a\sqrt{a}}$$

$$\int_0^\infty e^{-at}\sin mt\ dt \;=\; \frac{m}{a^2 + m^2}$$

$$\int_0^\infty e^{-at}\cos mt\ dt \;=\; \frac{a}{a^2 + m^2}$$

$$\int_0^\infty \frac{e^{-at} - e^{-bt}}{t} dt \;=\; \ln\frac{b}{a}$$

$$\int_0^\pi \sin nt \sin mt \, dt = 0 \quad (m \neq n)$$

$$\int_0^\pi \sin nt \sin mt \, dt = \frac{\pi}{2} \quad (m = n)$$

$$\int_0^\pi \cos nt \cos mt \, dt = 0 \quad (m \neq n)$$

$$\int_0^\pi \cos nt \cos mt \, dt = \frac{\pi}{2} \quad (m = n)$$

$$\int_0^\pi \sin nt \cos mt \, dt = 0 \quad (m = n)$$

$$\int_0^\pi \sin nt \cos mt \, dt = 0 \quad (m \neq n; (m + n) \text{ even})$$

$$\int_0^\pi \sin nt \cos mt \, dt = \frac{2m}{m^2 - n^2} \quad (m \neq n; (m + n) \text{ odd})$$

$$\int_0^\pi \sin^2 mt \, dt = \int_0^\pi \cos^2 mt \, dt = \frac{\pi}{2}$$

Trigonometric Functions

The following are some commonly used relations involving trigonometric functions.

$$\sin^2 A = 1 - \cos^2 A$$
$$\sec^2 A = 1 + \tan^2 A$$
$$\sin(-A) = -\sin A$$
$$\cos(-A) = \cos A$$
$$\sin(A + B) = \sin A \cos B + \cos A \sin B$$
$$\sin(A - B) = \sin A \cos B - \cos A \sin B$$
$$\cos(A + B) = \cos A \cos B - \sin A \sin B$$
$$\cos(A - B) = \cos A \cos B + \sin A \sin B$$
$$2 \sin A \cos B = \sin(A + B) + \sin(A - B)$$
$$2 \cos A \cos B = \cos(A + B) + \cos(A - B)$$
$$2 \sin A \sin B = \cos(A - B) - \cos(A + B)$$
$$\sin A + \sin B = 2 \sin \frac{1}{2}(A + B) \cos \frac{1}{2}(A - B)$$
$$\sin A - \sin B = 2 \sin \frac{1}{2}(A - B) \cos \frac{1}{2}(A + B)$$
$$\cos A + \cos B = 2 \cos \frac{1}{2}(A + B) \cos \frac{1}{2}(A - B)$$
$$\cos A - \cos B = 2 \sin \frac{1}{2}(A + B) \sin \frac{1}{2}(B - A)$$

$$p \cos A + q \sin A \;=\; r \sin(A + \theta)$$
$$\text{where} \qquad r = \sqrt{p^2 + q^2}, \;\; \sin \theta = p/r, \;\; \cos \theta = q/r$$
$$p \cos A + q \sin A \;=\; r \cos(A - \phi)$$
$$\text{where} \qquad r = \sqrt{p^2 + q^2}, \;\; \cos \phi = p/r, \;\; \sin \phi = q/r$$
$$\sin \frac{A}{2} \;=\; \sqrt{\frac{1}{2}(1 - \cos A)}$$
$$\cos \frac{A}{2} \;=\; \sqrt{\frac{1}{2}(1 + \cos A)}$$
$$\tan \frac{A}{2} \;=\; \sqrt{\frac{1 - \cos A}{1 + \cos A}}$$

Appendix G

Resistor Colour Codes

For low-power resistors, the resistance values are coded in colour bands printed around the surface of the resistors. Typical resistors contain three to five colour bands, each band representing a significant figure, a multiplier or a tolerance percentage. (Some have a sixth band to indicate the temperature coefficient.) The value represented by each colour is tabulated in Table G.1. Without ambiguity, we refer to the band nearest the end of a resistor as the first band, as shown in figure G.1.

For a three-band resistor (quite uncommon now), the tolerance is $\pm20\%$. The first two bands represent the significant figures, and the last band indicates the multiplier. For example, a $100\Omega \pm20\%$ resistor is marked with the following three bands:

brown–black–brown (100)

For a four-band resistor, the first two bands represent the significant figures, the third band represents the multiplier, and the last band indicates the tolerance. For example, a $47k\Omega \pm2\%$ resistor is marked with the following colour bands:

yellow–violet–orange–red (47k)

Finally, for a five-band resistor, the first three bands represent the significant figures, the fourth band represents the multiplier, and the last band indicates the tolerance. For example, a $249k\Omega \pm1\%$ resistor is marked with the following colour bands:

red–yellow–white–orange–brown (249k)

The following points regarding the resistor colour coding system are worth noting:

1. The tolerance band is a little thicker than the other bands.

2. Gold and silver do not appear in the figure bands. When a gold or silver band appears at the end of the resistor, it must be the tolerance band, and significant figures should be read from the other end.

3. If a sixth band exists, it indicates the *temperature coefficient* of the resistor. Yellow for 25ppm, orange for 15ppm, red for 50ppm, and brown for 100ppm.

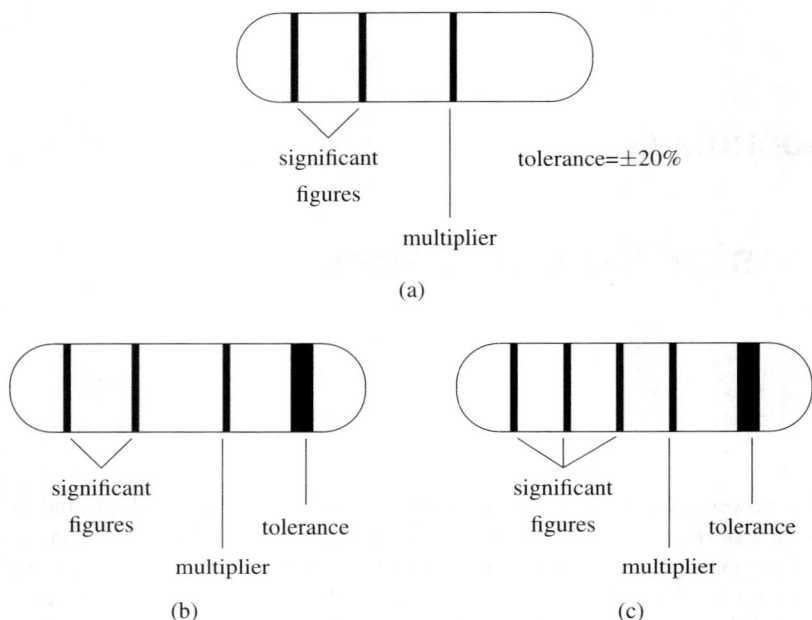

Figure G.1: Resistor colour bands

Colour	Significant figure	Multiplier	Tolerance
Silver	—	10^{-2}	±10%
Gold	—	10^{-1}	±5%
Black	0	1	—
Brown	1	10	±1%
Red	2	10^2	±2%
Orange	3	10^3	—
Yellow	4	10^4	—
Green	5	10^5	±0.5%
Blue	6	10^6	±0.25%
Violet	7	10^7	±0.1%
Grey	8	10^8	—
White	9	10^9	—
None	—	—	±20%

Table G.1: Values represented by colour bands

Answers to End-of-Chapter Problems

Chapter 1

1 0.086Ω, $34.88A$.

2 Circuits (b) and (d) are not solvable. (a) $V_x = 9V$ and $I_x = -3A$. (c) $V_x = -1V$ and $I_x = \frac{2}{3}A$.

3 (a) Voltage across $1\Omega = \frac{5}{3}V$, Voltage across $2\Omega = \frac{10}{3}V$, Voltage across $3\Omega = 5V$. (b) Current in $\frac{1}{2}\Omega = \frac{2}{9}A$, Current in $\frac{1}{3}\Omega = \frac{1}{3}A$, Current in $\frac{1}{4}\Omega = \frac{4}{9}A$.

4 (a) $R_1 + R_4 + \dfrac{R_2 R_3}{R_2 + R_3}$, (b) $R_6 + \dfrac{R_1\left(R_2 + R_3 + \frac{R_4 R_5}{R_4 + R_5}\right)}{R_1 + R_2 + R_3 + \frac{R_4 R_5}{R_4 + R_5}}$.

5 (a) $V_{R_1} = 4V$, $I_{R_2} = 2A$. (b) $V_{R_1} = \frac{25}{6}V$, $I_{R_2} = \frac{5}{3}A$.

6 $I_5 = \dfrac{R_2 R_4 - R_1 R_3}{R_5(R_1 + R_4)(R_2 + R_3) + R_1 R_4(R_2 + R_3) + R_2 R_3(R_1 + R_4)} V_0$.

7 (a) $2R$. (b) $(1 + \sqrt{5})R/2$.

8 (a) $I_x = 0.5A$, $V_x = -1V$. (b) $V_x = \frac{8}{3}V$, $I_x = 2A$.

10 $V_1 = \frac{8}{5}V$, $V_x = \frac{64}{5}V$.

Chapter 2

1 For each pair of v and i, the set of v and the set of i satisfy Kirchhoff's laws independently.

2 $\hat{v}_L = 1V$, $\hat{i}_L = \frac{1}{2}A$. $i_3 = -\frac{1}{2}A$, i.e., i_3 flows upwards. But yet nothing is wrong about this circuit because resistance values can be negative! No violation of assumption of reciprocity.

4 $V_x = 4V$.

5 (a) $V_T = -5.97V$, $I_N = -1.2A$, $R_T = R_N = 4.97\Omega$. (b) $V_T = 8000V$, $I_N = 8A$,

$R_T = R_N = 1000\Omega$. (c) $V_T = -6\text{V}$ (no Norton circuit). (d) $V_T = 3.33\text{V}$, $I_N = 5\text{A}$, $R_T = R_N = 0.67\Omega$. (e) $I_N = -5\text{A}$ (no Thévenin circuit). (f) $V_T = 582.15\text{V}$, $I_N = 64\text{A}$, $R_T = R_N = 9.09\Omega$.

7 (a) $V_T = 1.909\text{V}$, $I_N = 0.14\text{A}$, $R_T = R_N = 13.64\Omega$. (b) $V_T = 24.2\text{V}$, $I_N = 1.72\text{A}$, $R_T = R_N = 14\Omega$.

8 Maximum power transfer takes place when $R_L = 9.75\Omega$, and maximum power $P_{\max} = 0.314\text{W}$.

9 (a) 4.667Ω. (b) 1Ω.

10 $V_T = 582.15\text{V}$, $I_N = 64\text{A}$, $R_T = R_N = 9.09\Omega$.

Chapter 3

1 Circuit (b) is planar.

(a)
$$\begin{bmatrix} 3 & -2 & 0 & 0 \\ -2 & 5 & -3 & 0 \\ 0 & -3 & 5 & 0 \\ 0 & 0 & 0 & 1 \end{bmatrix} \begin{bmatrix} V_1 \\ V_2 \\ V_3 \\ V_4 \end{bmatrix} = \begin{bmatrix} 0 \\ 1 \\ -2 \\ 2 \end{bmatrix}$$

(b)
$$\begin{bmatrix} 5 & 0 & -2 \\ 0 & 2 & -1 \\ -2 & -1 & 3 \end{bmatrix} \begin{bmatrix} V_1 \\ V_2 \\ V_3 \end{bmatrix} = \begin{bmatrix} 1 \\ -1 \\ -2 \end{bmatrix}$$

(c)
$$\begin{bmatrix} 4 & -3 \\ -3 & 6 \end{bmatrix} \begin{bmatrix} I_1 \\ I_2 \end{bmatrix} = \begin{bmatrix} 2 \\ 1 \end{bmatrix}$$

(d)
$$\begin{bmatrix} 6 & -2 & 0 \\ -2 & 7 & -2 \\ 0 & -2 & 4 \end{bmatrix} \begin{bmatrix} I_1 \\ I_2 \\ I_3 \end{bmatrix} = \begin{bmatrix} 3 \\ 0 \\ 2 \end{bmatrix}$$

2 $V_x = 4.7\text{V}$.

$$\begin{bmatrix} \dfrac{1}{15} + \dfrac{1}{30} + \dfrac{1}{2} & \dfrac{-1}{30} & \dfrac{-1}{2} \\[2mm] \dfrac{-1}{30} & \dfrac{1}{25} + \dfrac{1}{30} + \dfrac{1}{10} & \dfrac{-1}{10} \\[2mm] \dfrac{-1}{2} & \dfrac{-1}{10} & \dfrac{1}{2} + \dfrac{1}{10} \end{bmatrix} \begin{bmatrix} V_1 \\ V_2 \\ V_3 \end{bmatrix} = \begin{bmatrix} -1 \\ 0 \\ 5 \end{bmatrix}$$

3 $I_y = -458.5\text{mA}$.

4 $I_x = 121.74\text{mA}$.

$$
\begin{bmatrix}
50 + 100 & -50 & -100 & 0 \\
-50 & 10 + 50 & -5 & 0 \\
-100 & -5 & 100 + 5 & 0 \\
0 & 0 & 0 & 10
\end{bmatrix}
\begin{bmatrix}
I_1 \\ I_2 \\ I_3 \\ I_4
\end{bmatrix}
=
\begin{bmatrix}
2 - 3 \\ 0 \\ -4 \\ 3 + 4
\end{bmatrix}
$$

5 $V_y = -76.65\text{V}$.

6 (a) $V_x = -10\text{V}$. (b) $I_y = -1.6\text{A}$.

7 (a) $V_x = -80.16\text{V}$. (b) $I_y = -0.3467\text{A}$.

8 m mesh equations in mesh method versus $n - 1$ equations in nodal method.

9 (a) $V_{o/c} = 80\text{V}$, $R_T = 36.7\Omega$. (b) $V_{o/c} = 10\text{V}$, $R_T = 7.5\Omega$. (c) $V_{o/c} = 2\text{V}$, $R_T = 4\Omega$. Thus, maximum power transfer occurs when $R_{ab} = 4\Omega$.

Chapter 4

1 $\frac{1}{2}C(10 - 3)^2 = 0.0098\text{J}$.

2 (Energy loss when charging from 3V to 8V) = (energy loss when charging from $t = 0$ to $\tau \ln \frac{7}{2}$) = 0.009J. Knowledge of resistor value is not needed.

3 $v(0^+) = 0$, $v(\infty) = IR_2$, $\tau = CR_2$, $v_C(t) = IR_2 \left(1 - e^{-t/\tau}\right)$.

4 $v(0^+) = \frac{R_2 U}{R_1 + R_2 + R_3}$, $v(\infty) = \frac{R_2 U}{R_1 + R_2}$, $\tau = C(R_1 \| R_2)$,

$$
v_C(t) = R_2 U \left[\frac{1}{R_1 + R_2 + R_3} + \left[\frac{1}{R_1 + R_2} - \frac{1}{R_1 + R_2 + R_3} \right] \left(1 - e^{-t/\tau}\right) \right]
$$

5 $v_{C1}(t) = 0.5E + 0.25(1 - e^{-t/\tau})E$, $v_{C2} = 0.25(1 - e^{-t/\tau})E$, where $\tau = R(C_1 \| C_2) = RC_1 C_2/(C_1 + C_2)$.

6 (a) $v_C(t) = 6 - e^{-t/8}$. (b) $v_C(t) = 5 + 3e^{-t/2}$. (c) $v_C(t) = 8 - 2e^{-t/4}$. Loss = $C_1 E^2/16$.

7 $i_{L1}(t) = -0.25I + 0.75e^{-t/\tau}I$, $i_{L2}(t) = 1.25I + 0.75Ie^{-t/\tau}$, where $\tau = (L_1 \| L_2)/R = L_1 L_2/(L_1 + L_2)R$.

8 (a) true. (b) false. (c) true. (d) false. (e) true. (f) true.

9 (a) $v_C(t) = 8.33 - 3.33e^{-t/\tau}$ V, $i_R(t) = 0.67 + 0.33e^{-t/\tau}$ A, where $\tau = 0.00333$s. (b) $i_L(t) = 3 - 2e^{-t/\tau}$ A, $v_R(t) = 2e^{-t/\tau}$ V, where $\tau = 0.1$s.

Chapter 5

1 The system of equations of n coupled inductors is

$$
L_{11}\frac{di_1}{dt} + L_{12}\frac{di_2}{dt} + \cdots + L_{1n}\frac{di_n}{dt} = v_1
$$

$$
L_{21}\frac{di_1}{dt} + L_{22}\frac{di_2}{dt} + \cdots + L_{2n}\frac{di_n}{dt} = v_2
$$

$$\cdots = \cdots$$

$$L_{n1}\frac{di_1}{dt} + L_{n2}\frac{di_2}{dt} + \cdots + L_{nn}\frac{di_n}{dt} = v_n$$

2 $v_1 = 2\cos t - 1.5\sin t$, $v_2 = 0.4\cos t - 0.8\sin t$.

3 $\kappa = 1$. The equivalent circuit as observed from the left terminals is a parallel combination of an inductor L and a resistor $R/4$.

4 $n = 14$. Power dissipation = 0.158 W.

5 $A = L_1 - M$, $B = L_2 - M$, $C = M$, $E = (L_1L_2 - 3M^2)/M$, $F = (L_1L_2 - 3M^2)/(L_2 - M)$, $G = (L_1L_2 - 3M^2)/(L_1 - M)$.

6 $i_1 = \frac{2}{3}$A, $v_1 = v_3 = -\frac{2}{3}$V, $i_2 = i_4 = -\frac{1}{3}$A, $v_2 = \frac{4}{3}$V, $v_4 =2$V, $i_3 =1$A.

7 $i_1(t) = 2.36e^{-0.5t} - 1.36e^{-6t} + 3$ A, $i_2(t) = 1.18e^{-0.5t} + 1.82e^{-6t}$ A. (Verify that $i_1(\infty) = 3$A, and $i_2(\infty) = 0$A.)

8 When $M = 2L$, magnetizing inductance = L (primary side) and leakage inductance = 0. When $M = L$, magnetizing inductance = L (primary side) and leakage inductance = $3L$ (secondary side).

10 0.138A (amplitude).

Chapter 6

1 (i) $v_{av} = 0$, $v_{rms} = 0.7071$. (ii) $v_{av} = 10$, $v_{rms} = 10.0995$. (iii) $v_{av} = 2$, $v_{rms} = 2.236$. (iv) $v_{av} = 0.5$, $v_{rms} = 0.7071$.

2 (a) Average = $\alpha(a + A)/2$, rms = $\sqrt{(a^2 + aA + A^2)\alpha/3}$. (b) Average = $A(1 - \cos\alpha)/\pi$, rms = $\sqrt{A^2(2\alpha - \sin 2\alpha)/4\pi}$.

3 Average power = $954.93 \times [1 - \cos 0.2\alpha\pi - \cos 0.2(1 + \alpha)\pi - \cos 0.2(2 + \alpha)\pi - \cos 0.2(3 + \alpha)\pi - \cos 0.2(4 + \alpha)\pi]$ W.

4 Sum = $15.2563\sin(\omega t + 101.75°)$.

5 Let $\omega = 2\pi/T$. (a) $\frac{4}{\pi}\sin\omega t + \frac{4}{3\pi}\sin 3\omega t + \frac{4}{5\pi}\sin 5\omega t + \cdots$,
(b) $-\frac{8}{\pi^2}\cos\omega t - \frac{8}{9\pi^2}\cos 3\omega t - \frac{8}{25\pi^2}\cos 5\omega t - \cdots$,
(c) $1.13\sin\frac{\pi t}{2} + 0.19\sin\frac{3\pi t}{2} + 0.08\sin\frac{5\pi t}{2} + 0.05\sin\frac{7\pi t}{2} + \cdots$,
(d) $\frac{2}{\pi}\sin\omega t - \frac{1}{\pi}\sin 2\omega t + \frac{2}{3\pi}\sin 3\omega t - \frac{1}{2\pi}\sin 4\omega t + \frac{2}{5\pi}\sin 5\omega t + \cdots$,
(e) $-\frac{1}{4} + (\frac{3}{\pi}\sin\omega t + \frac{3}{2\pi}\sin 2\omega t + \frac{1}{\pi}\sin 3\omega t + \frac{3}{5\pi}\sin 5\omega t + \frac{1}{\pi}\sin 6\omega t + \frac{3}{7\pi}\sin 7\omega t + \frac{1}{3\pi}\sin 9\omega t + \cdots) + (\frac{3}{\pi}\cos\omega t - \frac{1}{\pi}\cos 3\omega t + \frac{3}{5\pi}\cos 5\omega t - \frac{3}{7\pi}\cos 7\omega t + \cdots)$,
(f) $-\frac{1}{4} + (\frac{12}{\pi^2}\sin\omega t - \frac{4}{3\pi^2}\sin 3\omega t + \frac{12}{25\pi^2}\sin 5\omega t - \frac{12}{49\pi^2}\sin 7\omega t + \cdots) + (-\frac{6}{\pi^2}\cos 2\omega t - \frac{2}{3\pi^2}\cos 6\omega t - \frac{6}{25\pi^2}\cos 10\omega t - \cdots)$.

6 Let $T(t)$ be the triangular wave of figure 6.16 (b), and $R(t)$ be the rectangular wave of figure 6.16 (a). $T'(t) = \frac{2\omega}{\pi}R(t)$.

7 84.5W, 1.69W, 4225W.

8 (a) 0.93, 0.093W. (b) 0.33, 0.033W. (c) 0.66, 0.066W. (d) 0.3, 0.03W. (e) 1.57, 0.157W. (f) 1, 0.1W.

9 (a) 0. (b) 0. (c) 0. (d) 0. (e) –3W. (f) –3W.

Chapter 7

1 Top-left: $Z = R + j\omega L$, no singular point for real ω. Top-right: $Z = j\left(\omega L - \frac{1}{\omega C}\right)$, singular point at $\omega = 1/\sqrt{LC}$ where $Z = 0$. Lower-left: $Z = \frac{-j(1-\omega^2 LC_2)}{\omega(C_1+C_2)(1-\omega^2 LC')}$ where $C' = C_1 C_2/(C_1 + C_2)$, singular points at $\omega = 1/\sqrt{LC'}$ where $Z \Rightarrow \infty$ and at $\omega = 1/\sqrt{LC_2}$ where $Z = 0$. Lower-right: $Z = R + j\left(\omega L - \frac{1}{\omega C}\right)$, no singular point for real ω.

2 $v_o(t) = V_o + \tilde{v}_o(t)$, $V_o = 10$V, $\tilde{v}_o(t) = 9.89\sin(\omega t - 81.47°)$ mV at $\omega = 5000$rad/s, $\tilde{v}_o(t) = 61.7\sin(\omega t - 21.8°)$ mV at $\omega = 300$rad/s.

3 (a) $V_1 = 49.95\angle-89.9994°$ (top left node), $V_2 = 0.0499975\angle-89.43°$ (top right node). (b) $I_1 = 8.94\angle-21.02°$ (left mesh clockwise), $I_2 = 16.14\angle-23.79°$ (right mesh clockwise).

5 $S = 330$VA, p.f. $= 0.8939$, $\phi = 26.62°$.

6 $S = VI^* = 200\angle25°$VA, $P = 181.26$W, $Q = 84.52$var, $Z = 2\angle25° = 1.81 + j0.85\Omega$.

7 $V_s = 6.067$V, $I_s = 3.064$A rms, $P = 18$W, $Q = -4.66$var, $S = 18.59$VA, p.f. $= 0.97$.

8 Case 1: $I_{in} = 8.43$A rms, $P_{in} = 1247.7$W, p.f. $= 0.62$ lagging. Case 2: $I_{in} = 17.35$A rms, $P_{in} = 1477.49$W, p.f. $= 0.35$ lagging.

9 0.0733F.

Chapter 8

1 $s^2 + 2005s + 10000 = 0$, $\lambda = -2000, -5$, $v_o(t) = Ae^{-2000t} + Be^{-5t}$.

2 Zeros at $\pm j10$, poles at $-2000, -5$. $\lim_{s\to\infty} |V_o/V_i| = 10000$, $\lim_{s\to 0} = 100$.

3 $0 < \zeta < 1$.

5 $\lambda = -1000, 0, -2 \pm j\sqrt{5}$, $A + Be^{-1000t} + Ce^{-2t}\cos(\sqrt{5}t + \phi)$.

6 (a) Use G instead of R in the derivation.

$$\frac{V_o}{V_{in}} = \frac{G_1}{G_1 G_2 + s(G_2 G_1 L + C_2) + s^2(G_1 C_2 L + G_2 C_1 L + G_1 C_1 L) + s^3 L C_1 C_2}$$

$$Z_{in} = \frac{G_1 + G_2 + s(G_1 G_2 L + C_2) + s^2(G_1 C_2 L + G_2 C_1 L + G_1 C_1 L) + s^3 C_1 C_2 L}{G_1 G_2 + s(C_2 G_1 + C_1 G_2 + C_1 G_1) + s^2 C_1 C_2}$$

Characteristic equation is $2 + 2s + 3s^2 + s^3 = 0$. Poles are $-2.52, -0.24 \pm j0.86$, and zero at infinity.

(b)

$$\frac{V_o}{V_{in}} = \frac{s^2 LC}{1 + sLG + s^2 LC}$$

$$Z_{in} = \frac{s^2 LC + sLG + 1}{sC(sGL + 1)}$$

Characteristic equation is $1 + s + s^2 = 0$. Poles are $-0.5 \pm j0.866$, and zeros are $0, 0$.

8

$$F(s) = \frac{2000\pi \times 10^3 \left(1 + \dfrac{s}{600\pi}\right)}{\left(1 + \dfrac{s}{20\pi}\right)\left(1 + \dfrac{200\pi \times 10^3}{s}\right)\left(1 + \dfrac{s}{100\pi \times 10^6}\right)}$$

Mid-band gain is $2000\pi \times 10^3$ or 136dB.

9

$$F(s) = \frac{200\pi \left(1 + \dfrac{s}{600\pi}\right)}{\left(1 + \dfrac{20\pi}{s}\right)\left(1 + \dfrac{s}{200\pi \times 10^3}\right)\left(1 + \dfrac{s}{100\pi \times 10^6}\right)}$$

Chapter 9

1 (a) Cutsets: $\{a, c, e\}$, $\{a, b\}$, $\{b, d, f\}$, etc., KCL equations are $I_a + I_c + I_e = 0$, $I_a - I_b = 0$, $I_b + I_d + I_f = 0$. Loops: $\{a, b, d, c\}$, $\{a, b, f, e\}$, $\{e, c, d, f\}$, KVL equations are $V_a + V_b - V_d - V_C = 0$, $V_a + V_b - V_f - V_e = 0$, $-V_e + V_c + V_d - V_f = 0$.
(b) Cutsets: $\{a, b, e\}$, $\{a, b, f\}$, $\{d, c, f\}$, etc., KCL equations are $I_a - I_b - I_e = 0$, $I_a - I_b - I_f = 0$, $I_d + I_c - I_f = 0$. Loops: $\{a, b, c, d\}$, $\{a, e, f, g, d\}$, $\{b, e, f, g, c\}$. KVL equations are $V_a + V_b - V_c + V_d = 0$, $V_a + V_e + V_f + V_g + V_d = 0$. $-V_b + V_e + V_f + V_g + V_c = 0$.
(c) Cutsets: $\{a, b, d, f\}$, $\{c, f\}$, $\{c, d, e, g\}$, etc., KCL equations are $I_a + I_b + I_d + I_f = 0$, $I_c - I_f = 0$, $I_c + I_d - I_e - I_g = 0$. Loops: $\{a, b\}$, $\{c, d, f\}$, $\{b, e, d\}$, etc., KVL equations are $V_a - V_b = 0$, $V_c + V_f - V_d = 0$, $V_b - V_e - V_d = 0$.

2 (a) yes, (b) no, (c) no, (d) yes.

3 (a) Tree $\{a, b, c, e\}$, independent KVL: $V_d - V_b - V_a + V_c = 0$, $V_f - V_b - V_a + V_e = 0$, independent KCL: $I_a + I_d + I_f = 0$, $I_b + I_d + I_f = 0$, $I_c - I_d = 0$, $I_e - I_f = 0$.
(b) Tree $\{a, b, c, e, f\}$, independent KVL: $V_d + V_a + V_b - V_c = 0$, $V_g + V_c - V_b - V_e + V_f = 0$. independent KCL: $I_a - I_d = 0$, $I_b - I_d + I_g = 0$, $I_c + I_d - I_g = 0$, $I_e - I_g = 0$, $I_f - I_g = 0$.
(c) Tree $\{a, c, d, g\}$, independent KVL: $V_b - V_a = 0$, $V_e - V_a + V_d = 0$, $V_f - V_d + V_c = 0$, $V_h + V_g - V_a + V_d = 0$. independent KCL: $I_a + I_b + I_e + I_h = 0$, $I_c - I_f = 0$, $I_d + I_f - I_e - I_h = 0$, $I_g - I_h = 0$.

4 $I_x = -4\text{A}, I_y = -8\text{A}, I_z = -2\text{A}, V_x = 7\text{V}, V_y = 7\text{V}.$

5 (a)

$$Q = \begin{bmatrix} a & b & c & e & d & f \\ 1 & 0 & 0 & 0 & 1 & 1 \\ 0 & 1 & 0 & 0 & 1 & 1 \\ 0 & 0 & 1 & 0 & -1 & 0 \\ 0 & 0 & 0 & 1 & 0 & -1 \end{bmatrix}$$

$$B = \begin{bmatrix} a & b & c & e & d & f \\ -1 & -1 & 1 & 0 & 1 & 1 \\ -1 & -1 & 0 & 1 & 0 & 1 \end{bmatrix}$$

(b)

$$Q = \begin{bmatrix} a & b & c & e & f & d & g \\ 1 & 0 & 0 & 0 & 0 & -1 & 0 \\ 0 & 1 & 0 & 0 & 0 & -1 & 1 \\ 0 & 0 & 1 & 0 & 0 & 1 & -1 \\ 0 & 0 & 0 & 1 & 0 & 0 & -1 \\ 0 & 0 & 0 & 0 & 1 & 0 & -1 \end{bmatrix}$$

$$B = \begin{bmatrix} a & b & c & e & f & d & g \\ 1 & 1 & -1 & 0 & 0 & 1 & 0 \\ 0 & -1 & 1 & 1 & 1 & 0 & 1 \end{bmatrix}$$

(c)

$$Q = \begin{bmatrix} a & c & d & g & b & f & h \\ 1 & 0 & 0 & 0 & 1 & 0 & 1 \\ 0 & 1 & 0 & 0 & 0 & -1 & 0 \\ 0 & 0 & 1 & 0 & 0 & 1 & -1 \\ 0 & 0 & 0 & 1 & 0 & 0 & -1 \end{bmatrix}$$

$$B = \begin{bmatrix} a & c & d & g & b & f & h \\ -1 & 0 & 0 & 0 & 1 & 0 & 0 \\ 0 & 1 & -1 & 0 & 0 & 1 & 0 \\ -1 & 0 & 1 & 1 & 0 & 0 & 1 \end{bmatrix}$$

6

$$Q = \begin{bmatrix} E_C & E_4 & R_1 & R_2 & R_5 & I_3 & I_6 \\ 1 & 0 & 0 & 1 & 0 & 1 & 1 \\ 0 & 1 & 0 & 0 & 1 & -1 & -1 \\ 0 & 0 & 1 & -1 & 0 & -1 & 0 \end{bmatrix}$$

$$B = \begin{bmatrix} E_0 & E_4 & R_1 & R_2 & R_5 & I_3 & I_6 \\ -1 & 0 & 1 & 1 & 0 & 0 & 0 \\ 0 & -1 & 0 & 0 & 1 & 0 & 0 \\ -1 & 1 & 1 & 0 & 0 & 1 & 0 \\ -1 & 1 & 0 & 0 & 0 & 0 & 1 \end{bmatrix}$$

KCL equations $QI = 0$, KVL equations $BV = 0$.

7 $V(R_1) = 6V$, $V(R_2) = 4V$, $I(E_0) = -7A$, $I(E_4) = 1A$, $I(R_5) = 2A$, $I(R_1) = 6A$, $I(R_2) = 4A$, $V(I_3) = 2V$, $V(I_6) = 8V$.

10 Replace C by L, and R_C by $1/R_L$. $I_L(t) = I_i(1 - e^{-tR_L/L})$. $V = I_i R_L e^{-tR_L/L}$.

Chapter 10

1 $t = 9$, $l = 11$. Cutset-voltage method needs 6 equations, while loop-current method needs 9 equations. Use cutset-voltage method.

2 (i) not solvable, (ii) not solvable, (iii) solvable, (iv) solvable, (v) solvable.

3 (iii) 0 equations (inspection), (iv) 1 equation, loop-current method, (v) 1 equation, cutset-voltage method.

4 (a) Use loop-current method. Check answers by comparing node voltages: 0.8V (top left), 0.2V (top middle), 0.2V (top right), −0.2V (middle left), 0.6V (center), 0V (bottom middle as ref), −0.8V (bottom right). (b) Use loop-current method. Node voltages are −1V (top left), 1V (top middle), −2V (top right), 2V (middle left), 0V (center as ref), 5V (bottom middle).

5 Loop-current method. Check answers by comparing node voltages: 0V (bottom right as ref), −0.2V (top right), 0V (top left), −1V (middle left), 0V (center), −0.6V (middle right).

7 $g = 0.2$A/V. Power supplied by controlled source = 8W.

8 $R_{in} = 10\Omega$.

9 $V_{o/c} = -0.0227V$, $I_{s/c} = -0.0227A$, $R_T = 10\Omega$.

10 $i_1 = i_{s1}/2$, $i_2 = -i_{s1}/12$, $i_3 = 7i_{s1}/24$, $i_4 = -7i_{s1}/12$, $i_5 = 5i_{s1}/24$, $i_6 = i_{s1}/12$, $i_7 = i_{s1}/8$, $i_{v_{s1}} = i_6 = i_{s1}/12$.

Chapter 11

1 4-dim state vector, order = 4.

3 Number of independent storage elements = 7. A particular choice of tree is $\{u, C_1, C_2, C_3, C_4, R_1, L_2\}$, and the corresponding state vector is

$$x = \begin{bmatrix} v_{C1} \\ v_{C2} \\ v_{C3} \\ v_{C4} \\ i_{L1} \\ i_{L3} \\ i_{L4} \end{bmatrix}$$

The choice of tree not unique, hence state vector not unique. Circuit order = 7.

4 Assume that all branches are directed either downwards or to the right.

$$
Q =
\begin{array}{c}
\begin{array}{ccccccc|ccccc}
u & C_1 & C_2 & C_3 & C_4 & R_1 & L_2 & C_5 & R_2 & L_1 & L_3 & L_4 \\
1 & 0 & 0 & 0 & 0 & 0 & 0 & 0 & 0 & 1 & 0 & 0 \\
0 & 1 & 0 & 0 & 0 & 0 & 0 & 0 & 0 & -1 & 1 & 1 \\
0 & 0 & 1 & 0 & 0 & 0 & 0 & 0 & 0 & 0 & -1 & 0 \\
0 & 0 & 0 & 1 & 0 & 0 & 0 & 1 & 1 & 0 & 0 & -1 \\
0 & 0 & 0 & 0 & 1 & 0 & 0 & -1 & -1 & 0 & 0 & 0 \\
0 & 0 & 0 & 0 & 0 & 1 & 0 & 0 & 0 & -1 & 1 & 1 \\
0 & 0 & 0 & 0 & 0 & 0 & 1 & 0 & 0 & 0 & -1 & -1
\end{array}
\end{array}
$$

$$
B =
\begin{array}{c}
\begin{array}{ccccccc|ccccc}
u & C_1 & C_2 & C_3 & C_4 & R_1 & L_2 & C_5 & R_2 & L_1 & L_3 & L_4 \\
0 & 0 & 0 & -1 & 1 & 0 & 0 & 1 & 0 & 0 & 0 & 0 \\
0 & 0 & 0 & -1 & 1 & 0 & 0 & 0 & 1 & 0 & 0 & 0 \\
-1 & 1 & 0 & 0 & 0 & 1 & 0 & 0 & 0 & 1 & 0 & 0 \\
0 & -1 & 1 & 0 & 0 & -1 & 1 & 0 & 0 & 0 & 1 & 0 \\
0 & -1 & 0 & 1 & 0 & -1 & 1 & 0 & 0 & 0 & 0 & 1
\end{array}
\end{array}
$$

5 Particular choice of tree: $\{u_1, C_2, R_8, L_6\}$

$$
Q =
\begin{array}{c}
\begin{array}{cccc|ccccc}
u_1 & C_2 & R_8 & L_6 & C_4 & R_3 & R_5 & L_7 & i_9 \\
1 & 0 & 0 & 0 & 1 & 1 & 1 & 1 & 0 \\
0 & 1 & 0 & 0 & 1 & 1 & 1 & 1 & 0 \\
0 & 0 & 1 & 0 & 0 & 0 & -1 & 0 & -1 \\
0 & 0 & 0 & 1 & 0 & 0 & 0 & -1 & 1
\end{array}
\end{array}
$$

$$
B =
\begin{array}{c}
\begin{array}{cccc|ccccc}
u_1 & C_2 & R_8 & L_6 & C_4 & R_3 & R_5 & L_7 & i_9 \\
-1 & -1 & 0 & 0 & 1 & 0 & 0 & 0 & 0 \\
-1 & -1 & 0 & 0 & 0 & 1 & 0 & 0 & 0 \\
-1 & -1 & 1 & 0 & 0 & 0 & 1 & 0 & 0 \\
-1 & -1 & 0 & 1 & 0 & 0 & 0 & 1 & 0 \\
0 & 0 & 1 & -1 & 0 & 0 & 0 & 0 & 1
\end{array}
\end{array}
$$

State vector: $x = \begin{bmatrix} v_2 \\ i_7 \end{bmatrix}$.

$$
\dot{x} =
\begin{bmatrix}
\dfrac{-1}{C_2 + C_4}\left[\dfrac{1}{R_3 \| (R_5 + R_8)}\right] & \dfrac{1}{C_2 + C_4} \\[3mm]
\dfrac{-1}{L_6 + L_7} & 0
\end{bmatrix} x
$$
$$
+
\begin{bmatrix}
\dfrac{1}{C_2 + C_4}\left[\dfrac{1}{R_3 \| (R_5 + R_8)}\right] & \dfrac{-1}{C_2 + C_4}\left[\dfrac{R_8}{R_5 + R_8}\right] \\[3mm]
\dfrac{1}{L_6 + L_7} & 0
\end{bmatrix}
\begin{bmatrix} u_1 \\ i_9 \end{bmatrix}
$$

$$+ \begin{bmatrix} \dfrac{C_4}{C_2 + C_4} & 0 \\ 0 & \dfrac{L_6}{L_6 + L_7} \end{bmatrix} \begin{bmatrix} \dot{u}_1 \\ \dot{i}_9 \end{bmatrix}$$

Characteristic equation:

$$s^2 + \frac{s}{C_2 + C_4} \left[\frac{1}{R_3 \| (R_5 + R_8)} \right] + \frac{1}{(C_2 + C_4)(L_6 + L_7)} = 0$$

6 Transfer functions:

$$\frac{V_2(s)}{U_1(s)} = \frac{1 + 1.5s + 0.5s^2}{1 + 1.5s + s^2} \qquad \frac{I_7(s)}{U_1(s)} = \frac{-0.5s(1 - 2s)}{1 + 1.5s + s^2}$$

7 Time domain expressions for given inputs:

$$v_2(t) = \left(10 - 5e^{-0.75t} \cos \frac{\sqrt{7}}{4}t + \frac{15}{\sqrt{7}} e^{-0.75t} \sin \frac{\sqrt{7}}{4}t \right) u_s(t)$$

$$i_7(t) = \left(2.5 + \frac{20}{\sqrt{7}} e^{-7.5t} \sin \frac{\sqrt{7}}{4}t \right) u_s(t)$$

8 Tree: $\{u_8, C_5, R_3, R_6, L_2\}$ (not unique). State vector $x = [v_5 \ i_4]^T$.

$$\dot{x} = \begin{bmatrix} 0 & \dfrac{1}{C_5 + C_7} \\ \dfrac{-1}{L_2 + L_4} & \dfrac{-(R_3 + R_6 \| R_9)}{L_2 + L_4} \end{bmatrix} x$$

$$+ \begin{bmatrix} 0 & 0 \\ \dfrac{-R_6}{R_6 + R_9} \left(\dfrac{1}{L_2 + L_4} \right) & 0 \end{bmatrix} \begin{bmatrix} u_8 \\ i_1 \end{bmatrix}$$

$$+ \begin{bmatrix} \dfrac{-C_7}{C_5 + C_7} & 0 \\ 0 & \dfrac{L_2}{L_2 + L_4} \end{bmatrix} \begin{bmatrix} \dot{u}_8 \\ \dot{i}_1 \end{bmatrix}$$

9 Natural response: $Ae^{-0.382t} + Be^{-2.618t}$.
Complete solution: $x_1(t) = 9.7e^{-0.382t} + 0.5e^{-2.618t} + 0.0037 \sin 5t - 0.2 \cos 5t$, $x_2(t) = 3.707e^{-0.382t} + 1.312e^{-2.618t} - 0.03 \sin 5t - 0.019 \cos 5t$.
Check initial conditions: $x_1(0^+) = 10$, $x_2(0^+) = 5$.

10 Characteristic equation: $s^2 + 2s + 1 = 0$. $x_1(t) = -1 + 5te^{-t} + 3e^{-t}$, $x_2(t) = -1 + 10te^{-t} + e^{-t}$. Check initial conditions: $x_1(0^+) = 2$, $x_2(0^+) = 0$. Note that $x_1(0^-) = 1$ and a jump occurs at $t = 0$ due to the $\delta(t)$ function.

Chapter 12

1 Left: $y_{11} = (1/R + 1/2R), y_{12} = y_{21} = -1/2R, y_{22} = (1/R) + (1/2R), z_{11} = R\|3R, z_{12} = z_{21} = R/4, z_{22} = R\|3R$. Right: $y_{11} = 1/R, y_{12} = y_{21} = 0, y_{22} = 1/R, z_{11} = R, z_{12} = z_{21} = 0, z_{22} = R$.

2 Left: $g_{11} = (1/R) + (1/3R), g_{12} = -1/3, g_{21} = 1/3, g_{22} = 2R\|R, h_{11} = R\|2R, h_{12} = 1/3, h_{21} = -1/3, h_{22} = (1/R) + (1/3R)$. Right: $g_{11} = 1/R, g_{12} = 0, g_{21} = 0, g_{22} = R, h_{11} = R, h_{12} = 0, h_{21} = 0, h_{22} = 1/R$.

3 General expression for Z_L:

$$Z_L = \frac{-k_{22} + k_{11} \pm \sqrt{(k_{22} - k_{11})^2 + 4k_{12}k_{21}}}{2k_{21}}$$

Z_L exists if $k_{21} \neq 0$. Special case: $k_{11} = k_{22} \Rightarrow Z_L = \sqrt{k_{12}/k_{21}}$.

5

$$K_1 = \begin{bmatrix} 3 & 2 \\ 1 & 1 \end{bmatrix} \quad K_2 = \begin{bmatrix} 1 & 2 \\ 0 & 1 \end{bmatrix} \quad K_1 K_2 = \begin{bmatrix} 3 & 8 \\ 1 & 3 \end{bmatrix} \quad K_2 K_1 = \begin{bmatrix} 5 & 4 \\ 1 & 1 \end{bmatrix}$$

6 $[Y] = [Y_1] + [Y_2]$.

7 $[G] = [G_1] + [G_2]$.

8 No solution since $[Z_2]$ does not exist.

9 $[H] = [H_1] + [H_2]$.

Bibliography

Circuit Analysis

1. D.A. Bell, *Electric Circuits: Principles, Applications and Computer Analysis*, Englewood Cliffs NJ: Prentice Hall, 1995.
2. H.W. Bode, *Network Analysis and Feedback Amplifier Design*, Princeton NJ: Van Nostrand, 1945.
3. R.L. Boylestad, *Introductory Circuit Analysis*, Englewood Cliffs NJ: Prentice Hall, 1997.
4. A. Budak, *Circuit Theory Fundamentals and Applications*, Englewood Cliffs NJ: Prentice Hall, 1987.
5. H.J. Carlin and A.B. Giordano, *Network Theory: An Introduction to Reciprocal and Non-reciprocal Circuits*, Englewood Cliffs NJ: Prentice Hall, 1964.
6. W.K. Chen (Ed.), *The Circuits and Systems Handbook*, New York: CRC Press, 1996.
7. L.O. Chua, C.A. Desoer and E.S. Kuh, *Linear and Nonlinear Circuits*, New York: McGraw-Hill, 1987.
8. D.R. Cunningham and J.A. Stuller, *Circuit Analysis*, New York: Houghton Mifflin Company, 1995.
9. C.A. Desoer and E.S. Kuh, *Basic Circuit Theory*, New York: McGraw-Hill, 1969.
10. R.C. Dorf and J.A. Svoboda, *Introduction to Electric Circuits*, New York: Wiley, 1996.
11. J.A. Edminister, *Schaum's Outline of Theory and Problems of Electric Circuits*, New York: McGraw-Hill, 1983.
12. J.K. Fidler and L. Ibbotson, *Introductory Circuit Theory*, Singapore: McGraw-Hill, 1989.
13. T.L. Floyd, *Principles of Electric Circuits*, Upper Saddle River NJ: Prentice Hall, 1997.
14. S. Franco, *Electric Circuits Fundamentals*, Fort Worth: Saunders College, 1995.

15. B. Friedland, O. Wing and R.B. Ash, *Principles of Linear Networks*, New York: McGraw-Hill, 1961.

16. E.A. Guillemin, *Introductory Circuit Theory*, New York: Wiley, 1953.

17. W.H. Hayt and J.E. Kemmerly, *Engineering Circuit Analysis*, New York: McGraw-Hill, 1978.

18. A. Henderson, *Electrical Networks*, London: Edward Arnold, 1990.

19. L.P. Huelsman, *Basic Circuit Theory*, New York: Prentice Hall, 1992.

20. J.D. Irwin, *Basic Engineering Circuits*, Englewood Cliffs NJ: Prentice Hall, 1996.

21. D.E. Johnson, J.L. Hilburn, J.R. Johnson and P.D. Scott, *Basic Electric Circuit Analysis*, Englewood Cliffs NJ: Prentice Hall, 1995.

22. B.K. Kinariwala and F.F. Kuo, *Linear Circuits and Computation*, New York: Wiley, 1973.

23. F.F. Kuo, *Network Analysis and Synthesis*, New York: Wiley, 1962.

24. C.R. Paul, *Analysis of Linear Circuits*, New York: McGraw-Hill, 1989.

25. K.F. Sander, *Electric Circuit Analysis*, Reading MA: Addison-Wesley, 1992.

26. J.O. Scanlan and R. Levy, *Circuit Theory*, Edinburgh: Oliver & Boyd, 1973.

27. R.E. Scott, *Elements of Linear Circuits*, Reading MA: Addison-Wesley, 1965.

28. H.H. Skilling, *Electric Networks*, New York: Wiley, 1974.

29. K.C.A. Smith and R.E. Alley, *Electrical Circuits: An Introduction*, New York: Cambridge University Press, 1992.

30. W.D. Stanley, *Transform Circuit Analysis for Engineering and Technology*, Englewood Cliffs NJ: Prentice Hall, 1997.

31. G.C. Temes and J.W. LaPatra, *Introduction to Circuit Synthesis and Design*, New York: McGraw-Hill, 1977.

32. T.N. Trick, *Introduction to Circuit Analysis*, New York: Wiley, 1977.

33. M.E. Van Valkenburg and B.K. Binariwala, *Linear Circuits*, Englewood Cliffs NJ: Prentice Hall, 1982.

34. R. Yorke, *Electric Circuit Theory*, New York: Pergamon Press, 1986.

PSPICE Simulation

1. E. Brumgnach, *PSPICE for Windows*, New York: Delmar, 1994.

2. R.W. Goody, *PSPICE for Windows: A Circuit Simulation Primer*, New York: Prentice Hall, 1995.

3. P.W. Tuinenga, *SPICE. A Guide to Circuit Simulation and Analysis using PSPICE*, Englewood Cliffs NJ: Prentice Hall, 1995.

Linear Systems

1. C.T. Chen, *Linear System Theory and Design*, New York: Holt, Rinehart and Winston, 1970.
2. D.K. Cheng, *Theory of Linear Systems*, Reading MA: Addison-Wesley, 1959.
3. T. Kailath, *Linear Systems,* Englewood Cliffs NJ: Prentice Hall, 1980.
4. L.A. Zadeh and C.A. Desoer, *Linear System Theory*, New York: McGraw-Hill, 1963.

Engineering Mathematics

1. V.I. Arnold, *Ordinary Differential Equations*, Cambridge MA: MIT Press, 1983.
2. H.S. Carslaw, *Theory of Fourier Series and Integrals*, London: Macmillan, 1930.
3. R.V. Churchill, *Fourier Series and Boundary Value Problems*, New York: McGraw-Hill, 1963.
4. T. Croft, R. Davison and M. Hargreaves, *Engineering Mathematics: A Modern Foundation for Electronic, Electrical, and Control Engineers*, Reading MA: Addision-Wesley, 1992.
5. F.E. Hohn, *Elementary Matrix Algebra*, New York: Macmillan, 1958.
6. E. Kreyszig, *Advanced Engineering Mathematics*, New York: Wiley, 1983.
7. G. Strang, *Linear Algebra and its Applications*, New York: Academic Press, 1976.

Index

066|||426709|||410068||426709|||410066|||426709||| 426709|||410066||||426709|||410066||||426|